T0223479

Embedded Software Development

Development

The Open-Source Approach

Embedded Systems

Series Editor

Richard Zurawski
ISA Corporation, San Francisco, California, USA

Embedded Software Development

The Open-Source Approach

Ivan Cibrario Bertolotti

National Research Council of Italy
Turin, Italy

Tingting Hu

National Research Council of Italy
Politecnico di Torino
Turin, Italy

CRC Press
Taylor & Francis Group
Boca Raton London New York

CRC Press is an imprint of the
Taylor & Francis Group, an **informa** business

CRC Press
Taylor & Francis Group
6000 Broken Sound Parkway NW, Suite 300
Boca Raton, FL 33487-2742

First issued in paperback 2020

© 2016 by Taylor & Francis Group, LLC
CRC Press is an imprint of Taylor & Francis Group, an Informa business

No claim to original U.S. Government works

ISBN-13: 978-1-4665-9392-3 (hbk)
ISBN-13: 978-0-367-73743-6 (pbk)

Visit the Taylor & Francis Web site at
http://www.taylorandfrancis.com

and the CRC Press Web site at
http://www.crcpress.com

Ché nessun quaggiù lasciamo,
né timore, né desir
(For we leave no one behind us,
nor dreads, nor desires)
— *Ivan*

谨以此书献给我的外公陈光禄，外婆宋凤枝
(This book is dedicated to my grandfather Chen Guanglu,
my grandmother Song Fengzhi)
— *Tingting*

Contents

PART II   Advanced Topics

Foreword

Embedded software touches many aspects in our daily life by defining the behavior of the things that surround us, may it be a phone, a thermostat, or a car. Over time, these things have been made more capable by a combination of advances in hardware and software. The features and capabilities of these things are further augmented by connecting to other things or to external software services in the cloud, e.g., by a training watch connecting wirelessly to a heartrate monitor that provides additional sensor data and to a cloud service providing additional analysis and presentation capabilities. This path of evolution can be seen in most domains, including industrial embedded systems, and has over time added new layers of knowledge that are needed for development of embedded systems.

Embedded systems cover a large group of different systems. For instance, a phone is a battery-powered device with relatively high processing power and ample options for wireless connectivity, while a thermostat is likely to have scarce resources and limited connectivity options. A car on the other hand represents another type of embedded system, comprising a complex distributed embedded system that operates in harsh environments. Although the requirements and constraints on these systems are quite different, there are still strong commonalities among most types of embedded systems, considering the software layers that sit closest to the hardware, typically running on embedded or real-time operating systems that offer similar services.

Every new generation of a software-based system typically is more elaborate and operates in a more complex and dynamic environment than the previous one. Over the last four decades many things have changed with respect to software based systems in general; starting in the 1980s with the early desktop PC, like the IBM XT that I used at the university, which was similar to industrial and embedded systems with respect to operating system support and development environments. In the 1990s, real-time and embedded operating systems services remained mostly the same, while the evolution of general purpose operating systems witnessed many changes including graphical user interfaces. During the first decade of the new millennium we saw the introduction of smart phones, server virtualization, and data centers, and currently we see the emergence of large scale open-source software platforms for cloud computing and the Internet of Things (IoT).

Some embedded systems such as consumer electronics have been an integral part of these latest developments while industrial system adoption is slower and more careful. Although, the system context today is much more complex than thirty-five years ago and many software layers have been added, the foundation, the software that sits close to the hardware, is still much the same with the same requirements for timeliness and predictability.

As the software stack in embedded systems increases and as the functionality reaches beyond the embedded device through connectivity and collaboration with surrounding systems, the burden on the engineer potentially increases in order to master the entire scope of development. To some degree this can be addressed by using more powerful software platforms, better tools, and so on that alleviate the developer from handling some details. But unlike general purpose computing systems many of the embedded devices still have strict constraints with respect to memory and computational resources. For industrial embedded systems additional constraints related to safety, reliability, and availability apply. There are also aspects such as constraints on power consumption to enhance battery life or to facilitate fanless operation needed for high availability in some harsh or hard to reach environments. Therefore, detailed control of the lower levels of the software stack is still necessary for the development of many embedded systems.

The fast-paced development of consumer and general purpose technologies makes it difficult to directly adopt these technologies for industrial embedded systems. A common denominator for industrial systems within many different domains is that they are expected to be operational for a very long time. An example is the world's first electrical and microchip enabled industrial robot that was developed at the research department of Allmänna Svenska Elektriska Aktiebolaget (ASEA, which later became ABB) in the early 1970s. Today, some 40 years later, some of these robots are still in operation within Swedish industry. Moreover, even if individual equipment sometimes operate unchanged, the software platform of the specific product family often evolves over a long time, outliving several generations of hardware. This puts requirements on software portability and availability of tooling, third party support, drivers, etc. Hence, technology must be selected for stability over a long time.

An interesting aspect of the conflict between long-lived systems and fast-paced technology is what role open-source can play. Can open-source help in the provisioning of long-term software technologies? Certainly Linux® and the GCC compiler are examples of open-source platforms and tools heavily used for embedded systems and that have had long-term support and development. Perhaps some of the emerging open-source projects will play the same role for embedded systems in the era of IoT technologies.

From a setback in early 2000, following the dot-com crash, there was interest in embedded systems from new generations of young people through maker culture and easily accessible open-source operating systems like Linux and FreeRTOS™, and hardware projects like Raspberry Pi™, and Arduino™. Capturing this interest is a great opportunity for continued evolution of embedded systems not only in the consumer segment but also for industrial systems.

This book provides a comprehensive description covering many aspects of embedded systems development and explains their interrelationships, which is a very important complement to other in-depth literature on specific subjects such as operating systems. Today there are communities that provide a lot of very specific and detailed information related to embedded systems e.g., on setting up a specific package for an operating system.

What is lacking is continuous and coherent descriptions holding all fragments together presenting a holistic view on the fundamentals of the embedded system, the involved components, how they operate in general, and how they interact and depend on each other. As such, this book fills a gap in the literature and will prove valuable to students and professionals who need a single coherent source of information.

Kristian Sandström
ABB Corporate Research
Västerås, Sweden

Raspberry Pi^{TM} is a trademark of the Raspberry Pi Foundation.
Arduino^{TM} is the trademark of Arduino LLC.
Linux$^{®}$ is the registered trademark of Linus Torvalds in the U.S. and other countries.
FreeRTOS^{TM} is the trademark of Real Time Engineers Ltd.

Preface

This book is the outcome of the authors' research experience in the field of real-time operating systems and communications applied to control systems and other classes of embedded applications, often carried out in strict cooperation with industry. Through the years, we have been positively influenced by many other people with whom we came in contact.

They are too numerous to mention individually, but we are nonetheless indebted to them for their contribution to our professional growth. A special thank you goes to our industrial partners, who are constantly in touch with most of the topics presented in this book. Their questions, suggestions, and remarks that we collected along the way were helpful in making the book clearer and easier to read.

We would also like to express our appreciation to our coworkers for all their support and patience during the long months while we were busy with the preparation of the manuscript.

Last, but not least, we are grateful to Richard Zurawski, who gave us the opportunity to write this book, and Kristian Sandström for his insightful foreword. We are also indebted to CRC Press publishing, editorial, and marketing staff, Nora Konopka, John Gandour, Kyra Lindholm, Laurie Oknowsky, Karen Simon, Michele Smith, and Florence Kizza in particular. Without their help, the book would have probably never seen the light.

The Authors

Ivan Cibrario Bertolotti earned the Laurea degree (*summa cum laude*) in computer science from the University of Torino, Turin, Italy, in 1996. Since then, he has been a researcher with the National Research Council of Italy (CNR). Currently, he is with the Institute of Electronics, Computer, and Telecommunication Engineering (IEIIT) of CNR, Turin, Italy.

His research interests include real-time operating system design and implementation, industrial communication systems and protocols, and formal methods for vulnerability and dependability analysis of distributed systems. His contributions in this area comprise both theoretical work and practical applications, carried out in cooperation with leading Italian and international companies.

Dr. Cibrario Bertolotti taught several courses on real-time operating systems at Politecnico di Torino, Turin, Italy, from 2003 until 2013, as well as a PhD degree course at the University of Padova in 2009. He regularly serves as a technical referee for the main international conferences and journals on industrial informatics, factory automation, and communication. He has been an IEEE member since 2006.

Tingting Hu earned a master's degree in computer engineering and a PhD degree in control and computer engineering, both from Politecnico di Torino, Turin, Italy. Since 2010, she has been a research fellow with the National Research Council of Italy (CNR). Currently, she is with the Institute of Electronics, Computer, and Telecommunication Engineering (IEIIT) of CNR, Turin, Italy.

Her main research interests are the design and implementation of real-time operating systems and communication protocols, focusing on deterministic and flexible execution and communication for distributed real-time embedded systems. A significant amount of her research activities are carried out in strict collaboration with industry.

Dr. Hu is actively involved in several regional and national industrial research projects in the context of the Italian "Factory of the Future" framework program. Moreover, in 2014 she taught a postgraduate-level course about real-time operating systems and open-source software for embedded applications aimed at company technical managers. She has been an IEEE member since 2011 and serves as technical referee for several primary conferences in her research area.

List of Figures

List of Tables

1 Introduction

The goal of this book is to introduce readers to *embedded software development*. This is a very broad subject and encompasses several important topics in computer science and engineering—for instance, operating systems, concurrent programming, compilers, code optimization, hardware interfaces, testing, and verification.

Each topic is quite complex by itself and very well deserves one whole textbook, or probably more, to be thoroughly discussed. Unfortunately, since most of those textbooks concentrate on one specific topic, their styles of writing are usually far away from each other.

For instance, concurrent programming is often presented using a formal approach based on an abstract mathematical model for describing a computer system and the concurrent activities it must execute, along with their data dependencies and synchronization constraints.

Such an approach is perfectly fine, provided that the reader has enough available time to fully grasp the model and enough experience to realize how a real system and its requirements fit into the model itself.

On the opposite side, standalone descriptions of how embedded hardware devices interface with a microcontroller are usually compiled by hardware engineers. As a consequence, although they are perfectly correct for what concerns the content, their point of view may easily be quite far away from what a programmer expects.

As a consequence, readers are left with the hard task of mixing and matching all these valuable sources of information, which sometimes use their own nomenclature and jargon, in order to build their own unified, consistent knowledge about embedded software development in their minds.

This might hardly be feasible when the reader is just entering the embedded software development domain as a novice. The problem is further compounded by the fact that the embedded software development process is quite different from its general-purpose counterpart. Hence, previous experience in writing, for instance, PC-based applications, may not be easy to leverage in the new domain.

This book takes a rather different path. In fact, all basic topics presented in the first part of the book are discussed by following a *unified* approach and consistently emphasizing their *relationships* and mutual dependencies, rather than isolating them in their own world.

For instance, the description of a central topic like *interrupt handling* is carried out starting from the hardware event that may trigger an interrupt request, and then guiding the reader through the whole process—partly performed by hardware, and partly by software—that eventually culminates in the execution of the corresponding interrupt handling routine.

In this way, readers gain a complete understanding of several aspects of interrupt handling that are often kept "behind the scenes" in general-purpose programming, but are indeed of utmost importance to guarantee that embedded systems work reliably and efficiently.

Furthermore, virtually all chapters in the first part of the book make explicit reference to real fragments of code and, when necessary, to a specific operating system, that is, FREERTOS. This also addresses a shortcoming of other textbooks, that is, the lack of practical programming examples presented and commented on in the main text (some provide specific examples in appendices).

From this point of view FREERTOS is an extremely useful case study because it has a very limited memory footprint and execution-time overhead, but still provides a quite comprehensive range of primitives.

In fact, it supports a multithreaded programming model with synchronization and communication primitives that are not far away from those available on much bigger, and more complex systems. Moreover, thanks to its features, is has successfully been used in a variety of real-world embedded applications.

At the same time, it is still simple enough to be discussed in a relatively detailed way within a limited number of pages and without overwhelming the reader with information. This also applies to its hardware abstraction layer—that is, the code module that layers the operating system on top of a specific hardware architecture—which is often considered an "off-limits" topic in other operating systems.

As a final remark, the book title explicitly draws attention to the *open-source* approach to software development. In fact, even though the usage of open-source components, at least in some scenarios (like industrial or automotive applications) is still limited nowadays, it is the authors' opinion that open-source solutions will enjoy an ever-increasing popularity in the future.

For this reason, *all* the software components mentioned and taken as examples in this book—from the software development environment to the real-time operating system, passing through the compiler and the software analysis and verification tools—are themselves open-source.

—

The first part of the book guides the reader through the key aspects of embedded software development, spanning from the peculiar requirements of embedded systems in contrast to their general-purpose counterparts, without forgetting network communication, implemented by means of open-source protocol stacks.

In fact, even though the focus of this book is mainly on software development techniques, most embedded systems are nowadays *networked* or *distributed*, that is, they consist of a multitude of nodes that cooperate by communicating through a network. For this reason, embedded programmers must definitely be aware of the opportunities and advantages that adding network connectivity to the systems they design and develop may bring.

The chapters of the first part of the book are:

- Chapter 2, *Embedded Applications and Their Requirements*. This chapter outlines the role and purpose of embedded systems, introducing readers to their internal structure and to the most common way they are interfaced to software development tools. The chapter also gives a first overview of the embedded software development process, to be further expanded in the following.

- Chapter 3, GCC-*Based Software Development Tools*. The main topic of this chapter is a thorough description of the GNU compiler collection (GCC)-based software development system, or *toolchain*, which is arguably the most popular open-source product of this kind in use nowadays. The discussion goes through the main toolchain components and provides insights on their inner workings, focusing in particular on the aspects that may affect programmers' productivity,

- Chapter 4, *Execution Models for Embedded Systems*. This chapter and the next provide readers with the necessary foundations to design and implement embedded system software. Namely, this chapter presents in detail two different software execution models that can be profitably applied to model and express the key concept of *concurrency*, that is, the parallel execution of multiple activities, or tasks, within the same software system.

- Chapter 5, *Concurrent Programming Techniques*. In this chapter, the concept of execution model is further expanded to discuss in detail how concurrency must be managed to achieve correct and timely results, by means of appropriate *concurrent programming* techniques.

- Chapter 6, *Scheduling Algorithms and Analysis*. After introducing some basic nomenclature, models, and concepts related to task-based *scheduling algorithms*, this chapter describes the most widespread ones, that is, rate monotonic (RM) and earliest deadline first (EDF). The second part of the chapter briefly discusses *scheduling analysis*, a technique that allows programmers to predict the worst-case timing behavior of their systems.

- Chapter 7, *Configuration and Usage of Open-Source Protocol Stacks*. In recent years, many embedded systems rapidly evolved from centralized to *networked* or *distributed* architectures, due to the clear advantages this approach brings. In this chapter, we illustrate how an open-source protocol stack—which provides the necessary support for inter-node communication—can easily be integrated in an embedded software system and how it interfaces with other software components.

- Chapter 8, *Device Driver Development*. Here, the discourse goes from the higher-level topics addressed in the previous three chapters to a greater level of detail, concerning how software manages and drives *hardware devices*. This is an aspect often neglected in general-purpose software development, but of utmost importance when embedded systems are considered, because virtually all of them are strongly tied to at least some dedicated hardware.

- Chapter 9, *Portable Software*. While the previous chapters set the stage for effective embedded software development, this chapter outlines the all-important trade-off between code execution efficiency and *portability*, that is, easiness of migrating software from one project to another. This aspect is becoming more and more important nowadays due to the ever-increasing need to reduce time to market and to implement quick prototyping.

- Chapter 10, *The* FREERTOS *Porting Layer*. This technically oriented chapter scrutinizes the main components of the FREERTOS porting layer, diving into the implementation details of key operating system concepts, for instance, *context switching* and the implementation of low-level critical regions by disabling interrupts.
- Chapter 11, *Performance and Footprint at the Toolchain Level*. This chapter illustrates, by means of a running real-world example for the most part, how a modern high-level software development toolchain can be used to achieve a satisfactory level of optimization, for what concerns both execution speed and memory footprint, without resorting to assembly-level programming or other non-portable software development techniques.
- Chapter 12, *Example: A* MODBUS *TCP Device*. To conclude the first part of the book, this chapter contains a comprehensive example of how the software development techniques described in the previous chapters can be applied to design and build a simple, but complete embedded system.

In the second part, the book presents a few advanced topics, focusing mainly on improving software *quality* and *dependability*. The importance of these goals is ever increasing nowadays, as embedded systems are becoming commonplace in critical application areas, like anti-lock braking system (ABS) and motion control.

In order to reach the goal, it is necessary to adopt a range of different techniques, spanning from the software design phase to runtime execution, passing through its implementation. In particular, this book first presents the basic principles of formal verification through *model checking*, which can profitably be used since the very early stages of algorithm and software design.

Then, a selection of *runtime* techniques to prevent or, at least, detect memory corruption are discussed. Those techniques are useful to catch any software error that escaped verification and testing, and keep within bounds the damage it can do to the system and its surroundings.

Somewhat in between these two extremes, *static code analysis* techniques are useful, too, to spot latent software defects that may escape manual code inspection. With respect to formal verification, static code analysis techniques have the advantage of working directly on the actual source code, rather than on a more abstract model.

Moreover, their practical application became easier in recent years because several analysis tools moved away from being research prototypes and became stable enough for production use. The book focuses on one of them as an example.

The second part of the book consists of the following chapters:

- Chapter 13, *Model Checking of Distributed and Concurrent Systems*. This chapter describes how formal verification, and model checking in particular, can be profitably used to improve software quality and even prove that it satisfies some critical requirements. The practical advantages of this technique become evident as embedded software complexity grows, thus limiting the effectiveness of more traditional testing techniques.

- Chapter 14, *Model Checking: An Example.* Building upon the general concept of model checking introduced previously, this chapter shows the application of model checking to a case study of practical, industrial interest. It demonstrates how model checking can be used to identify low-probability defects that may never be observed during testing, and how counterexamples found by the model checker can be helpful to correct them.
- Chapter 15, *Memory Protection Techniques.* Since it is usually impossible to *design* and *develop* perfect software, even after extensive testing and formal verification, another way to enhance software dependability is to introduce various *runtime* checks. This chapter focuses on several specific kinds of check, aimed at preventing or detecting memory corruption.
- Chapter 16, *Security and Dependability Aspects.* This chapter describes how software dependability can be further enhanced by means of sophisticated *static code analysis* techniques, which go beyond what an ordinary compiler is able to do. Furthermore, the same techniques also enhance software quality from what concerns *security* attacks, an aspect of utmost importance as embedded systems are nowadays often connected to public communication networks.

The bibliography at the end of the book has been kept rather short because it has been compiled with software practitioners in mind. Hence, instead of providing an exhaustive and detailed list of references that would have been of interest mainly to people willing to dive deep into the theoretical aspects of embedded real-time systems, we decided to highlight a smaller number of additional sources of information.

In this way, readers can more effectively use the bibliography as a starting point to seek further knowledge on this rather vast field, without getting lost. Within the works we cite, readers will also find further, more specific pointers to pursue their quest.

Part I

Basics of Embedded Software Development

2 Embedded Applications and Their Requirements

CONTENTS

This chapter outlines the central role played by embedded software in a variety of contemporary appliances. At the same time, it compares embedded and general-purpose computing systems from the hardware architecture and software development points of view. This is useful to clarify and highlight why embedded software development differs from application software development for personal computers most readers are already familiar with.

2.1 ROLE AND PURPOSE OF EMBEDDED SYSTEMS

Nowadays, embedded systems are becoming more and more popular and widespread. They are omnipresent from our daily life to different sections of industry. For example, even simple consumer appliances, such as microwave ovens, washing machines, fridges and so on, include several microcontrollers to perform predefined sets of functions/procedures.

Embedded systems also play a significant role in the development of smart homes and building automation. Embedded nodes are installed to assess changes in operational environment—for instance, to detect that nobody is in a room—and carry out corresponding actions—for instance, turn off all the lights automatically.

This kind of automation is quite helpful as saving energy is becoming a bigger concern day by day. The same embedded computer may also be used to complete other kinds of activities, for example, personnel access control to buildings based on digital identification like badges.

What's more, embedded systems are deeply involved in the transportation industry, especially automotive. It is quite common to find that even inexpensive cars are equipped with more than 10 embedded nodes, including those used for anti-lock braking system (ABS). For what concerns industry automation, embedded systems are deployed in production lines to carry out all sorts of activities, ranging from motion control to packaging, data collection, and so on.

The main concerns of embedded systems design and development are *different* from general-purpose systems. Embedded systems are generally equipped with limited resources, for instance, small amount of memory, low clock frequency, leading to the need for better code optimization strategies. However, fast CPUs sometimes simply cannot be adopted in industrial environments because they are supposed to work within a much narrower range of temperature, for instance $[0, 40]$ degrees Celsius.

Instead, industrial-grade microprocessors are generally assumed to perform sustainable correct functioning even up to 85 degrees. This leads to one main concern of embedded systems, that is *reliability*, which encompasses hardware, software, and communication protocol design. In addition, processor heat dissipation in such environments is another issue if they are working at a high frequency.

All these differences bring unavoidable changes in the way embedded software is developed, in contrast with the ordinary, general-purpose software development process, which is already well known to readers. The main purpose of this chapter is to outline those differences and explain how they affect programmers' activities and way of working.

In turn, this puts the focus on how to make the best possible use of the limited resources available in an embedded system by means of code optimization. As outlined above, this topic is of more importance in embedded software development with respect to general-purpose development, because resource constraints are usually stronger in the first case.

Moreover, most embedded systems have to deal with real-world events, for instance, continuously changing environmental parameters, user commands, and others. Since in the real world events inherently take place independently from each other and in parallel, it is natural that embedded software has to support some form of parallel, or *concurrent* execution. From the software development point of view, this is generally done by organizing the code as a set of activities, or *tasks*, which are carried out concurrently by a real-time operating system (RTOS).

The concept of task—often also called *process*—was first introduced in the seminal work of Dijkstra [48]. In this model, any concurrent application, regardless of its nature or complexity, is represented by, and organized as, a set of tasks that, conceptually, execute in parallel.

Each task is autonomous and holds all the information needed to represent the evolving execution state of a sequential program. This necessarily includes not only the program instructions but also the state of the processor (program counter, registers) and memory (variables).

Informally speaking, each task can be regarded as the execution of a sequential program by "its own" conceptual processor even though, in a single-processor system the RTOS will actually implement concurrent execution by switching the physical processor from one task to another when circumstances warrant.

Therefore, thoroughly understanding the details of how tasks are executed, or *scheduled*, by the operating system and being aware of the most common *concurrent programming* techniques is of great importance for successful embedded software development. This is the topic of Chapters 4 through 6.

In the past, the development of embedded systems has witnessed the evolution from *centralized* to *distributed* architectures. This is because, first of all, the easiest way to cope with the increasing need for more and more computing power is to use a larger number of processors to share the computing load. Secondly, as their complexity grows, centralized systems cannot scale up as well as distributed systems.

A simple example is that, in a centralized system, one more input point may require one more pair of wires to bring data to the CPU for processing. Instead, in a distributed system, many different Inputs/Outputs values can be transmitted to other nodes for processing through the same shared communication link. Last but not the least, with time, it becomes more and more important to integrate different subsystems, not only horizontally but also vertically.

For example, the use of buses and networks at the factory level makes it much easier to integrate it into the factory management hierarchy and support better business decisions. Moreover, it is becoming more and more common to connect embedded systems to the Internet for a variety of purposes, for instance, to provide a web-based user interface, support firmware updates, be able to exchange data with other equipment, and so on.

For this reason *protocol stacks*, that is, software components which implement a set of related communication protocols—for instance, the ubiquitous TCP/IP protocols—play an ever-increasing role in all kinds of embedded systems. Chapter 7 presents in detail how a popular open-source TCP/IP protocol stack can be interfaced, integrated, and configured for use within an embedded system.

For the same reason Chapter 8, besides illustrating in generic terms how software and hardware shall be interfaced in an embedded system—by means of suitable *device drivers*—also shows an example of how protocol stacks can be interfaced with the network hardware they work with.

Another major consideration in embedded systems design and development is *cost*, including both hardware and software, as nowadays software cost is growing and becomes as important as hardware cost. For what concerns software, if existing applications and software modules can be largely reused, this could significantly save time and effort, and hence, reduce cost when integrating different subsystems or upgrading the current system with more advanced techniques/technologies.

An important milestone toward this goal, besides the adoption of appropriate software engineering methods (which are outside the scope of this book), consists of writing *portable* code. As discussed in Chapters 9 and 10, an important property of portable code is that it can be easily compiled, or *ported*, to diverse hardware architectures with a minimum amount of change.

This property, besides reducing software development time and effort—as outlined previously—also brings the additional benefit of improving software reliability because less code must be written anew and debugged when an application is moved from one architecture to another. In turn, this topic is closely related to code optimization techniques, which are the topic of Chapter 11.

Generally, embedded systems also enforce requirements on *real-time* performance. It could be *soft* real-time in the case of consumer appliances and building automation, instead of *hard* real-time for critical systems like ABS and motion control.

More specifically, two main aspects directly related to real-time are *delay* and *jitter*. Delay is the amount of time taken to complete a certain job, for example, how long it takes a command message to reach the target and be executed. Delay variability gives rise to jitter. For example, jobs are completed sometimes sooner, sometimes later.

Hard real-time systems tolerate much less jitter than their counterparts and they require the system to behave in a more *deterministic* way. This is because, in a hard real-time system, deadlines must always be met and any jitter that results in missing the deadline is unacceptable. Instead, this is allowed in soft real-time systems, as long as the probability is sufficiently small. This also explains why jitter is generally more of concern. Several of these important aspects of software development will be considered in more detail in Chapter 12, within the context of a real-world example.

As we can see, in several circumstances, embedded systems also have rather tight *dependability* requirements that, if not met, could easily lead to *safety* issues, which is another big concern in some kinds of embedded systems. Safety is not only about the functional correctness (including the timing aspect) of a system but also related to the security aspect of a system, especially since embedded systems nowadays are also network-based and an insecure system is often bound to be unsafe.

Therefore, embedded software must often be tested more accurately than general-purpose software. In some cases, embedded software correctness must be ensured not only through careful testing, but also by means of formal verification. This topic is described in Chapter 13 from the theoretical point of view and further developed in Chapter 14 by means of a practical example. Further information about security and dependability, and how they can be improved by means of automatic software analysis tools, is contained in Chapter 16.

A further consequence, related to both code reliability and security, of the limited hardware resources available in embedded architectures with respect to their general-purpose counterparts, is that the hardware itself may provide very limited support to detect software issues as early as possible and before damage is done to the system.

From this point of view, the software issue of most interest is *memory corruption* that takes place when part of a memory-resident data structure is overwritten with inconsistent data, often due to a wayward task that has no direct relationship with the data structure itself. This issue is also often quite difficult to detect, because the effects of memory corruption may be subtle and become manifest a long time after the corruption actually took place.

In order to tackle memory corruption, general-purpose processors are almost invariably equipped with sophisticated hardware components logically interposed between the processor itself and main memory, which are usually called memory management units (MMUs). Besides providing other useful memory-management features, a MMU allows the RTOS to specify which areas or memory a certain task is allowed to access and in which way (for instance, read-only data access, read-write data access, or instruction execution).

The MMU is able to trap any attempt made by a task to access a forbidden memory area or the use of a forbidden access mode before the access actually takes place, and hence, before memory is corrupted or, for read accesses, before the task gains access to information that does not pertain to it. When the MMU detects an illegal access, the offending task stops executing, the RTOS takes control of the processor and initiates a recovery action, which often involves the termination of the task itself.

To limit their cost and power consumption, embedded processors rarely implement a full-fledged MMU. Instead, some of them do provide a scaled-down version of this component, which can perform only memory protection, often in a simplified way with respect to an ordinary MMU. For instance, in the ARM Cortex family of processor cores [8] this feature is optionally provided by a component called memory protection unit (MPU), which is part of a broader protected memory system architecture (PMSA).

However, more often than not, low-cost microcontrollers offer no hardware-based support for memory protection. As a consequence, programmers shall resort to software-based protection techniques when required, even though those techniques are merely able to *detect* memory corruption rather than *prevent* it. A more detailed discussion of the different classes of memory protection usually available on embedded systems will be the topic of Chapter 15.

2.2 MICROCONTROLLERS AND THEIR INTERNAL STRUCTURE

The recent advances in microcontroller design and implementation made their internal structure quite complex and very similar to the structure of a whole computer system built twenty years ago. As a result, even inexpensive microcontrollers nowadays embed a variety of other elements besides the processor itself.

As shown in Figure 2.1, a microcontroller typically contains:

- One or more processor cores, which are the functional units responsible for program execution.
- Multiple internal memory banks, with different characteristics regarding capacity, speed, and volatility.
- Optionally, one or more memory controllers to interface the microcontroller with additional, external memory.
- A variety of input–output controllers and devices, ranging from very simple, low-speed devices like asynchronous serial receivers/transmitters to very fast and complex ones, like Ethernet and USB controllers.

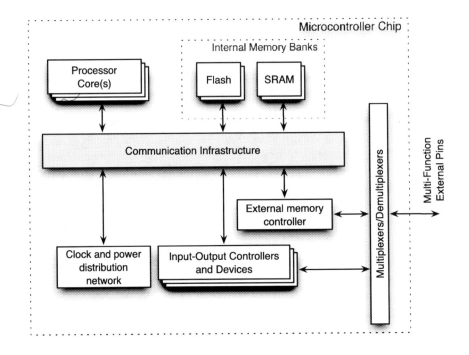

Figure 2.1   Internal structure of a microcontroller (simplified).

- A bank of multiplexers and demultiplexers to route the on-chip input and output signals associated with the memory controllers and the various internal devices toward the microcontroller external input–output ports.
- A communication infrastructure to let all the above-mentioned units exchange data and synchronization information.
- One or more clock generators, along with a clock and power distribution network. Both networks are used to provide an appropriate clock and power supply to the microcontroller units in a selective way. This is useful in order to save power, for instance, by removing clock and power from units that are unused in a certain application.

Is is therefore evident that, even though most of this book will focus on the processing capabilities of microcontrollers, in terms of program execution, it is extremely important to consider the microcontroller architecture as a whole during component selection, as well as software design and development.

For this reason, this chapter contains a brief overview of the major components outlined above, with special emphasis on the role they play from the programmer's perspective. Interested readers are referred to more specific literature [50, 154] for detailed information. After a specific microcontroller has been selected for use, its hardware data sheet of course becomes the most authoritative reference on this topic.

Arguably the most important component to be understood and taken into account when designing and implementing embedded software is *memory*. In fact, if it is true that processor cores are responsible for instruction *execution*, those instructions are stored in memory and must be continuously retrieved, or *fetched*, from memory to be executed.

Moreover, program instructions heavily refer to, and work on, *data* that are stored in memory, too. In a similar way, processor cores almost invariably make use of one or more memory-resident *stacks* to hold arguments, local variables, and return addresses upon function calls.

This all-important data structure is therefore referenced implicitly upon each function call (to store input arguments and the return address into it), during the call itself (to retrieve input arguments, access local variables, and store function results), and after the call (to retrieve results).

As a consequence, the performance and determinism of a certain program is deeply affected, albeit indirectly, by the exact location of its instructions, data, and stacks in the microcontroller's memory banks. This dependency is easily forgotten by just looking at the program source code because, more often than not, it simply does not contain this kind of information.

This is because most programming languages (including the C language this book focuses on) do allow the programmer to specify the abstract *storage class* of a variable (for instance, whether a variable is local or global, read-only or read-write, and so on) but they do not support any standard mechanisms that allow programmers to indicate more precisely *where* (in which memory bank) that variable will be allocated.

As will be better described in Chapters 3 and 9, this important goal must therefore be pursued in a different way, by means of other components of the software development toolchain. In this case, as shown in Figure 2.2, the *linker* plays a central role because it is the component responsible for the final allocation of all memory-resident objects defined by the program (encompassing both code and data), to fit the available memory banks.

Namely, compiler-dependent extensions of the programming language are first used to *tag* specific functions and data structures in the source code, to specify where they should be allocated in a symbolic way. Then, the linker is instructed to pick up all the objects tagged in a certain way and allocate them in a specific memory bank, rather than using the default allocation method.

An example of this strategy for a GCC-based toolchain will be given in Chapters 8 and 11, where it will be used to distribute the data structures needed by the example programs among the memory banks available on the microcontroller under study. A higher-level and more thorough description of the technique, in the broader context of software portability, will be given in Chapter 9 instead.

2.2.1 FLASH MEMORY

The two most widespread kinds of internal memory are *flash memory* and *static random access memory (SRAM)*. Flash memory is *non-volatile*, that is, it does not

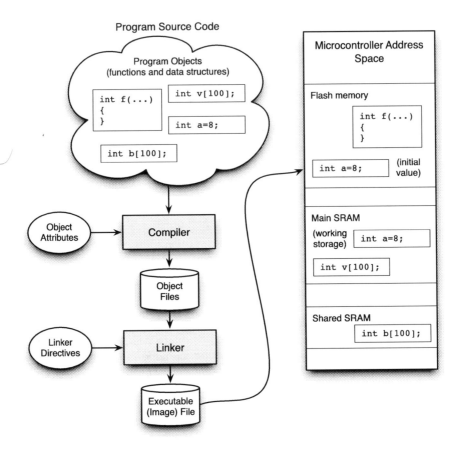

Figure 2.2   Allocation of program objects in different memory banks.

lose its contents when power is removed from the microcontroller. On the other hand, it works as a *read-only* memory during normal use.

Write operations into flash memory are indeed possible, but they require a special procedure (normal store operations performed by the processor are usually not adequate to this purpose), are relatively slow (several orders of magnitude slower than read operations) and, in some cases, they can only be performed when the microcontroller is put into a special operating mode by means of dedicated tools.

Flash memory is usually used to store the program code (that, on recent architectures, is read-only by definition), constant data, and the initial value of global variables.

In many cases, flash memory is not as fast as the processor. Therefore, it may be unable to sustain the peak transfer rate the processor may require for instruction and data access, and it may also introduce undue delays or *stalls* in processing activities if those accesses are performed *on demand*, that is, when the processor asks for them.

For this reason, units of various complexity are used to try and predict the next flash memory accesses that will be requested by the processor and execute them in advance, while the processor is busy with other activities.

For instance, both the NXP LPC24xx and LPC17xx microcontroller families [126, 128] embed a *Memory accelerator module (MAM)* that works in combination with the flash memory controller to accelerate both code and data accesses to flash memory.

In this way, if the prediction is successful, the processor is never stalled waiting for flash memory access because processor activities and flash memory accesses proceed concurrently.

On the other hand, if the prediction is unsuccessful, a processor stall will definitely occur. Due to the fact that prediction techniques inherently work on a statistical basis, it is often very hard or impossible to predict exactly *when* a stall will occur during execution, and also *for how long* it will last. In turn, this introduces a certain degree of non-determinism in program execution timings, which can hinder its real-time properties in critical applications.

In those cases, it is important that programmers are aware of the existence of the flash acceleration units just described and are able to turn them off (partially or completely) in order to change the trade-off point between *average performance* and execution *determinism*.

2.2.2 STATIC RANDOM ACCESS MEMORY (SRAM)

Static random access memory (SRAM), unlike flash memory, is *volatile*, that is, it loses its value when power is removed from the microcontroller. As a consequence, it cannot be used as a permanent storage location for code and data.

On the other hand, internal SRAM can be directly read and written by the processor and it is most often the fastest memory available in the whole system. As such, it is able to feed instructions and data to the processor in a fully deterministic way, without ever introducing any delay.

As a consequence, SRAM is commonly used to store global read-write variables and to host the processor stacks. Moreover, it can also be used to store portions of code for which execution speed and determinism are of utmost importance. However, SRAM code execution brings a couple of side effects that should be carefully evaluated and taken into account.

1. Being SRAM volatile, it cannot be used to permanently retain the code, which would otherwise be lost as soon as power is removed from the system. Hence, it is necessary to store the code elsewhere, in a non-volatile memory (usually flash memory) and copy it into SRAM at system startup, bringing additional *complexity* to system initialization.

    Furthermore, after the code has been copied, its *execution address* (that is, the address at which it is executed by the processor) will no longer be the same as its *load address* (the address of the memory area that the linker originally assigned to it).

This clearly becomes an issue if the code contains absolute memory addresses to refer to instructions within the code itself, or other forms of position-dependent code, because these references will no longer be correct after the copy and, if followed, they will lead code execution back to flash memory.

This scenario can be handled in two different ways, at either the compiler or linker level, but both require additional care and configuration instructions to those toolchain components, as better described in Chapter 3. Namely:

- It is possible to configure the compiler—for instance, by means of appropriate command-line options and often at the expense of performance—to generate *position-independent code (PIC)*, that is, code that works correctly even though it is moved at will within memory.
- Another option is to configure the linker so that it uses one base address (often called *load memory address (LMA)*) as its target to store the code, but uses a different base address (the *virtual memory address (VMA)*) to calculate and generate absolute addresses.

2. When the code is spread across different memory banks—for instance, part of it resides in flash memory while other parts are in SRAM—it becomes necessary to jump from one bank to another during program execution—for instance, when a flash-resident function calls a SRAM-resident function. However, memory banks are often mapped far away from each other within the microcontroller's address range and the relative displacement, or *offset*, between addresses belonging to different banks is large.

Modern microcontrollers often encode this offset in jump and call instructions to locate the target address and, in order to reduce code size and improve performance, implement several different instruction variants that support different (narrower or wider) offset ranges. Compilers are unaware of the target address when they generate jump and call instructions—as better explained in Chapter 3, this is the linker's responsibility instead. Hence they, by default, choose the instruction variant that represents the best trade-off between addressing capability and instruction size.

While this instruction variant is perfectly adequate for jumps and calls among functions stored in the same memory bank, the required address offset may not fit into it when functions reside in different memory banks. For the reasons recalled above, the compiler cannot detect this issue by itself. Instead, it leads to (generally rather obscure) link-time errors and it may be hard for programmers to track these errors back to their original cause.

Although virtually all compilers provide directives to force the use of an instruction variant that supports larger offsets for function calls, this feature has not been envisaged in most programming language standards, including the C language [89]. Therefore, as discussed in Chapter 9, compiler-dependent language extensions have to be used to this purpose, severely impairing code portability.

3. When SRAM or, more in general, the same bank of memory is used by the processor for more than one kind of access, for instance, to access both instructions and data, memory *contention* may occur.

In fact, processors are nowadays typically designed according to a Harvard-style architecture [108] and have got *multiple* memory and peripheral access buses, which can operate in parallel to enhance performance.

For instance, even the relatively simple and inexpensive NXP LPC17xx microcontroller family [128] incorporates an ARM Cortex-M3 processor core [9] that is equipped with three separate buses, one for instructions, another for data, and a third one for peripheral access.

However, memory banks almost invariably provide a *single* access port, which can accommodate only one read or write transaction at any given time, for the bank as a whole. As a consequence, if the processor generates more than one access request toward the same bank at the same time, albeit on separate buses, these requests will no longer be satisfied in parallel because they will be serialized at the memory access port level.

This means that the potential parallelism at the processor bus level can no longer be exploited in this case and execution may be slowed down by the extra memory access latency that comes as a consequence.

It should also be noted that further limitations to parallelism may also come as a side effect of the on-chip interconnection architecture, to be described in Section 2.2.4.

4. Last but not least, due to hardware-related considerations (for instance, chip area and power consumption) SRAM is in scarce supply on most microcontrollers, especially low-end ones. On the other hand, code size is relatively large, given the growing trend to adopt reduced instruction set computing (RISC) processor architectures rather than complex instruction set computing (CISC) ones. Therefore, it's essential to carefully scrutinize which portions of code shall indeed go into SRAM to avoid depleting it.

2.2.3   EXTERNAL MEMORY

On some microcontrollers, it is also possible to add *external* memory, that is, memory implemented by additional hardware components external to the microcontroller and connected to it by a suitable bus. The external bus is managed by means of one or more dedicated memory controllers resident on the microcontroller itself. The same technique can also be used to connect high-speed external peripherals.

For what concerns memory, three main kinds of memory can be incorporated in a system in this way:

- *Flash memory*
- *SRAM*
- *Dynamic Random Access Memory (DRAM)*

On them, external flash memory and SRAM share the same general characteristics as their on-chip counterparts. Instead, DRAM is rarely found in commerce as a kind of on-chip microcontroller memory, due to difficulties to make the two chip production processes coexist. It is worth mentioning the main DRAM properties here because they have some important side effects on program execution, especially in an embedded real-time system. Interested readers are referred to [155] for more detailed information about this topic.

DRAM, like SRAM, is a random access memory, that is, the processor can freely and directly read from and write into it, without using any special instructions or procedures. However, there are two important differences concerning access *delay* and *jitter*:

1. Due to its internal architecture, DRAM usually has a much higher capacity than SRAM but read and write operations are slower. While both on-chip and external SRAM are able to perform read and write operations at the same speed as the processor, DRAM access times are one or two orders of magnitude higher in most cases.

   To avoid intolerable performance penalties, especially as processor speed grows, DRAM access requires and relies on the interposition of another component, called *cache*, which speeds up read and write operations. A cache is basically a fast memory of limited capacity (much smaller than the total DRAM capacity), which is as fast as SRAM and holds a copy of DRAM data recently accessed by the processor.

   The caching mechanism is based on widespread properties of programs, that is, their memory access *locality*. Informally speaking, the term locality means that, if a processor just made a memory access at a certain address, its next accesses have a high *probability* to fall in the immediate vicinity of that address.

   To persuade ourselves of this fact, by intuition, let us consider the two main kinds of memory access that take place during program execution:

   - *Instruction fetch.* This kind of memory access is inherently sequential in most cases, the exception being jump and call instructions. However, they represent a relatively small fraction of program instruction and many contemporary processors provide a way to avoid them completely, at least for very short-range conditional jumps, by means of conditionally executed instructions. A thorough description of how conditionally executed instructions work is beyond the scope of this book. For instance, Reference [8] discusses in detail how they have been implemented on the ARM Cortex family of processor cores.
   - *Data load and store.* In typical embedded system applications, the most commonly used memory-resident data structure is the array, and arrays elements are quite often (albeit not always) accessed within loops by means of some sort of sequential indexing.

   Cache memory is organized in fixed-size blocks, often called cache *lines*, which are managed as an indivisible unit. Depending on the device, the block size usually is a power of 2 between 16 and 256 bytes. When the processor initiates a transaction to read or write data at a certain address, the cache controller checks whether or not a line containing those data is currently present in the cache.

   - If the required data are found in the cache a fast read or write transaction, involving only the cache itself and not memory, takes place. The fast transaction is performed at the processor's usual speed and does not introduce any extra delay. This possibility is called cache *hit*.

- Otherwise, the processor request gives origin to a cache *miss*. In this case, the cache controller performs two distinct actions:
  a. It selects an empty cache line. If the cache is completely full, this may entail storing the contents of a full cache line back into memory, an operation known as *eviction*.
  b. The cache line is filled with the data block that surrounds the address targeted by the processor in the current transaction.

In the second case, the transaction requested by the processor finishes only after the cache controller has completed both actions outlined previously. Therefore, a cache miss entails a significant performance penalty from the processor's point of view, stemming from the extra time needed to perform memory operations.

On the other hand, further memory access transactions issued by the processor in the future will likely hit the cache, due to memory access locality, with a significant reduction in data access time.

From this summary description, it is evident that cache performance heavily depends on memory access locality, which may vary from one program to another and even within the same program, depending on its current activities.

An even more important observation, from the point of view of real-time embedded systems design, is that although it is quite possible to satisfactorily assess the *average* performance of a cache from a *statistical* point of view, the exact cache behavior with respect to a specific data access is often hard to predict.

For instance, it is impossible to exclude scenarios in which a sequence of cache misses occurs during the execution of a certain section of code, thus giving rise to a worst-case execution time that is much larger than the average one.

To make the problem even more complex, cache behavior also depends in part on events external to the task under analysis, such as the allocation of cache lines—and, consequently, the eviction of other lines from the cache—due to memory accesses performed by other tasks (possibly executed by other processor cores) or interrupt handlers.

2. Unlike SRAM, DRAM is unable to retain its contents indefinitely, even though power is continuously applied to it, unless a periodic *refresh* operation is performed. Even though a detailed explanation of the (hardware-related) reasons for this and of the exact procedure to be followed to perform a refresh cycle are beyond the scope of this book, it is useful anyway to briefly recall its main consequences on real-time code execution.

Firstly, it is necessary to highlight that during a refresh cycle DRAM is unable to perform regular read and write transactions, which must be postponed, unless advanced techniques such as *hidden refresh* cycles are adopted [155]. Therefore, if the processor initiates a transaction while a refresh cycle is in progress, it will incur additional delay—beyond the one needed for the memory access itself—unless the transaction results in a cache hit.

Secondly, the exact time at which a refresh cycle starts is determined by the memory controller (depending on the requirements of the DRAM connected to it) and not by the processor.

As a consequence, even though the time needed to perform a refresh cycle is small (of the same order of magnitude as a regular memory read or write operation), from the processor's point of view refresh cycles are seen as seemingly unpredictable events that unexpectedly delay memory transactions and, more generally, program execution.

A further consideration concerning the external bus is about its *speed* and *width* limitations, especially on low-end systems. Those limitations do not depend on the devices attached to them and mainly depend on signal routing considerations, both internal and external to the microcontroller chip.

- For what concerns internal signal routing, we already mentioned that typical microcontrollers have a limited number of external pins, which is not big enough to route all internal input–output signals to the printed circuit board (PCB).

  The use of an external (parallel) bus is likely to consume a significant number of pins and make them unavailable for other purposes. A widespread workaround for this issue is to artificially limit the external bus width to save pins.

  For instance, even tough 32-bit microcontrollers support an external 32-bit bus, hardware designers may limit its width to 16 or even 8 bits.

  Besides the obvious advantage in terms of how many pins are needed to implement the external bus, a negative side-effect of this approach is that more bus cycles become necessary to transfer the same amount of data. Assuming that the bus speed is kept constant, this entails that more time is needed, too.

- To simplify external signal routing hardware designers may also keep the external bus speed slower than the maximum theoretically supported by the microcontroller, besides reducing its width. This is beneficial for a variety of reasons, of which only the two main ones are briefly presented here.

  - It makes the system more tolerant to signal propagation time skews and gives the designer more freedom to route the external bus, by means of longer traces or traces of different lengths.
  - Since a reduced number of PCB traces is required to connect the microcontroller to the external components, fewer routing conflicts arise against other parts of the layout.

A rather extreme, but representative, example of how the number of external wires needed to connect an external memory—flash memory in this case—can be reduced to a minimum is represented by the recently introduced serial peripheral interface (SPI) flash interface. This interface was first introduced in the personal computer domain and is nowadays gaining popularity in the embedded system world, too.

It can be configured as either a serial or a low-parallelism interface (from 1 to 4 data bits can be transferred in parallel) that needs only from a minimum of 3 to a maximum of 6 wires, plus ground, to be implemented. These figures are much lower than what a parallel bus, even if its width is kept at 8 bits, requires.

As it is easy to imagine, the price to be paid is that this kind of interface is likely unable to sustain the peak data transfer rate a recent processor requires, and the interposition of a cache (which brings all the side effects mentioned previously) is mandatory to bring average performance to a satisfactory level.

Regardless of the underlying reasons, those design choices are often bound to cap the performance of external memory below its maximum. From the software point of view, in order to assess program execution performance from external memory, it is therefore important to evaluate not only the theoretical characteristics of the memory components adopted in the system, but also the way they have been connected to the microcontroller.

2.2.4 ON-CHIP INTERCONNECTION ARCHITECTURE

Besides the external connections mentioned previously, contemporary microcontrollers incorporate a significant number of *internal* connections, among functional units residing on the same chip.

The overall on-chip interconnection architecture therefore plays a very significant role in determining and, sometimes, limiting microcontroller performance and determinism, as the external bus does with respect to external memory and devices.

Earlier system's internal interconnections were based on a simple, single-bus structure supporting only one data transfer operation, or transaction, at a time. To deal with the growing complexity of microcontrollers, which nowadays embed internal components of diverse speed and characteristics, and to achieve better modularity, performance, and flexibility, bus-based architectures soon evolved in an interconnection infrastructure incorporating multiple heterogeneous buses, which communicate by means of *bridges*.

For instance, as summarized in Figure 2.3, the LPC2468 microcontroller [125, 126] is equipped with the following internal buses:

- A *local* bus that connects the processor to the main SRAM bank and to the fast general-purpose input-output (GPIO) ports.
- Two *advanced high-performance bus (AHB)* buses, hosting one high-performance input-output controller each, namely Ethernet and USB, as well as the two additional SRAM banks used by those controllers for data storage. One of the AHB buses also hosts the external memory controller, through which external memory and devices can be accessed, as explained in Section 2.2.3.
- One *advanced peripheral bus (APB)* bus shared among all the other, lower-performance peripherals like, for instance, CAN controllers.

The components connected to a bus can be categorized into two different classes:

1. Bus *masters*, which are able to initiate a read or write transaction on the bus, targeting a certain slave.

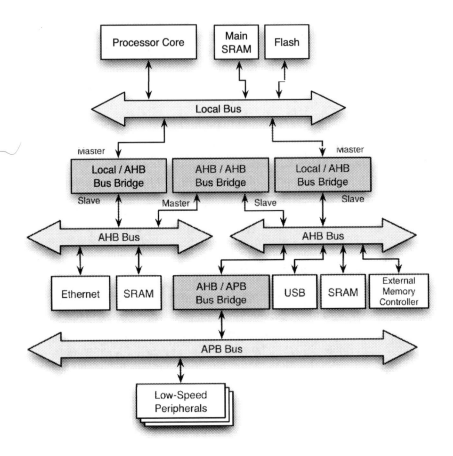

Figure 2.3 Bus- and bridge-based on-chip interconnection.

2. Bus *slaves*, which can respond to transactions initiated by masters, but cannot initiate a transaction on their own.

A typical example of bus master is, of course, the processor. However, peripheral devices may act as bus masters, too, when they are capable of autonomous direct memory access (DMA) to directly retrieve data from, and store them to, memory without processor intervention.

A different approach to DMA, which does not require devices to be bus masters, consists of relying on a general-purpose *DMA controller*, external to the devices. The DMA controller itself is a bus master and performs DMA by issuing two distinct bus transactions, one targeting the device and the other one targeting memory, on behalf of the slaves.

For instance, to transfer a data item from a device into memory, the DMA controller will first wait for a trigger from the device, issue a read transaction targeting the device (to read the data item from its registers), and then issue a write transaction targeting memory (to store the data item at the appropriate address). This approach simplifies device implementation and allows multiple devices to share the same DMA hardware, provided they will not need to use it at the same time.

In order to identify which slave is targeted by the master in a certain transaction, masters provide the target's *address* for each transaction and slaves respond to a unique range of addresses within the system's address space. The same technique is also used by bridges to recognize when they should forward a transaction from one bus to the other.

Each bus supports only one ongoing transaction at a time. Therefore all bus masters connected to the same bus must compete for bus access by means of an arbitration mechanism. The bus arbiter chooses which master can proceed and forces the others to wait when multiple masters are willing to initiate a bus transaction at the same time. On the contrary, several transactions can proceed in parallel on different buses, provided those transactions are *local* to their bus, that is, no bridges are involved.

As shown in Figure 2.3, buses are interconnected by means of a number of bridges, depicted as gray blocks.

- Two bridges connect the local bus to the AHB buses. They let the processor access the controllers connected there, as well as the additional SRAM banks.
- One additional bridge connects one of the AHB buses to the APB bus. All transactions directed to the lower-performance peripherals go through this bridge.
- The last bridge connects the two AHB buses together, to let the Ethernet controller (on the left) access the SRAM bank residing on the other bus, as well as external memory through the external memory controller.

The role of a bridge between two buses A and B is to allow a bus master M residing, for instance, on bus A to access a bus slave S connected to bus B. In order to do this, the bridge plays two different roles on the two buses at the same time:

- On bus A, it works as a bus slave and responds *on behalf of S* to the transaction initiated by M.
- On bus B, it works as a bus master, performing *on behalf of M* the transaction directed to S.

The kind of bridge described so far is the simplest one and works in an *asymmetric* way, that is, is able to forward transactions initiated on bus A (where its master port is) toward bus B (where the slave port is), but not vice versa. For instance, referring back to Figure 2.3, the AHB/AHB bridge can forward transactions initiated by a master on the left-side AHB toward a slave connected to the right-side AHB, but not the opposite.

Other, more complex bridges are *symmetric* instead and can assume both master and slave roles on both buses, albeit not at the same time. Those bridges can also be seen as a pair of asymmetric bridges, one from A to B and the other from B to A, and they are often implemented in this way.

As outlined above, bus arbitration mechanisms and bridges play a very important role to determine the overall performance and determinism of the on-chip interconnection architecture. It is therefore very important to consider them at design time and, especially for what concerns bridges, make the best use of them in software.

When the processor (or another bus master) crosses a bridge to access memory or a peripheral device controller, both the processor itself and the system as a whole may incur a *performance* and *determinism* degradation. Namely:

- The processor clearly incurs a performance degradation when accessing lower-speed peripheral buses, which may not be able to sustain the processor's peak data transfer rate. Moreover, depending on their internal construction, bridges themselves may introduce additional delay besides the time strictly required to perform the two bus transactions they are coordinating.

  The overall performance of the system may degrade, too, because a single transaction spanning one or more bridges keeps multiple buses busy at the same time and those buses are unable to support other simultaneous transactions.

  At the same time—unless the system implements more sophisticated bus access protocols supporting split transactions—a transaction going from a faster bus to a slower bus keeps the faster bus occupied for the time needed to complete the transaction on the slower bus.

- Even more importantly for a real-time system, memory or peripheral access times become less *deterministic* because, informally speaking, the more bridges the transaction has to cross to reach its target, the more likely it becomes that bus arbitration mechanisms introduce delay somewhere along the path because of bus contention.

In order to evaluate the exact amount of contention to expect, it is important to highlight once more that, although the processor can be considered the most important bus master in the system, some DMA-capable peripheral devices can act as masters, too. Fast devices, such as the Ethernet controller, may indeed generate a significant amount of bus traffic. In these cases, it is important to configure the bus arbiter to give priority to the masters that are more critical from the real-time performance point of view, if possible.

One last aspect worth mentioning is that bridges also play a crucial role in determining slave *accessibility* from a certain master. In fact, in order for a master to be able to successfully target a certain slave in a transaction it must be possible to build a complete path from the master to the slave within the interconnection system, possibly passing through one or more bridges.

If building such a path is impossible, then the bus master simply cannot access the slave. This situation is possible especially for bus masters other than the processor, and is indeed common especially on low-cost microcontrollers. As a consequence,

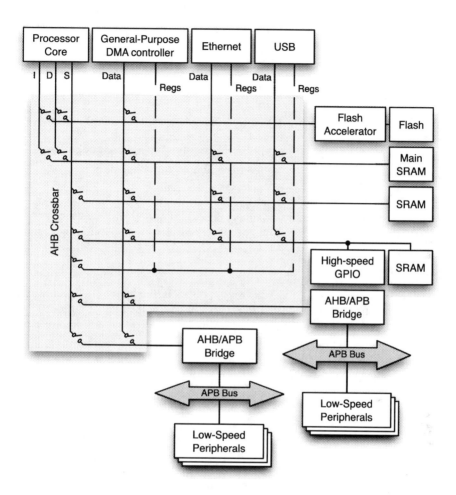

Figure 2.4 Crossbar-based on-chip interconnection.

especially when designing and implementing software drivers for DMA-capable devices, it is very important to ensure that the data structures to be used by the device are indeed accessible to it. In turn, as will be better explained in Chapters 3 and 9, this requires additional instructions to the linker in order to force those data structures to be allocated in the intended memory bank.

In any case, as the number of on-chip peripherals and their data bandwidth requirements grow, bridge-based interconnection shows its limits in terms of achievable parallelism and flexibility. For this reason, as shown in Figure 2.4, recent microcontrollers are shifting from the bus- and bridge-based interconnection discussed previously to a *crossbar-based* interconnection, even in low-cost implementations.

As shown in the figure, which depicts in a simplified way the internal interconnections of the LPC1768 microcontroller [127, 128], a crossbar resembles a matrix that is able to connect a number of masters to a number of slave components at the same time and mostly independently from each other. For this reason, it is also called *interconnection matrix* in some technical literature.

As an example, in the specific microcontroller being considered, the four bus masters are:

- The processor itself, which is able to manage as a master three outstanding transactions at a time, one on the *instruction* bus (I in the figure), another on the *data* bus (D in the figure), and a third one on the *system* bus (S in the figure), which is mainly devoted to peripheral device access.
- The Ethernet and USB controllers, each managing as masters their own data transfer bus (Data in the figure).
- The general-purpose direct memory access (DMA) controller, which provides memory data transfer services to peripherals that, on their own, are unable to become masters. The general-purpose DMA controller bus is also labeled Data in the figure.

All these buses are shown as solid vertical lines in the figure. It should also be remarked that the Ethernet and USB controllers, as well as the general-purpose DMA controller, also work as slaves on a separate bus and allow the processor to read and write their internal registers through it. These buses are labeled as Regs in the figure and shown as dashed vertical lines. They will be discussed separately in the following because in a way they represent a deviation with respect to the general structure of the crossbar.

Referring back to Figure 2.4, slaves are shown on the right and at the bottom. Besides the usual ones—namely, flash memory, high-speed GPIO, and a couple of SRAM banks—two bridges also work as slaves on the crossbar side, in order to grant access to low-speed peripherals. The bus segments they are connected to are shown as horizontal lines in the figure.

At this level, since those peripherals are not performance-critical and do not have important data bandwidth requirements, the interconnection infrastructure can still be bridge-based without adverse side-effects on performance.

Informally speaking, the crossbar itself works like a matrix of switches that can selectively connect one vertical bus segment (leading to a master) to one horizontal segment (leading to a slave). Since multiple switches can be active at the same time, multiple transactions can pass through the switch at the same time and without blocking each other, provided that they all use distinct bus segments.

As said previously, the bus segments shown as dashed vertical lines somewhat deviate from the regular structure of the crossbar and are used by components (for instance, the Ethernet controller) that can be both bus masters and slaves. This kind of behavior is needed in some cases because, continuing the example concerning the Ethernet controller,

- It must be able to autonomously read from memory and write into memory the frames to be transmitted and being received, respectively, and also

Table 2.1

Microcontrollers versus General-Purpose Processors

Feature	Microconrollers	General-purpose processors
On-chip peripherals	Yes	No (or a few)
On-chip memory	Yes	No (except caches)
External high-speed buses	Optional	Mandatory
External I/O connections	Limited	Full

- It must allow the processor to read and write its internal registers, to access status information and control its behavior, respectively.

Accordingly, these devices are connected to a vertical bus segment, on which they operate as masters, as well as a horizontal bus segment, on which they work as slaves and are targeted by other masters for register access.

Although a crossbar-based interconnection increases the degree of bus transaction parallelism that can be achieved in the system, it should also be noted that, on the other hand, it does not completely remove all bus-related timing dependencies from the system because blocking still occurs when two masters intend to access the same slave at the same time.

For instance, referring again to Figure 2.4, regardless of the presence of a crossbar-based interconnection, the processor core and the Ethernet controller will still block each other if they try to access the main SRAM bank together. Moreover— as also shown in the figure—in order to save complexity and cost, the switch matrix within the crossbar may not be complete, thus introducing additional constraints about which bus segments can, or cannot, be connected.

Another extremely important on-chip network, which has not yet been discussed so far, is responsible for conveying *interrupt* requests from most other functional units to the processor. The performance of this network is critically important for embedded systems. As for the data-transfer interconnection just discussed, several architectural variants are possible and are in use nowadays. They will be described in more detail in Chapter 8, along with the software development methods and techniques used to properly handle interrupts.

2.3 GENERAL-PURPOSE PROCESSORS VERSUS MICROCONTROLLERS

After summarizing the internal structure of recent microcontrollers in Section 2.2, the goal of this section is to highlight some high-level differences between those microcontrollers with respect to the processors used in general-purpose computing. The main ones are summarized in Table 2.1.

The first difference worth mentioning is about the number and variety of on-chip peripheral components that are included in a microcontroller with respect to a general-purpose processor.

Since their main role is to provide computing power to a more complex system, made of several distinct components, general-purpose processors usually embed only a very limited number of on-chip peripherals.

Two notable exceptions to this general trend are extremely high-speed peripherals, like peripheral component interconnect (PCI) express and direct media interface (DMI) bus controllers, with their associated DMA controller blocks, as well as integrated graphics processing units (GPUs). The implementation of all lower-speed peripherals (for instance, Ethernet and USB controllers) is left to other chips, historically known as "south bridges" and connected to the processor chip by means of a high-speed interface, like DMI.

On the contrary, microcontrollers are invariably equipped with a large variety of on-chip peripherals, which brings two contrasting consequences on embedded system design and software development. Namely:

- On the positive side, microcontroller-based systems can often be designed and built around a single chip, that is, the microcontroller itself. Besides the obvious benefits in terms of reducing the system's physical size and power consumption, this fact brings an additional advantage, too.

  In fact, microcontrollers are able to keep most high-speed interconnections between the processor and peripheral components—namely, the ones described in Section 2.2.4—*within* the microcontroller chip. Only lower-speed buses must be routed external to the chip. In turn, this greatly simplifies printed circuit board (PCB) design and debugging, while also reducing its cost and manufacturing time.

- On the negative side, designing a microcontroller-based system is often less flexible because the choice of the microcontroller—to be done once and for all near the beginning of the design—deeply affects which peripherals will be available in the system and it may be hard to change it at a later time.

  Indeed, most of the differences between members of the same microcontroller family concern on-chip peripherals whereas other characteristics, for instance, the kind of processor and its instruction set, are generally uniform across the whole family.

A very similar thought is also valid for on-chip memory. General-purpose processors usually don't have any kind of on-chip memory and rely on an external memory bus to connect the processor to a bank of DRAM. To achieve an adequate transfer speed, this bus often operates at an extremely high frequency. Hence, it becomes hard to lay out and route, due to severe constraints on the maximum length of the PCB traces between the processor and the memory chips or cards, as well as the maximum length difference, or *skew*, between them.

The only kind of internal memory usually found in a general-purpose processor is cache memory, mentioned in Section 2.2.3 and, in some cases and especially on complex processors, a small ROM used to facilitate the processor startup sequence.

On the contrary, in most cases the amount of internal memory (both flash and RAM) available in a microcontroller is big enough to accommodate the application without the need of any additional, external components. Indeed, many microcontrollers do not implement any form of external memory bus, so it is plainly impossible to add more memory in this way.

As explained previously, the lack of external memory simplifies and speeds up hardware design, but makes system memory harder or impossible to expand at a later time. As a consequence, picking the microcontroller with the correct amount of internal memory becomes more important to avoid memory shortages during software development.

In summary, as also highlighted in Table 2.1, the use of one or more external high-speed buses is unavoidable when using a general-purpose processor whereas it is optional, and unlikely, when a suitable microcontroller is adopted.

Another peculiarity of microcontrollers that is worth mentioning here is that, due to the high number of peripheral devices embedded in the chip and the need of keeping package cost under control, it is usually *impossible* to use all peripherals at the same time.

This is because, unlike general-purpose processors, the number of physical I/O pins available on a microcontroller exceeds the number of logical I/O functions made available by the on-chip peripherals. As a consequence, a single pin is shared among multiple logical functions belonging to different peripheral devices and it can, of course, only support one at a time.

The exact function that a pin will correspond to must be chosen by the programmer, usually at system initialization time, by programming a (usually rather complex) network of multiplexers and demultiplexers that connect each I/O pin to the corresponding internal I/O points. As an example, even on the relatively simple LPC1768 microcontroller [127, 128], there are about 150 multiplexers and demultiplexers, which are controlled by a group of 11 32-bit registers (from PINSEL0 to PINSEL10).

Programming the previously mentioned multiplexers in the correct way is often not a trivial task because, first of all, it is an extremely low-level activity with respect to the level of abstraction application programmers are used to work at. For this reason, the corresponding code fragments are often inherited from one project to another and it is easy to neglect to update them when the new peripherals are needed.

Secondly, it is usually possible to route a logical function of a certain peripheral to more than one physical I/O pin. In this case, the best internal (on-chip) routing choice depends not only on which other peripherals need to be used in the project, but may also depend on some aspects of the hardware design, for instance, the (off-chip) routing of the corresponding traces through the PCB. In turn, this requires a cooperation between hardware and software designers that may not be easy to achieve, especially when the design group grows.

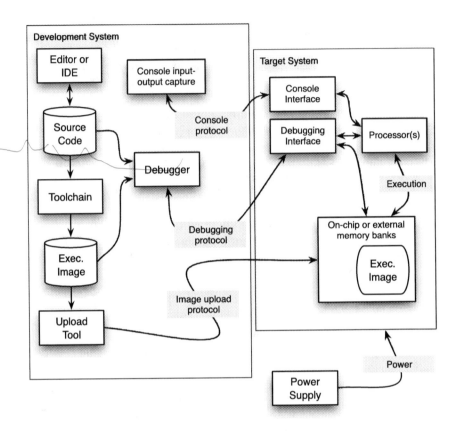

Figure 2.5   Summary of the embedded software development process.

2.4   EMBEDDED SOFTWARE DEVELOPMENT PROCESS

The process followed to develop software for an embedded system departs significantly from what is usually done for general-purpose software. Many of the differences stem from the fact that in general-purpose software development the computer used by the programmer to develop the new software is most often the same one that runs the software itself.

Instead, as shown in Figure 2.5, in the embedded software development process there is usually a sharp distinction between software development and execution. Namely:

1. A *development system*, shown on the left of the figure and often consisting of a personal computer, hosts and executes the programs used to develop the new software and compile, deploy, and debug it.
2. A *target system*, shown on the right, executes the new software under development, after it has been uploaded into its memory banks.

Within the development system, several software components still closely resemble their counterpart used for general-purpose software development. Namely:

- An *editor* helps programmers create, modify and, more in general, manage source code modules. Depending on its level of complexity and sophistication, the editor's role may span from a mere programmer's aid to "put text" into source code modules to a full-fledged integrated development environment (IDE) able to perform the following functions (among others).
  - Syntax coloring and, in some cases, other operations that require knowledge of the language syntax and part of its semantics. Among these, probably the most useful ones concern the capability of automatically retrieving the definition of, and references to, a variable or function given its name.
  Other useful features provided by the most sophisticated IDEs also include automatic code refactoring. By means of code refactoring it becomes possible, for instance, to extract part of the code of a function and create a new, standalone function from it.
  - Automatic integration with software versioning and revision control systems, for instance, concurrent version system (CVS) [28] or Apache subversion (SVN) [117]. In short, a software versioning and revision control system keeps track of all changes made by a group of programmers to a set of files, and allows them to collaborate even though they are physically separated in space and time.
  - Interaction with the toolchain, which allows programmers to rebuild their code directly from the IDE, without using a separate command shell or other interfaces. Most IDEs are also able to parse the toolchain output and, for instance, automatically open a source file and draw the programmer's attention to the line at which the compiler detected an error.
  - Ability to interact with the debugger and use its facilities directly from the IDE interface. In this way, it becomes possible to follow program execution from the source code editing window, explore variable values, and visualize the function call stack directly.
  Although none of the previously mentioned IDE features are strictly necessary for successful software development, their favorable effect on programmers' productivity increases with the size and complexity of the project. Therefore, for big and complex projects, where several programmers are involved concurrently, it may therefore be wise to adopt a suitable IDE right from the beginning, rather than retrofit it at a later time.
  A thorough discussion of editors and IDEs is beyond the scope of this book. Staying within the open-source software domain, probably the most widespread products are Emacs [156] and Eclipse [54]. Interested readers may refer to their documentation for further information.
- A *toolchain* is responsible for transforming a set of source code modules into an executable image, containing the program code (in machine lan-

guage form) and data (suitably allocated in the target system memory). The only difference with respect to general-purpose software development is that, instead of producing executable code for the development system itself, the toolchain produces code for the target system, which may use a different processor with a different instruction set.

Since, by intuition, the toolchain plays a central role in the software development process, detailed information about the toolchain will be given in Chapter 3. Moreover, Chapters 9 and 11 contain further information on how to profitably use the toolchain to attain specific goals, such as portability or code optimization from different points of view.

On the other hand, other components are either unique to the embedded software development process or they behave differently than their general-purpose counterparts. In particular:

- The *debugger* must interact with the processor on the target board in order to perform most of its functions, for instance, single-step execution or display the current value of a variable stored in the target system memory. To this purpose, it must implement an appropriate debugging protocol, also understood by the target system, and communicate with the target through a suitable debugging interface, for instance, JTAG [80].
- The toolchain stores the executable image of the program on the development system, as a disk file. To execute the program, its image must first of all be moved into the memory banks of the target system by means of an *upload* tool. Communication between the development and the target system for image upload may require the conversion of the executable image into a format understood by the target system and is controlled by a communication protocol that both sides must implement. For instance, many members of the NXP LPC microcontroller family, like the LPC1768 [127, 128], support the in-system programming (ISP) protocol to upload the executable image into the on-chip flash memory.
- The target system may be unable to autonomously manage console input–output, which is often used to print out information and debugging messages, as well as support simple interaction with the embedded software during development. This is especially true during the early stages of software development, when software to drive many of the on-board peripheral devices of the target system may not be available yet.
  For this reason, most target systems provide a minimal console input–output capability through a very simple hardware interface, often as simple as an asynchronous serial port made visible to the development system as a USB device. In this case, the development system must be able to connect to this interface and make it available to the programmer, for instance, by means of a terminal emulator.
- Last, but not least, since the target system is totally separate from the development system, it becomes necessary to supply *power* to it.

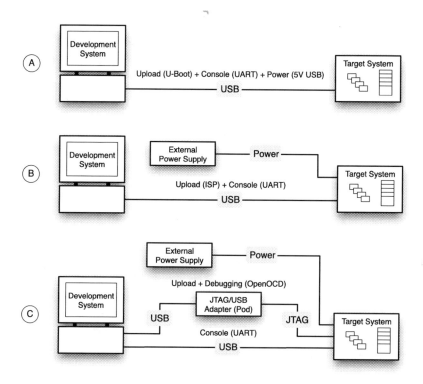

Figure 2.6 Possible ways of connecting the development system with the target board.

Accordingly, supporting all the above-mentioned features requires the establish-ment of several logical connections between the development system and the target board, namely:

1. If the target system offers debugging capabilities, the debugger must be connected to the corresponding debugging interface.
2. A suitable connection is needed to upload the executable image into the target system.
3. Console input–output data reach the development system through their own connection.
4. Finally, the target system must be connected to a power source.

As illustrated by the examples shown in the Figure 2.6, depending on the tar-get board configuration and complexity, the previously mentioned logical connec-tions may be realized by means of different combinations of physical connections. Namely:

- Relatively simple development boards support all the previously mentioned logical connections by means of a single physical connection, as shown in Figure 2.6 (a). In this case, a universal serial bus (USB) connection is often used to power the target system and make one of the microcontroller's asynchronous serial ports available to the development system through the USB itself.

In turn, the serial port is used to upload the code using a bootloader that has been preloaded into the development board by the manufacturer and supports a serial connection, for instance, U-Boot [46]. After the program under development has been started, the same serial port is also used to interact with the program, by means of a terminal emulator on the development system side.

This is the case, for instance, of the Embedded Artist EA LPC2468 OEM board [56].

- Custom-made boards may need a separate physical connection for the power supply in order to work around the power limitations of USB.

  Moreover, it may be inconvenient to fit a full-fledged bootloader like U-Boot in a production system. In this case, the USB connection is still used to interact with the program under development as before, but executable image upload is done in a different way.

  For instance, as shown in Figure 2.6 (b), many recent microcontrollers have an internal read-only memory (ROM) with basic bootloading capabilities. This is the case, for instance, of the NXP LPC1768 microcontroller [127, 128], whose ROM implements an in-system programming (ISP) protocol to upload an executable image into the on-chip flash memory.

  Of course, besides being useful by itself, the ISP protocol can also be used for the initial upload of a more sophisticated bootloader, like the previously mentioned U-Boot, into a portion of the microcontroller flash memory reserved to this purpose. Then, subsequent uploads can be done as in the previous case.

- A possible issue with the previous two ways of connecting the development system with the target board is the relatively limited support for *debugging*. It is indeed possible to support remote debugging through a serial port, for instance, by means of the remote debugging mode of the GNU debugger GDB [158]. However, some software (called *remote stub* in the GDB nomenclature) needs to be executed on the development board to this purpose.

  As a consequence, this debugging method works only as long as the target board is still "healthy enough" to execute some code and, depending on the kind of issues the software being developed encounters, this may or may not be true. It is therefore possible to lose debugging capabilities exactly when they are most needed.

  In order to address this inconvenience, some microcontrollers offer a lower-level code uploading and debugging interface, most often based on a JTAG [80] interface. The main difference is that this kind of debugging interface is hardware-based, and hence, it works in any case.

Table 2.2

Embedded versus General-Purpose Software Development Environments

Component	Development environment	
	Embedded	General-purpose
Editor or IDE	Standard	Standard
Software versioning & revision control	Standard	Standard
Compiler	Cross	Native
Debugger interface	Remote	Local
Debugging capabilities	Limited	Full-fledged
Code upload	Required	Not required
Code execution	Target board	Development system
I/O from program under development	Remote connection	Local

As shown in Figure 2.6 (c), a third connection allows the development system to interact with the target board for debugging purposes using JTAG. Since PCs usually don't support this kind of interface natively, the connection is usually made by means of a JTAG–USB converter, often called *pod*. For the software point of view, many pods support the open on-chip debugger (OpenOCD) [134]. It is an open-source project aiming at providing a free and open platform for on-chip debugging, in-system programming, and boundary-scan testing. As such, the JTAG connection also supports code uploading to the on-chip and external memory banks. In this case, the serial port is used exclusively for program-controlled console input–output.

Since, as can be perceived from the few examples discussed previously, there is no established, single standard for this kind of connection, it is very important that programmers thoroughly understand the kinds of connections their development system, toolchain, and development board support, in order to choose and use the right tools that, as explained, differ case by case. Further information about the topic can be retrieved from the reference material about the tools involved, cited in the text.

2.5 SUMMARY

Embedded systems are nowadays used in all areas of industry and permeate our daily life even though, due to their nature, people may not even appreciate their existence. As a consequence, an ever-increasing amount of application and system software has to be developed for them, often satisfying quality and dependability requirements that exceed the ones of general-purpose software.

As shown in Table 2.2, significant differences exist between embedded software development environments with respect to the ones typically used for general-purpose computing. These differences become more significant as we move away from the "surface" of the software development environment—that is, the editor or

IDE that programmers use to write their code—and get closer to the point where code is executed.

At the same time, the microcontrollers most commonly used in embedded systems are also significantly different from general-purpose processors, from the hardware point of view. At least some of these differencies have important repercussions on software development, and hence, programmers cannot be unaware of them.

Both aspects have been discussed in this chapter, with the twofold goal of setting the foundation for the more specific topics to be discussed in this book and provide readers with at least a few ideas and starting points for further learning.

3 GCC-Based Software Development Tools

CONTENTS

This chapter presents the main components of a software development toolchain exclusively based on the GNU compiler collection (GCC) and other open-source projects. All of them are currently in widespread use, especially for embedded software development, but their documentation is sometimes highly technical and not easy to understand, especially for novice users.

After introducing the main toolchain components, the chapter goes into more detail on how they work and how they are used. This information has therefore the twofold goal of providing a short reference about the toolchain, and also of giving interested readers a sound starting point to help them move to the reference toolchain documentation in a smoother way. More detailed information about the compiler, mainly focusing on software portability aspects, will be given in Chapter 9.

The very last part of the chapter also gives an overview of how individual toolchain components are configured and built for a specific target architecture. A practical example of toolchain construction will be given in Chapter 12.

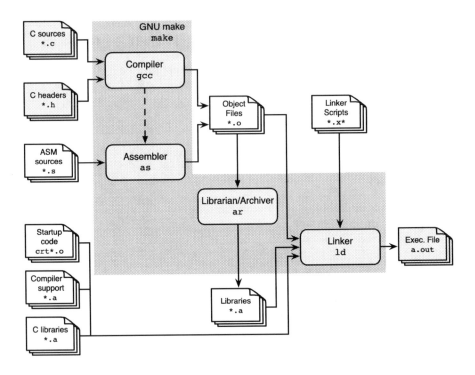

Figure 3.1   Overview of the toolchain workflow.

3.1   OVERVIEW

Generally speaking a toolchain is a complex set of software components. Collectively, they translate source code into executable machine code. Figure 3.1 outlines the general, theoretical toolchain workflow. As shown in the figure, its main components are:

1. The GCC *compiler* translates a C source file (that may include other headers and source files) and produces an object module. The translation process may require several intermediate code generation steps, involving the assembly language. In the last case, the compiler implicitly invokes the assembler as [55], too.

   The gcc program is actually a *compiler driver*. It is a programmable component, able to perform different actions by appropriately invoking other toolchain components, depending on the input file type (usually derived from its filename extension). The actions to be performed by gcc, as well as other options, are configured by means of a *specs* string or file. Both the compiler driver itself and its specs string are discussed in more detail in Section 3.2.

2. The ar *librarian* collects multiple object modules into a library. It is worth recalling that, especially in the past, the librarian was often called *archiver*, and the name of the toolchain component still derives from this term. The same tool also

performs many other operations on a library. For instance, it is able to extract or delete a module from it.

Other tools, like nm and objdump, perform more specialized operations related to object module contents and symbols. All these modules will not be discussed further in the following, due to space limitations. Interested readers may refer to their documentation [138] for more information.

3. The *linker* LD presented in Section 3.4, links object modules together and against libraries guided by one or more *linker scripts*. It resolves cross references to eventually build an *executable image*. Especially in small-scale embedded systems, the linking phase usually brings "user" and "system" code together into the executable image.

4. There are several categories of system code used at link time:

   - The *startup* object files crt*.o contain code that is executed first, when the executable image starts up. In standalone executable images, they also include system and hardware initialization code, often called *bootstrap* code.
   - The *compiler support library* libgcc.a contains utility functions needed by the code generator but too big/complex to be instantiated inline. For instance, integer multiply/divide or floating-point operations on processors without hardware support for them.
   - *Standard C libraries*, libc.a and libm.a, to be briefly discussed in Section 3.5.
   - Possibly, the *operating system* itself. This is the case of most small-scale operating systems. FREERTOS, the real-time operating system to be discussed and taken as a reference in Chapter 5, belongs to this category.

5. Last, but not least, another open-source component, *GNU make*, is responsible for coordinating the embedded software build process as a whole and automating it. In Figure 3.1 it is shown as a dark gray background that encompasses the other toolchain components and it will be the subject of Section 3.7.

An interesting twist of open-source toolchains is that the toolchain components are themselves written in a high-level language and are distributed as *source code*. For instance, the GNU compiler for the C programming language GCC is itself written in C. Therefore, a working C compiler is required in order to build GCC. There are several different ways to solve this "chicken and egg" problem, often called *bootstrap problem*, depending on the kind of compiler and build to be performed.

However, in embedded software development, the kind of toolchain most frequently used is the *cross-compilation* toolchain. As outlined in Chapter 2, cross compilation is the process of generating code for a certain architecture (the *target*) by means of a special toolchain, which runs on a different architecture (the development system, or *host*). This is because, due to limitations concerning their memory capacity and processor speed, embedded systems often lack the capability of compiling their own code.

The *bootstrap problem* is somewhat simpler to solve in this case, because it is possible to use a native toolchain on the host to build the cross-compilation toolchain.

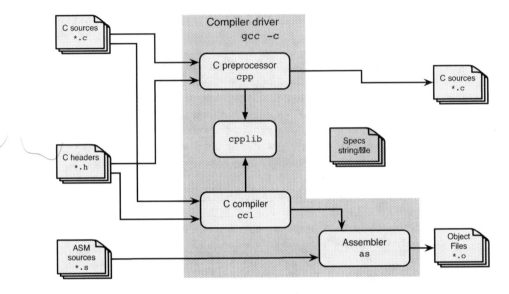

Figure 3.2   Overview of the compiler driver workflow.

The availability of a native toolchain is usually not an issue because most open-source operating system distributions already provide one ready-to-use.

3.2   COMPILER DRIVER WORKFLOW

The general outline of the compiler driver workflow is shown in Figure 3.2. According to the abstract view given in the previous section, when the compiler driver is used to compile a C source file, in theory it should perform the following steps:

1. preprocess the source code,
2. compile the source code into assembly code, and
3. assemble it to produce the output object file.

These steps are implemented by separate components, that is:

1. the C preprocessor cpp, better presented in Section 3.3,
2. the C compiler proper cc1, and
3. the assembler as.

The internal structure of the compiler deserves a chapter by itself due to its complexity. It will be more deeply discussed in Chapter 11, with particular emphasis on how it is possible to tune its workflow, in order to improve the performance and footprint of the code it generates.

In the following, we will focus instead on some peculiarities of the GCC-based toolchain that deviate from the abstract view just presented, as well as on the compiler driver itself.

First of all, as shown in the figure, the preprocessor is *integrated* in the C compiler and both are implemented by `cc1`. A *standalone preprocessor* `cpp` does exist, but it is not used during normal compilation.

In any case, the behavior of the standalone preprocessor and the one implemented in the compiler is consistent because both make use of the same preprocessing library, `cpplib`, which can also be used directly by application programs as a general-purpose macro expansion tool.

On the contrary, the assembler is implemented as a separate program, `as`, that is not part of the GCC distribution. Instead, it is distributed as part of the binary utilities package BINUTILS [138]. It will not be further discussed here due to space constraints.

One aspect peculiar to the GCC-based toolchain is that the compiler driver is programmable. Namely, it is driven by a set of *rules*, contained in a "specs" string or file. The specs string can be used to customize the behavior of the compiler driver. It ensures that the compiler driver is as flexible as possible, within its design envelope.

In the following, due to space constraints, we will just provide an overview of the expressive power that specs strings have, and illustrate what can be accomplished with their help, by means of a couple of examples. A thorough documentation of specs string syntax and usage can be found in [159].

First of all, the rules contained in a specs string specify which sequence of programs the compiler driver should run, and their arguments, depending on the kind of file provided as input. A default specs string is built in the compiler driver itself and is used when no custom specs string is provided elsewhere.

The sequence of steps to be taken in order to compile a file can be specified depending on the *suffix* of the file itself.

Other rules, associated with some command-line options may change the arguments passed by the driver to the programs it invokes.

```
*link: %{mbig-endian:-EB}
```

For example, the specs string fragment listed above specifies that if the command-line option `-mbig-endian` is given to the compiler driver, then the linker must be invoked with the `-EB` option.

Let us now consider a different specs string fragment:

```
*startfile: crti%O%s crtbegin%O%s new_crt0%O%s
```

In this case, the specs string specifies which object files should be unconditionally included at the start of the link. The list of object files is held in the `startfile` variable, mentioned in the left-hand part of the string, while the list itself is in the right-hand part, after the colon (`:`). It is often useful to modify the default set of objects in order to add language-dependent or operating system-dependent files without forcing programmers to mention them explicitly whenever they link an executable image.

More specifically:

- The `*startfile:` specification *overrides* the internal specs variable `startfile` and gives it a new value.

- `crti%O%s` and `crtbegin%O%s` are the standard initialization object files typical of C language programs. Within these strings,
  - `%O` represents the default suffix of object files. By default it is expanded to `.o` on Linux-based development hosts.
  - `%s` specifies that the object file is a system file and shall be located in the system search path rather than in user-defined directories.
- `new_crt0%O%s` replaces one of the standard initialization files of the C compiler to provide, as said previously, operating-system, language or machine-specific initialization functions.

3.3   C PREPROCESSOR WORKFLOW

The preprocessor, called `cpp` in a GNU-based toolchain, performs three main activities that, at least conceptually, take place *before* the source code is passed to the compiler proper. They are:

1. File inclusion, invoked by the `#include` directive.
2. Macro definition, by means of the `#define` directive, and expansion.
3. Conditional inclusion/exclusion of part of the input file from the compilation process, depending on whether or not some macros are defined (for instance, when the `#ifdef` directive is used) and their value (`#if` directive).

As it is defined in the language specification, the preprocessor works by *plaintext substitution*. However, Since the preprocessor and the compiler grammars are the same at the *lexical (token) level*, as an optimization, in a GCC-based toolchain the preprocessor also performs tokenization on the input files. Hence, it provides tokens instead of plaintext to the compiler.

A token [1] is a data structure that contains its *text* (a sequence of characters), some information about its *nature* (for instance whether the token represents a number, a keyword, or an identifier) and *debugging information* (file and line number it comes from).

Therefore, a token conveys additional information with respect to the portion of plaintext it corresponds to. This information is needed by the compiler anyway to further process its input, and hence, passing tokens instead of plaintext avoids duplicated processing.

Figure 3.3 contains a simplified view of the preprocessor workflow. Informally speaking, as it divides the input file into tokens, the preprocessor checks all of them and carries out one of three possible actions.

1. When the input token is a *preprocessor keyword*, like `#define` or `#include`, the preprocessor analyzes the tokens that follow it to build a complete statement and then obeys it.
   For example, after `#define`, it looks for a macro name, followed by the (optional) macro body. When the macro definition statement is complete, the preprocessor records the association *name* → *body* in a *table* for future use.

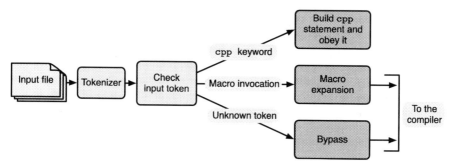

Figure 3.3   Simplified C preprocessor workflow.

It is useful to remark that, in this case, neither the preprocessor keyword, nor the following tokens (macro name and body in this example) are forwarded to the compiler. The macro *body* will become visible to the compiler only if the macro will be expanded later.

The *name* → *body* table is initialized when preprocessing starts and *discarded* when it ends. As a consequence, macro definitions are not kept across multiple compilation units. Initially, the table is not empty because the preprocessor provides a number of predefined macros.

2. When a macro is invoked, the preprocessor performs *macro expansion*. In the simplest case—that is, for *object-like* macros—macro expansion is triggered by encountering a macro name in the source code.

The macro is expanded by replacing its name with its body. Then, the result of the expansion is examined again to check whether or not further macro expansions can be done. When no further macro expansions can be done, the sequence of tokens obtained by the preprocessor as a result is forwarded to the compiler *instead of* the tokens that triggered macro expansion.

3. Tokens unknown to the preprocessor are simply passed to the compiler without modification. Since the preprocessor's and compiler's grammars are very different at the *syntax level*, many kinds of token known to the compiler have no meaning to the preprocessor, even though the latter is perfectly able to build the token itself.

For instance, it is obvious that type definitions are extremely important to the compiler, but they are completely transparent to the preprocessor.

The syntax of preprocessor keywords is fairly simple. In fact, they always start with a *sharp character* (#) in column one. Spaces are allowed between # and the rest of the keyword. The main categories of keyword are:

- Macro definition: `#define`.
- File inclusion: `#include`.
- Conditional compilation: `#ifdef`, `#ifndef`, `#if`, `#else`, `#elif`, and `#endif`.
- Other, for instance: `#warning` and `#error`.

Interested readers may refer to the full preprocessor documentation—that comes together with the GCC compiler documentation [160]—for further information about the other categories. In the following, we will mainly focus on the macro definition and expansion process.

There are two kinds of macros: *object-like* and *function-like* macros. Object-like macros are the simplest and their handling by the preprocessor can be summarized in two main points.

- The name of the macro is replaced by its body when it is encountered in the source file. The result is reexamined after expansion to check whether or not other macros are involved. If this is the case, they are expanded as well.
- Macro expansion does *not* take place when a macro is defined, but only when the macro is used. Hence, it is possible to have *forward references* to other macros within macro definitions.

For example, it is possible to define two macros, A and B, as follows.

```
#define B A+3
#define A 12
```

Namely, the definition of macro B does not produce any error although A has not been defined yet. When B is encountered in the source file, after the previously listed definitions, it is expanded as: B → A+3 → 12+3.

Since no other macro names are present in the intermediate result, macro expansion ends at this point and the three tokens 12, +, and 3 are forwarded to the compiler.

Due to the way of communication between the preprocessor and the compiler, explained previously and outlined in Figure 3.3, the compiler does not know how tokens are obtained. For instance, it cannot distinguish between tokens coming from macro expansion and tokens taken directly from the source file.

As a consequence, if B is used within a more complex expression, the compiler might get confused and interpret the expression in a counter-intuitive way. Continuing the previous example, the expression B*5 is expanded by the preprocessor as B*5 → A+3*5 → 12+3*5.

When the compiler parses the result, the evaluation of 3*5 is performed *before* +, due to the well-known precedence rules of arithmetic operators, although this may not be the behavior the programmer expects.

To solve this problem, it is often useful to put additional parentheses around macro bodies, as is shown in the following fragment of code.

```
#define B (A+3)
#define A 12
```

In this way, when B is invoked, it is expanded as: B → (A+3) → (12+3). Hence, the expression B*5 is seen by the compiler as (12+3)*5. In other words, the additional pair of parentheses coming from the expansion of macro B establishes the boundaries of macro expansion and overrides the arithmetic operator precedence rules.

The second kind of macros is represented by function-like macros. The main differences with respect to object-like macros just discussed can be summarized as follows.

- Function-like macros have a list of *parameters*, enclosed between parentheses (), after the macro name in their definition.
- Accordingly, they must be invoked by using the macro name followed by a list of *arguments*, also enclosed between ().

To illustrate how function-like macro expansion takes place, let us consider the following fragment of code as an example.

```
#define F(x, y) x*y*K
#define K 7
#define Z 3
```

When a function-like macro is invoked, its arguments are completely macro-expanded *first*. Therefore, for instance, the first step in the expansion of F(Z, 6) is: F(Z, 6) → F(3, 6).

Then, the parameters in the macro body are replaced by the corresponding, expanded arguments. Continuing our example:

- parameter x is replaced by argument 3, and
- y is replaced by 6.

After the replacement, the body of the macro becomes 3*6*K. At this point, the modified body replaces the function-like macro invocation. Therefore, F(3, 6) → 3*6*K.

The final step in function-like macro expansion consists of re-examining the result and check whether or not other macros (either object-like or function-like macros) can be expanded. In our example, the result of macro expansion obtained so far still contains the object-like macro name K and the preprocessor expands it according to its definition: 3*6*K → 3*6*7.

To summarize, the complete process of macro expansion when the function-like macro F(Z, 6) is invoked is

F(Z, 6) → F(3, 6)   (argument expansion)
→ 3*6*K   (parameter substitution in the macro body)
→ 3*6*7   (expansion of the result)

As already remarked previously about object-like macros, parentheses may be useful around parameters in function-like macro bodies, too, for the same reason.

For instance, the expansion of F(Z, 6+9) proceeds as shown below and clearly produces a counter-intuitive result, if we would like to consider F to be akin to a mathematical function.

$$F(Z, \ 6+9) \rightarrow F(3, \ 6+9) \qquad \text{(argument expansion)}$$
$$\rightarrow 3*6+9*K \qquad \text{(parameter substitution in the macro body)}$$
$$\rightarrow 3*6+9*7 \qquad \text{(expansion of the result)}$$

It is possible to work around this problem by defining `F(x, y)` as `(x)*(y)*(K)`. In this way, the final result of the expansion is:

$$F(Z, \ 6+9) \rightarrow \cdots \rightarrow (3)*(6+9)*(7)$$

as intended.

3.4 THE LINKER

The main purpose of the linker, usually called LD in the GNU toolchain and fully described in [34], is to *combine* a number of object files and libraries and place them into an *executable image*. In order to do this, the linker carries out two main activities:

1. it *resolves* inter-module symbol references, and
2. it *relocates* code and data.

The GNU linker may be invoked directly as `<cross>-ld`, where `<cross>` is the prefix denoting the target architecture when cross-compiling. However, as described in Section 3.2, it is more often called by the compiler driver `<cross>-gcc` automatically as required. In both cases, the linking process is driven by a *linker script* written in the Link Editor Command Language.

Before describing in a more detailed way how the linker works, let us briefly recall a couple of general linker options that are often useful, especially for embedded software development.

- The option `-Map=<file>` writes a *link map* to `<file>`. When the additional `--cref` option is given, the map also includes a cross reference table. Even though no further information about it will be given here, due to lack of space, the link map contains a significant amount of information about the outcome of the linking process. Interested readers may refer to the linker documentation [34] for more details about it.
- The option `--oformat=<format>` sets the format of the output file, among those recognized by `ld`. Being able to precisely control the output format helps to upload the executable image into the target platform successfully. Reference [34] contains the full list of supported output formats, depending on the target architecture and linker configuration.
- The options `--strip` and `--strip-debug` remove symbolic information from the output file, leaving only the executable code and data. This step is sometimes required for executable image upload tools to work correctly, because they might not handle any extra information present in the image properly.

Symbolic information is mainly used to keep debugging information, like the mapping between source code line numbers and machine code. For this reason, it is anyway unnecessary to load them into the target memory.

When `ld` is invoked through the compiler driver, linker options must be preceded by the escape sequence `-Wl` to distinguish them from options directed to the compiler driver itself. A comma is used to separate the escape sequence from the string to be forwarded to the linker and no intervening spaces are allowed. For instance, `gcc -Wl,-Map=f.map -o f f.c` compiles and links `f.c`, and gives the `-Map=f.map` option to the linker.

As a last introductory step, it is also important to informally recall the main differences between *object modules*, *libraries*, and *executable images* as far as the linker is concerned. These differences, outlined below, will be further explained and highlighted in the following sections.

- Object files are *always* included in the final executable image, instead the object modules found in libraries (and also called library modules) are used only *on demand*.
- More specifically, library modules are included by the linker only if they are needed to resolve pending symbol references, as will be better described in the following section.
- A library is simply a collection of unmodified object modules put together into a single file by the archiver or librarian `ar`.
- An executable image is formed by binding together object modules, either standalone or from libraries, by the linker. However, it is not simply a collection, like a library is, because the linker performs a significant amount of work in the process.

3.4.1 SYMBOL RESOLUTION AND RELOCATION

Most linker activities revolve around *symbol* manipulation. Informally speaking, a symbol is a convenient way to refer to the address of an object in memory in an *abstract* way (by means of a human-readable name instead of a number) and even before its exact *location* in memory is known.

The use of symbols is especially useful to the compiler during code generation. For example, when the compiler generates code for a backward jump at the end of a loop, two cases are possible:

1. If the processor supports *relative* jumps—that is, a jump in which the target address is calculated as the sum of the current program counter plus an offset stored in the jump instruction—the compiler may be able to generate the code completely and automatically by itself, because it knows the "distance" between the jump instruction and its target. The linker is not involved in this case.
2. If the processor only supports *absolute* jumps—that is, a jump in which the target address is directly specified in the jump instruction—the compiler must leave a

"blank" in the generated code, because it does not know where the code will eventually end up in memory. As will be better explained in the following, this blank will be filled by the linker when it performs symbol resolution and relocation.

Another intuitive example, regarding *data* instead of *code* addresses, is represented by global variables accessed by means of an extern declaration. Also in this case, the compiler needs to refer to the variable *by name* when it generates code, without knowing its memory address at all. Also in this case, the code that the compiler generates will be incomplete because it will include "blanks," in which symbols are referenced instead of actual memory addresses.

When the linker collects object files, in order to produce the executable image, it becomes possible to associate symbol *definitions* and the corresponding *references*, by means of a name-matching process known as *symbol resolution* or (according to an older nomenclature) *snapping*.

On the other hand, symbol values (to continue our examples, addresses of variables, and the exact address of machine instructions) become known when the linker *relocates* object contents in order to lay them out into memory. At this point, the linker can "fill the blanks" left by the compiler.

As an example of how symbol resolution takes place for data, let us consider the following two, extremely simple source files.

f.c

```
extern int i;

void f(void) {
    i = 7;
}
```

g.c

```
int i;
```

In this case, symbol resolution proceeds as follows.

- When the compiler generates code for f() it does not know where (and if) variable i is defined. Therefore, in f.o the address of i is left blank, to be filled by the linker.
- This is because the compiler works on exactly *one* source file at a time. When it is compiling f.c it does not consider g.c in any way, even though both files appear together on the command line.
- During symbol resolution, the linker observes that i is defined in g.o and associates the definition with the reference made in f.o.
- After the linker relocates the contents of g.o, the address of i becomes known and can eventually be used to complete the code in f.o.

It is also useful to remark that initialized data need a special treatment when the initial values must be in non-volatile memory. In this case, the linker must cooperate with the startup code (by providing memory layout information) so that the initialization can be performed correctly. Further information on this point will be given in Section 3.4.3.

3.4.2 INPUT AND OUTPUT SEQUENCES

As mentioned before, the linking process is driven by a set of commands, specified in a linker script. A linker script can be divided into three main parts, to be described in the following sections. Together, these three parts fully determine the overall behavior of the linker, because:

1. The *input and output* part picks the input files (object files and libraries) that the linker must consider and directs the linker output where desired.
2. The *memory layout* part describes the position and size of all memory banks available on the target system, that is, the space the linker can use to lay out the executable image.
3. The *section and memory mapping* part specifies how the input files contents must be mapped and relocated into memory banks.

If necessary, the linker script can be split into multiple files that are then bound together by means of the INCLUDE <filename> directive. The directive takes a file name as argument and directs the linker to include that file "as if" its contents appeared in place of the directive itself. The linker supports nested inclusion, and hence, INCLUDE directives can appear both in the main linker script and in an included script.

This is especially useful when the linker script becomes complex or it is convenient to divide it into parts for other reasons, for instance, to distinguish between architecture or language-dependent parts and general parts.

Input and output linker script commands specify:

- Which *input* files the linker will operate on, either object files or libraries. This is done by means of one or more INPUT() commands, which take the names of the files to be considered as arguments.
- The *sequence* in which they will be scanned by the linker, to perform symbol resolution and relocation. The sequence is implicitly established by the order in which input commands appear in the script.
- The special way a specific file or group of files will be handled. For instance, the STARTUP() command labels a file as being a startup file rather than a normal object file.
- Where to look for *libraries*, when just the library name is given. This is accomplished by specifying one or more search paths by means of the SEARCH_DIR() command.
- Where the *output*—namely, the file that contains the executable image—goes, through the OUTPUT() command.

Most of these commands have a command-line counterpart that, sometimes, is more commonly used. For instance, the -o command-line option acts the same as

OUTPUT() and mentioning an object file name on the linker command line has the same effect as putting it in an INPUT() linker script command.

The *entry point* of the executable image—that is, the instruction that shall be executed first—can be set by means of the ENTRY(<symbol>) command in the linker script, where <symbol> is a symbol.

However, it is important to remark that the *only* effect of ENTRY is to *keep a record* of the desired entry point and store it into the executable image itself. Then, it becomes the responsibility of the *bootloader* mentioned in Chapter 2—often simply called *loader* in linker's terminology—to obey what has been requested in the executable image.

When no loader is used—that is, the executable image is uploaded by means of an upload tool residing on the development host, and then runs on the target's "bare metal"—the entry point is defined by *hardware*. For example, most processors start execution from a location indicated by their *reset vector* upon powerup. Any entry point set in the executable image is ignored in this case.

All together, the *input* linker script commands eventually determine the linker *input sequence*. Let us now focus on a short fragment of a linker sequence that contains several input commands and describe how the input sequence is built from them.

```
INPUT(a.o, b.o, c.o)
INPUT(d.o, e.o)
INPUT(libf.a)
```

Normally, the linker scans the input files *once* and in the order established by the input sequence, which is defined by:

- The order in which files appear *within* the INPUT() command. In this case, b.o follows a.o in the input sequence and e.o follows d.o.
- If there are multiple INPUT() commands in the linker script, they are considered in the same sequence as they appear in the script.

Therefore, in our example the linker scans the files in the order: a.o, b.o, c.o, d.o, e.o, and libf.a.

As mentioned previously, object files can also be specified on the linker command line and become part of the input sequence, too. In this case:

- The command line may also include an option (-T) to refer to the linker script.
- The input files specified on the command line are combined with those mentioned in the linker script depending on *where* the linker script has been referenced.

For instance, if the command line is gcc ... a.o -Tscript b.o and the linker script script contains the command INPUT(c.o, d.o), then the input sequence is: a.o, c.o, d.o, and b.o.

As mentioned previously the *startup file* is a special object file because it contains low-level hardware initialization code. For example, it may set the CPU clock source

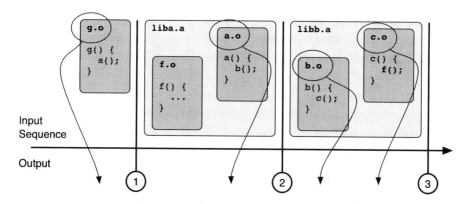

Figure 3.4   Linker's handling of object files and libraries in the input sequence.

and frequency. Moreover, it sets up the execution environment for application code. For instance, it is responsible of preparing initialized data for use. As a consequence, its position in memory may be constrained by the hardware startup procedure.

The STARTUP(<file>) command forces <file> to be the very first object file in the input sequence, regardless of where the command is. For example, the linker script fragment

```
INPUT(a.o, b.o)
STARTUP(s.o)
```

leads to the input sequence s.o, a.o, and b.o, even though s.o is mentioned last.

Let us now mention how the linker transforms the input sequence into the *output sequence* of object modules that will eventually be used to build the executable image. We will do this by means of an example, with the help of Figure 3.4. In our example, the input sequence is composed of an object file g.o followed by two libraries, liba.a and libb.a, in this order. They are listed at the top of the figure, from left to right. For clarity, libraries are depicted as lighter gray rectangles, while object files correspond to darker gray rectangles. In turn, object files contain function definitions and references, as is also shown in the figure.

The construction of the output sequence proceeds as follows.

- Object module g.o is unconditionally placed in the output sequence. In the figure, this action is represented as a downward-pointing arrow. As a consequence the symbol a, which is referenced from the body of function g(), becomes undefined at point ①.
- When the linker scans liba.a, it finds a definition for a in module a.o and resolves it by placing a.o into the output. This makes b undefined at point ②, because the body of a contains a reference to it.

- Since only a is undefined at the moment, only module a.o is put in the output. More specifically, module f.o is not, because the linker is not aware of any undefined symbols related to it.
- When the linker scans libb.a, it finds a definition of b and places module b.o in the output. In turn, c becomes undefined. Since c is defined in c.o, that is, another module within the *same library*, the linker places this object module in the output, too.
- Module c.o contains a reference to f, and hence, f becomes undefined. Since the linker scans the input sequence only once, it is unable to refer back to liba.a at this point. Even though liba.a defines f, that definition is not considered. At point ③ f is still undefined.

According to the example it is evident that the linker implicitly handles libraries as *sets*. Namely, the linker picks up object modules from a set *on demand* and places them into the output. If this action introduces additional undefined symbols, the linker looks into the set *again*, until no more references can be resolved. At this time, the linker moves to the next object file or library.

As also shown in the example, this default way of scanning the input sequence is problematic when libraries contain *circular cross references*. More specifically, we say that a certain library A contains a circular cross-reference to library B when one of A's object modules contains a reference to one of B's modules and, symmetrically, one of B's modules contains a reference back to one module of library A.

When this occurs, regardless of the order in which libraries A and B appear in the input sequence, it is always possible that the linker is unable to resolve a reference to a symbol, even though one of the libraries indeed contains a definition for it. This is what happens in the example for symbol f.

In order to solve the problem, it is possible to *group* libraries together. This is done by means of the command GROUP(), which takes a list of libraries as argument. For example, the command GROUP(liba.a, libb.a) groups together libraries liba.a and libb.a and instructs the linker to handle both of them as a *single set*.

Going back to the example, the effect of GROUP(liba.a, libb.a) is that it directs the linker to look back into the set, find the definition of f, and place module f.o in the output.

It is possible to mix GROUP() and INPUT() within the input sequence to transform just *part* of it into a set. For example, given the input sequence listed below:

```
INPUT(a.o, b.o)
GROUP(liba.a, libb.a)
INPUT(libc.a)
```

the linker will first examine a.o, and then b.o. Afterwards, it will handle liba.a and libb.a as a single set. Last, it will handle libc.a on its own.

Moreover, as will become clearer in the following, the use of GROUP() makes sense only for libraries, because object files are handled in a different way in the first place. Figure 3.5 further illustrates the differences. In particular, the input sequence

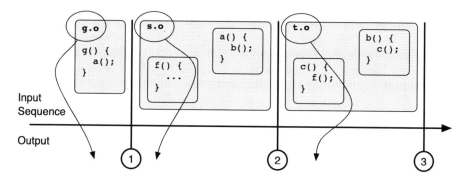

Figure 3.5   Object files versus libraries.

shown in Figure 3.5 is identical to the one previously considered in Figure 3.4, with the only exception that libraries have been replaced by object modules.

The input sequence of Figure 3.5 is processed as follows:

- Object module g.o is placed in the output and symbol a becomes undefined at point ①.
- When the linker scans s.o it finds a definition for a and places *the whole object module* in the output.
- This provides a definition of f even though it was not called for at the moment and makes b undefined at point ②.
- When the linker scans t.o it finds a definition of b and it places the whole module in the output. This also provides a definition of c.
- The reference to f made by c can be resolved successfully because the output already contains a definition of f.

As a result, there are no unresolved symbols at point ③. In other words, circular references between object files are resolved automatically because the linker places them into the output as a whole.

3.4.3   MEMORY LAYOUT

The MEMORY { ... } command is used to describe the memory layout of the target system as a set of *blocks* or *banks*. For clarity, command contents are usually written using one line of text for each block. Its general syntax is:

```
MEMORY
{
    <name> [(<attr>)] : ORIGIN = <origin>, LENGTH = <len>
    ...
}
```

where:

- ORIGIN and LENGTH are keywords of the linker script.
- <name> is the name given to the block, so that the other parts of the linker script can refer to that memory block by name.
- <attr> is optional. It gives information about the type of memory block and affects which kind of information the linker is allowed to allocate into it. For instance, R means read-only, W means read/write, and X means that the block may contain executable code.
- <origin> is the starting address of the memory block.
- <len> is the length, in *bytes*, of the memory block.

For example, the following MEMORY command describes the Flash memory bank and the main RAM bank of the LPC1768, as defined in its user manual [128].

```
MEMORY
{
    rom (rx)  : ORIGIN = 0x00000000, LENGTH = 512K
    ram (rwx) : ORIGIN = 0x10000000, LENGTH =  32K
}
```

To further examine a rather common issue that may come even from the seemingly simple topic of memory layout, let us now consider a simple definition of an initialized, global variable in the C language, and draw some comments on it. For example,

$$int\ a\ =\ 3;$$

- After the linker allocates variable a, it resides somewhere in RAM memory, for instance, at address 0x1000. RAM memory is needed because it must be possible to modify the value of a during program execution.
- Since RAM memory contents are not preserved when the system is powered off, its initial value 3 must be stored in nonvolatile memory instead, usually within a Flash memory bank. Hence, the initial value is stored, for instance, at address 0x0020.
- In order to initialize a, the value 3 must be copied from Flash to RAM memory. In a standalone system, the copy must be performed by either the bootloader or the startup code, but in any case *before* the main C program starts. This is obviously necessary because the code generated by the compiler *assumes* that initialized variables indeed contain their initial value. The whole process is summarized in Figure 3.6.
- To setup the initialized global variable correctly, the linker associates *two* memory addresses to initialized global data. Address 0x0020 is the *Load Memory Address* (LMA) of a, because this is the memory address where its *initial contents* are stored.
- The second address, 0x1000 in our example, is the *virtual memory address* (VMA) of a because this is the address used by the processor to refer to a at *runtime*.

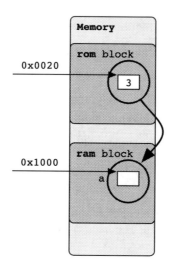

Figure 3.6 Initialized variables setup, VMA and LMA.

Often, the VMA and LMA of an object are *the same*. For example, the address where a function is stored in memory is the same address used by the CPU to call it. When they are not, a copy is necessary, as illustrated previously.

This kind of copy can *sometimes* be avoided by using the `const` keyword of the C language, so that read-only data are allocated only in ROM. However, this is not strictly guaranteed by the language specification, because `const` only determines the data *property* at the language level, but does not necessarily affect their *allocation* at the memory layout level.

In other words, data properties express how they can be manipulated in the program, which is not directly related to where they are in memory. As a consequence, the relationship between these two concepts may or may not be kept by the toolchain during object code generation.

From the practical point of view, it is important to remark that the linker follows the *same order* when it allocates memory for initialized variables in RAM and when it stores their initial value in ROM. Moreover, the linker does not interleave any additional memory object in either case. As a consequence, the layout of the ROM area that stores initial values and of the corresponding RAM area is the same. Only their starting addresses are different.

In turn, this implies that the *relative position* of variables and their corresponding initialization values within their areas is the same. Hence, instead of copying variable by variable, the startup code just copies the whole area in one single sweep.

The base addresses and size of the RAM and ROM areas used for initialized variables are provided to the startup code, by means of symbols defined in the linker script as described in Section 3.4.4.

As a final note for this section, it is worth remarking that there is an unfortunate clash of terminology between virtual memory addresses as they are defined in the linker's nomenclature and virtual memory addresses in the context of virtual memory systems, outlined in Chapter 15.

3.4.4   LINKER SCRIPT SYMBOLS

As described previously, the concept of *symbol* plays a central role in linker's operations. Symbols are mainly *defined* and *referenced* by object files, but they can also be defined and referenced in a *linker script*. There is just one "category" of symbols. Namely:

- A symbol defined in an object module can be referenced by the linker script. In this way, the object module can modify the inner workings of the linker script itself and affect section mapping and memory layout, just by defining symbols appropriately. For example, it is possible to set the *stack size* of the executable image from one of the object modules.
- Symmetrically, a symbol defined in the linker script can be referenced by an object module, and hence, the linker script can determine some aspects of the object module's behavior. For example, as mentioned in Section 3.4.3, the linker script can communicate to the startup code the base addresses and size of the RAM and ROM areas used for initialized variables.

An *assignment*, denoted by means of the usual = (equal sign) operator, gives a value to a symbol. The value is calculated as the result of an expression written on the right-hand side of the assignment. The expression may contain most C-language arithmetic and Boolean operators. It may involve both constants and symbols.

The result of an expression may be *absolute* or *relative* to the beginning of an output section, depending on the expression itself (mainly, the use of the ABSOLUTE() function) and also *where* the expression is in the linker script.

The special (and widely used) symbol . (dot) is the *location counter*. It represents the absolute or relative output location (depending on the context) that the linker is about to fill while it is scanning the linker script—and its input sequence in particular—to lay out objects into memory. With some exceptions, the location counter may generally appear whenever a normal symbol is allowed. For example, it appears on the right-hand side of an assignment in the following example.

```
__stack = .
```

This assignment sets the symbol __stack to the value of the location counter. Assigning a value to . moves the location counter. For example, the following assignment:

```
. += 0x4000
```

allocates 0x4000 bytes starting from where the location counter currently points and moves the location counter itself after the reserved area.

An assignment may appear in three different positions in a linker script and its position partly affects how the linker interprets it.

1. *By itself.* In this case, the assigned value is absolute and, contrary to the general rule outlined previously, the location counter . cannot be used.
2. As a statement *within a* SECTIONS *command.* The assigned value is *absolute* but, unlike in the previous case, the use of . is allowed. It represents an *absolute* location counter.
3. Within an *output section description*, nested in a SECTIONS command. The assigned value is *relative* and . represents the *relative* value of the location counter with respect to the beginning of the output section.

As an example, let us consider the following linker script fragment. More thorough and formal information about output sections is given in Section 3.4.5.

```
SECTIONS
{
    .   = ALIGN(0x4000);
    . += 0x4000;
    __stack = .;
}
```

In this example:

- The first assignment aligns the location counter to a multiple of 16Kbyte (0x4000).
- The second assignment moves the location counter forward by 16Kbyte. That is, it allocates 16Kbyte of memory for the stack.
- The third assignment sets the symbol __stack to the top of the stack. The startup code will refer to this symbol to set the initial stack pointer.

3.4.5 SECTION AND MEMORY MAPPING

The contents of each input object file are divided by the *compiler* (or assembler) into several categories according to their characteristics, like:

- code (.text),
- initialized data (.data),
- uninitialized data (.bss).

Each category corresponds to its own *input section* of the object file, whose name has also been listed above. For example, the object code generated by the C compiler is placed in the .text section of the input object files. Libraries follow the same rules because they are just collections of object files.

The part of linker script devoted to *section mapping* tells the linker how to fill the memory image with *output sections*, which are generated by collecting *input sections*. It has the following syntax:

```
SECTIONS
{
    <sub-command>
    . . .
}
```

where:

- The `SECTIONS` command encloses a sequence of *sub-commands*, delimited by braces.
- A sub-command may be:
    - an `ENTRY` command, used to set the initial entry point of the executable image as described in Section 3.4.2,
    - a symbol assignment,
    - an overlay specification (seldom used in modern programs),
    - a *section mapping* command.

A section mapping command has a relatively complex syntax, illustrated in the following.

```
<section> [<address>] [(<type>)] :
  [<attribute> ...]
  [<constraint>]
  {
    <output-section-command>
    ...
  }
  [> <region>] [AT> <lma_region>]
  [: <phdr> ...] [= <fillexp>]
```

Most components of a section mapping command, namely, the ones shown within brackets (`[]`), are optional. Within a section mapping command, an *output section command* may be:

- a *symbol assignment*, outlined in Section 3.4.4,
- *data values* to be included directly in the output section, mainly used for padding,
- a special *output section keyword*, which will not be further discussed in this book,
- an *input section* description, which identifies which input sections will become part of the output section.

An input section description must be written according to the syntax indicated below.

```
<filename> ( <section_name> ... )
```

It indicates which input sections must be mapped into the output section. It consists of:

- A `<filename>` specification that identifies one or more object files in the input sequence. Some wildcards are allowed, the most common one is `*`, which matches *all* files in the input sequence. It is also possible to *exclude* some input files, by means of `EXCLUDE_FILE(...)`, where `...` is the list of files to be excluded. This is useful in combination with wildcards, to refine the result produced by the wildcards themselves.

- One or more `<section_name>` specifications that identify which input sections, within the files indicated by `<filename>`, we want to refer to.

The order in which input section descriptions appear is important because it sets the order in which input sections are placed in the output sections.

For example, the following input section description:

```
*  ( .text .rodata )
```

places the `.text` and `.rodata` sections of all files in the input sequence in the output section. The sections appear in the output in the same order as they appear in the input.

Instead, these slightly different descriptions:

```
*  ( .text )
*  ( .rodata )
```

first places all the `.text` sections, and then all the `.rodata` sections.

Let us now examine the other main components of the section mapping command one by one. The very first part of a section mapping command specifies the output section *name*, *address*, and *type*. In particular:

- `<section>` is the name of the output section and is mandatory.
- `<address>`, if specified, sets the *VMA* of the output section. When it is not specified, the linker sets it *automatically*, based on the output memory block `<region>`, if specified, or the current location counter. Moreover, it takes into account the strictest alignment constraint required by the input sections that are placed in the output sections and the output sections alignment itself, which will be specified with an `[<attribute>]` and will be explained later.
- The most commonly used special output section `<type>` is NOLOAD. It indicates that the section shall not be loaded into memory when the program is run. When omitted, the linker creates a normal output section specified with the section name, for instance, `.text`.

Immediately thereafter, it is possible to specify a set of output section *attributes*, according to the following syntax:

```
[AT( <lma> )]
[ALIGN( <section_align> )]
[SUBALIGN( <subsection_align> )]
[<constraint>]
```

- The AT attribute sets the *LMA* of the output section to address `<lma>`.
- The ALIGN attribute specifies the alignment of the output section.
- The SUBALIGN attribute specifies the alignment of the input sections placed in the output section. It overrides the "natural" alignment specified in the input sections themselves.

- `<constraint>` is normally empty. It may specify under which constraints the output sections must be created. For example, it is possible to specify that the output section must be created only if all input sections are read-only [34].

The *memory block mapping* specification is the very last part of a section mapping command and comes after the list of output section commands. It specifies in which memory block (also called *region*) the output section must be placed. Its syntax is:

```
[> <region>] [AT> <lma_region>]
[: <phdr> ...] [= <fillexp>]
```

where:

- `> <region>` specifies the memory block for the output section *VMA*, that is, where it will be referred to by the processor.
- `AT> <lma_region>` specifies the memory block for the output section *LMA*, that is, where its contents are loaded.
- `<phdr>` and `<fillexp>` are used to assign the output section to an *output segment* and to set the *fill pattern* to be used in the output section, respectively.

Segments are a concept introduced by some executable image formats, for example the executable and linkable format (ELF) [33] format. In a nutshell, they can be seen as groups of sections that are considered as a single unit and handled all together by the loader.

The fill pattern is used to fill the parts of the output section whose contents are not explicitly specified by the linker script. This happens, for instance, when the location counter is moved or the linker introduces a gap in the section to satisfy an alignment constraint.

3.5   THE C RUNTIME LIBRARY

Besides defining the language itself, the international standards concerning the C programming language [87, 88, 89, 90] also specify a number of *library functions*. They range from very simple ones, like memcpy (that copies the contents of one memory area of known size into another) to very complex ones, like printf (that converts and print out its arguments according to a rich format specification).

The implementation of these functions is not part of the compiler itself and is carried out by another toolchain component, known as *C runtime library*. Library functions are themselves written in C and distributed as source code, which is then compiled into a library by the compiler during toolchain generation. Another example is the *C math library* that implements all floating-point functions specified by the standards except basic arithmetic.

Among the several C runtime libraries available for use for embedded software development, we will focus on NEWLIB [145], an open-source component widely used for this purpose.

The most important aspect that should be taken into account when porting this library to a new architecture or hardware platform, is probably to configure and provide an adequate *multitasking support* to it. Clearly, this is of utmost importance to ensure that the library operates correctly when it is used in a multitasking environment, like the one provided by any real-time operating system.

In principle, NEWLIB is a *reentrant* C library. That is, it supports multiple tasks making use of and calling it *concurrently*. However, in order to do this, two main constraints must be satisfied:

1. The library must be able to maintain some *per-task information*. This information is used, for instance, to hold the per-task `errno` variable, I/O buffers, and so on.
2. When multiple tasks share information through the library, it is necessary to implement critical regions within the library itself, by means of appropriate synchronization points. This feature is used, for instance, for dynamic memory allocation (like the `malloc` and `free` functions) and I/O functions (like `open`, `read`, `write`, and others).

It is well known that, in most cases, when a C library function fails, it just returns a Boolean error indication to the caller or, in other cases, a NULL pointer. Additional information about the reason for the failure is stored in the `errno` variable.

In a *single-task* environment, there may only be up to one pending library call at any given time. In this case, `errno` can be implemented as an ordinary global variable. On the contrary, in a *multitasking* environment, multiple tasks may make a library call concurrently. If `errno` were still implemented as a global variable, it would be impossible to know which task the error information is for. In addition, an error caused by a task would overwrite the error information used by all the others.

The issue cannot be easily solved because, as will be better explained in Chapters 4 and 5, in a multitasking environment the task execution order is nondeterministic and may change from time to time.

As a consequence, the `errno` variable must hold per-task information and must not be shared among tasks. In other words, one instance of the `errno` variable must exist for each thread, but it must still be "globally accessible" in the same way as a global variable. The last aspect is important mainly for backward compatibility. In fact, many code modules were written with single-thread execution in mind and preserving backward compatibility with them is important.

By "globally accessible," we mean that a given thread must be allowed to refer to `errno` from *anywhere* in its code, and the expected result must be provided.

The method adopted by NEWLIB to address this problem, on single-core systems, is depicted in Figure 3.7 and is based on the concept of *impure pointer*.

Namely, a *global* pointer (`impure_ptr`) allows the library to refer to the *per-task data structure* (`struct _reent`) of the running task. In order to do this, the library relies on appropriate support functions (to be provided in the *operating system support layer*), which:

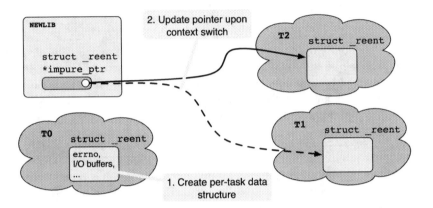

Figure 3.7   Handling of the NEWLIB impure pointer on single-core systems.

1. Create and destroy per-task data structures at an appropriate time, that is, task creation and deletion. This is possible because—even though the internal structure of the struct _reent is opaque to the operating system—its size is known.
2. Update impure_ptr upon every context switch, so that it always points to the per-task data structure corresponding to the running task.

In addition, some library modules need to maintain data structures that are shared among all tasks. For instance, the dynamic memory allocator maintains a single memory pool that is used by the library itself (on behalf of tasks) and is also made available to tasks for direct use by means of malloc(), free(), and other functions.

The formal details on how to manage concurrent access to shared resources, data structures in this case, will be given in Chapter 5. For the time being, we can consider that, by intuition, it is appropriate to let only one task at a time use those data structures, to avoid corrupting them. This is accomplished by means of appropriate *synchronization points* that force tasks to wait until the data structures can be accessed safely.

As shown in the left part of Figure 3.8, in this case, the synchronization point is *internal* to the library. The library implements synchronization by calling an operating system support function at critical region boundaries. For instance, the dynamic memory allocation calls __malloc_lock() before working on the shared memory pool, and __malloc_unlock() after it is done.

Therefore, a correct synchronization is a shared responsibility between the library and the operating system, because:

- the library "knows" where critical regions are and what they are for, and
- the operating system "knows" how to synchronize tasks correctly.

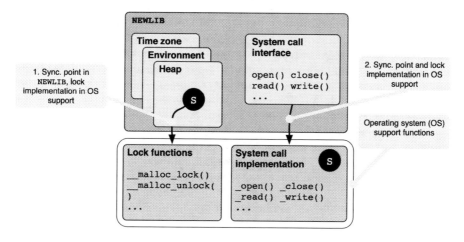

Figure 3.8   NEWLIB synchronization points.

The operating system support functions, to be implemented when the library is ported to a new architecture, have the crucial role of *gluing* these two parts together.

As shown in the right side of Figure 3.8, synchronization is also needed when tasks share other kinds of resource, like *files* and *devices*. For those resources, which historically were not managed by the library itself, resource management (and therefore synchronization) is completely *delegated* to other components, like the operating system itself or another specialized library that implements a filesystem.

Accordingly, for functions like `read()` and `write()`, the library only encapsulates the implementation provided by other components without adding any additional feature. Since these functions are often implemented as system calls, NEWLIB refers to them by this name in the documentation, even though this may or may not be true in an embedded system.

For instance, in Chapter 12 we will show an example of how to implement a USB-based filesystem by means of an open-source filesystem library and other open-source components. In that case, as expected and outlined here, the filesystem library requires support for synchronization to work correctly when used in a multitasking context.

3.6   CONFIGURING AND BUILDING OPEN-SOURCE SOFTWARE

It is now time to consider how individual toolchain components are built from their source code. In order to obtain a working toolchain component, three main phases are needed:

1. The *configuration* phase sets up all the parameters that will be used in the build phase and automatically generates a `Makefile`. As will be explained in Section 3.7, this file contains key information to automate the build process.

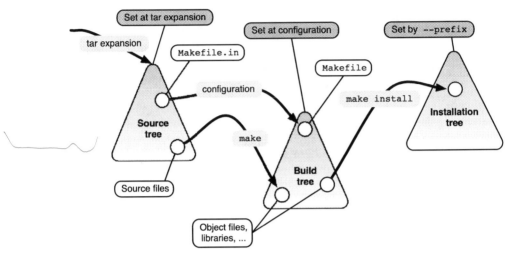

Figure 3.9   Summary of the toolchain components installation process.

Examples of parameters provided during configuration are the architecture that the toolchain component must target, where the component shall be installed on the development host, which extra libraries shall be used, and so on.

2. During the *build* phase, orchestrated by GNU make, intermediate files and byproducts as well as the final results are generated from source files.

3. After a successful build, the results are moved to their final location during the *installation* phase, started manually. In this way, it is possible to rebuild the toolchain multiple times without overwriting installed versions of the same component.

As shown in Figure 3.9, it is crucial to distinguish among three different kinds of directories that will be used during the build of each toolchain component:

- The *source* directory tree is the place where the source code of the component is stored, when its distribution archive is unpacked.
- The *build* directory tree is used in the build phase. The build phase is started by running GNU make. As a result, the corresponding `make` command must be launched from there. In turn, `make` uses the `Makefile` generated during the configuration phase. In this case, the configuration must also be run in the same place.
- The *installation* directory tree is used to install the toolchain component. It can be specified during the configuration phase, by means of the `--prefix` argument. Installation is started explicitly by means of the `make install` command.

The first preliminary step required to build a toolchain component is to create the source tree in the current directory, starting from the source package of the component, as shown in the following.

```
tar xf <source_package>
```

Then, we can create the build directory, change into it and configure the component, specifying where the installation tree is, as follows.

```
mkdir <build_dir>
cd <build_dir>
<path_to_source_tree>/configure \
    --prefix=<installation_tree> \
    ...<other_options>...
```

Finally, we can build and (if the build was successful) install the component.

```
make
make install
```

Each toolchain component must know its own installation prefix, that is, the root of the directory tree into which they will be installed. As shown previously, this information must be provided to the component at configuration time.

Although it is possible, in principle, to install each toolchain component in a different location, it is recommended to use a single installation prefix for all of them. A convenient way to do it is to set the environment variable PREFIX to a suitable path, and use it when configuring the toolchain components.

Moreover, most toolchain components need to locate and run executable modules belonging to other components, not only when they are used normally, but also while they are being built. For example, the build process of GCC also builds the C runtime library. This requires the use of the assembler, which is part of BINUTILS.

As toolchain components are built, their executable files are stored into the bin subdirectory under the installation prefix. The value of environment variable PATH must be extended to include this new path. This is because, during the build process, executable files are located (as usual) by consulting the PATH environment variable.

In the following, we will give more detailed information on how to build a working toolchain component by component. It is assumed that both the source code packages have already been downloaded into the current working directory. Moreover, if any toolchain component requires architecture or operating system-dependent patches, it is assumed that they are applied according to the instructions provided with them.

The first component to be built is BINUTILS, which contains the GNU binary utilities, by means of the following sequence of commands.

```
tar xjf binutils-<version>.tar.bz2

mkdir binutils-<version>-b

cd binutils-<version>-b

../binutils-<version>/configure --target=<target> \
    --prefix=$PREFIX --enable-multilib --disable-nls

make
```

In the previous listing:

- The most important configuration parameter is the intended target architecture, indicated by `<target>`.
- Similarly, `<version>` indicates the version of the BINUTILS package we intend to build.
- The `--prefix` argument specifies where BINUTILS will be installed.
- `--enable-multilib` activates support for multiple architecture and variant-dependent libraries.
- Finally, `--disable-nls` disables the national language support to make the build process somewhat faster.

The `--disable-werror` option allows the build process to continue after a warning. It may be useful when the host toolchain is much more recent than the toolchain being built, and hence, it produces unforeseen warnings. However, it should be used with care because it implicitly applies to the whole build. Therefore, it may mask real issues with the build itself.

The next two components to be configured and built are GCC and NEWLIB, by means of the following sequence of commands.

```
tar xjf gcc-<version>.tar.bz2
tar xzf newlib-<version>.tar.gz

cd gcc-<version>
tar xzf ../gmp-<version>.tar.gz
mv gmp-<version> gmp
tar xjf ../mpfr-<version>.tar.bz2
mv mpfr-<version> mpfr
ln -s ../newlib-<version>/newlib .
cd ..

mkdir gcc-<version>-b
cd gcc-<version>-b
../gcc-<version>/configure \
    --target=<target> --prefix=$PREFIX \
    --with-gnu-as --with-gnu-ld \
    --with-newlib \
    --enable-languages="c,c++" \
    --enable-multilib \
    --disable-shared --disable-nls

make
make install
```

In this sequence of commands:

- Like before, the GCC and NEWLIB source packages are expanded into two separate directories.

- In order to build GCC, two extra libraries are required: `gmp` (for multiple-precision integer arithmetic) and `mpfr` (for multiple-precision floating-point arithmetic). They are used at *compile-time*—that is, the executable image is *not* linked against them—to perform all arithmetic operations involving *constant* values present in the source code. Those operations must be carried out in extended precision to avoid round-off errors. The required version of these libraries is specified in the GCC installation documentation, in the `INSTALL/index.html` file within the GCC source package. The documentation also contains information about where the libraries can be downloaded from.

- The GCC configuration script is also able to configure `gmp` and `mpfr` automatically. Hence, the resulting `Makefile` builds all three components together and statically links the compiler with the two libraries. This is done if their source distributions are found in two subdirectories of the GCC source tree called `gmp` and `mpfr`, respectively, as shown in our example.

- Similarly, both GCC and NEWLIB can be built in a single step, too. To do this, the source tree of NEWLIB must be linked to the GCC source tree. Then, configuration and build proceed in the usual way. The `--with-newlib` option informs the GCC configuration script about the presence of NEWLIB.

- The GCC package supports various programming languages. The `--enable-languages` option can be used to restrict the build to a subset of them, only C and C++ in our example.

- The `--disable-shared` option disables the generation of *shared libraries*. Shared libraries are nowadays very popular in general-purpose operating systems, as they allow multiple applications to share the same memory-resident copy of a library and save memory. However, they are not yet commonly used in embedded systems, especially small ones, due to the significant amount of runtime support they require to dynamically link applications against shared libraries when they are launched.

3.7 BUILD PROCESS MANAGEMENT: GNU MAKE

The GNU make tool [65], fully described in [157], manages the *build process* of a software component, that is, the execution of the correct sequence of commands to transform its source code modules into a library or an executable program as efficiently as possible.

Since, especially for large components, it rapidly becomes unfeasible to rebuild the whole component every time, GNU make implements an extensive inference system that allows it to:

1. decide which parts of a component shall be rebuilt after some source modules have been updated, based on their *dependencies*, and then
2. automatically execute the appropriate sequence of *commands* to carry out the rebuild.

Both dependencies and command sequences are specified by means of a set of *rules*, according to the syntax that will be better described in the following. These rules can be defined either *explicitly* in a GNU make input file (often called `Makefile` by convention), and/or *implicitly*. In fact, `make` contains a rather extensive set of predefined, built-in rules, which are implicitly applied unless overridden in a `Makefile`.

GNU make retrieves the explicit, user-defined rules from different sources. The main ones are:

- One of the files `GNUmakefile`, `makefile`, or `Makefile` if they are present in the current directory. The first file found takes precedence on the others, which are silently neglected.
- The file specified by means of the `-f` or `--file` options on the command line.

It is possible to include a `Makefile` into another by means of the `include` directive. In order to locate the file to be included, GNU make looks in the current directory and any other directories mentioned on the command line, using the `-I` option. As will be better explained in the following, the `include` directive accepts any file name as argument, and even names computed on the fly by GNU make itself. Hence, it allows programmers to use any arbitrary file as (part of) a `Makefile`.

Besides options, the command line may also contain additional *arguments*, which specify the *targets* that GNU make must try to update. If no targets are given on the command line, GNU make pursues the *first* target explicitly mentioned in the `Makefile`.

It is important to remark that the word *target*, used in the context of GNU make, has a different meaning than the same word in the context of open-source software configuration and build, described in Section 3.6. There, target is the target architecture for which a software component must be built, whereas here it represent a goal that GNU make is asked to pursue.

3.7.1   EXPLICIT RULES

The general format of an *explicit* rule in a `Makefile` is:

```
<target> ... : <prerequisites> ...
        <command line>
        ...
```

In an explicit rule:

- The `target` is usually a file that will be (re)generated when the rule is applied.
- The `prerequisites` are the files on which `target` depends and that, when modified, trigger the regeneration of the target.
- The sequence of `command lines` are the actions that GNU make must perform in order to regenerate the target, in shell syntax.

Table 3.1

GNU make Command Line Execution Options

Option	Description
@	Suppress the automatic *echo* of the command line that GNU make normally performs immediately before execution.
–	When this option is present, GNU make ignores any *error* that occurs during the execution of the command line and continues.

- Every command line must be preceded by a *tab* and is executed in *its own* shell.

It is extremely important to pay attention to the last aspect of command line execution, which is often neglected, because it may have very important consequences on the effects commands have.

For instance, the following rule does *not* list the contents of directory somewhere.

```
all:
        cd somewhere
        ls
```

This is because, even though the cd command indeed changes current directory to somewhere within the shell it is executed by, the ls command execution takes place in a new shell, and the previous notion of current directory is lost when the new shell is created.

As mentioned previously, the prerequisites list specifies the *dependencies* of the target. GNU make looks at the prerequisites list to deduce *whether or not* a target must be regenerated by applying the rule. More specifically, GNU make applies the rule when one or more prerequisites are *more recent* than the target. For example, the rule:

```
kbd.o : kbd.c defs.h command.h
        cc -c kbd.c
```

specifies that the object file kbd.o (target) must be regenerated when at least one file among kbd.c defs.h command.h (prerequisites) has been modified. In order to regenerate kbd.o, GNU make invokes cc -c kbd.c (command line) within a shell.

The shell, that is, the command line interpreter used for command line execution is by default /bin/sh on unix-like systems, unless the Makefile specifies otherwise by setting the SHELL variable. Namely, it does not depend on the user login shell to make it easier to port the Makefile from one user environment to another.

Unless otherwise specified, by means of one of the command line execution options listed in Table 3.1, commands are *echoed* before execution. Moreover, when an *error* occurs in a command line, GNU make abandons the execution of the current rule and (depending on other command-line options) may stop completely. Command line execution options must appear at the very beginning of the command line, before the text of the command to be executed. In order to do the same things in a systematic way GNU make supports options like `--silent` and `--ignore`, which apply to all command lines or, in other words, change the default behavior of GNU make.

3.7.2 VARIABLES

A variable is a name defined in a `Makefile`, which represents a text string. The string is the *value* of the variable. The value of a certain variable VAR—by convention, GNU make variables are often written in all capitals—is usually retrieved and used (that is, *expanded*) by means of the construct `$(VAR)` or `${VAR}`.

In a `Makefile`, variables are expanded "on the fly," while the file is being read, except when they appear within a command line or on the right-hand part of variable assignments made by means of the assignment operator "=". The last aspect of variable expansion is important and we will further elaborate on it in the following, because the behavior of GNU make departs significantly from what is done by most other language processors, for instance, the C compiler.

In order to introduce a dollar character somewhere in a `Makefile` without calling for variable expansion, it is possible to use the escape sequence `$$`, which represents *one* dollar character, `$`.

Another difference with respect to other programming languages is that the `$()` operators can be nested. For instance, it is legal, and often useful, to state `$($(VAR))`. In this way, the *value* of a variable (like VAR) can be used as a variable *name*. For example, let us consider the following fragment of a `Makefile`.

```
MFLAGS = $(MFLAGS_$(ARCH))
MFLAGS_Linux  = -Wall -Wno-attributes -Wno-address
MFLAGS_Darwin = -Wall
```

- The variable `MFLAGS` is set to different values depending on the contents of the `ARCH` variable. In the example, this variable is assumed to be set elsewhere to the host operating system name, that is, either `Linux` or `Darwin`.
- In summary, this is a *compact* way to have different, operating system-dependent compiler flags in the variable `MFLAGS` without using conditional directives or write several separate `Makefiles`, one for each operating system.

A variable can get a value in several different, and rather complex ways, listed here in order of decreasing priority.

Table 3.2

GNU make Assignment Operators

Operator	Description
VAR = ...	Define a recursively-expanded variable
VAR := ...	Define a simply-expanded variable
VAR ?= ...	Defines the recursively-expanded variable VAR only if it has not been defined already
VAR += ...	Append . . . to variable VAR (see text)

1. As specified when GNU make is *invoked*, by means of an assignment statement put directly on its command line. For instance, the command make VAR=v invokes GNU make with VAR set to v.
2. By means of an *assignment* in a Makefile, as will be further explained in the following.
3. Through a shell *environment* variable definition.
4. Moreover, some variables are set *automatically* to useful values during rule application.
5. Finally, some variables have an *initial* value, too.

When there are multiple assignments to the same variable, the highest-priority one *silently* prevails over the others.

GNU make supports two kinds, or *flavors* of variables. It is important to comment on this difference because they are *defined* and *expanded* in different ways.

1. *Recursively-expanded* variables are defined by means of the operator =, informally mentioned previously. The evaluation of the right-hand side of the assignment, as well as the expansion of any references to other variables it may contain, are delayed until the variable being defined is itself expanded. The evaluation and variable expansion then proceed recursively.
2. *Simply-expanded* variables are defined by means of the operator :=. The value of the variable is determined once and for all when the assignment is executed. The expression on the right-hand side of the assignment is evaluated immediately, expanding any references to other variables.

Table 3.2 lists all the main assignment operators that GNU make supports. It is worth mentioning that the "append" variant of the assignment preserves (when possible) the kind of variable it operates upon. Therefore:

- If VAR is undefined it is the same as =, and hence, it defines a recursively expanded variable.

- If VAR is already defined as a simply expanded variable, it immediately expands the right-hand side of the assignment and appends the result to the previous definition.
- If VAR is already defined as a recursively expanded variable, it appends the right-hand side of the assignment to the previous definition without performing any expansion.

In order to better grasp the effect of delayed variable expansion, let us consider the following two examples.

```
X = 3
Y = $(X)
X = 8
```

In this first example, the value of Y is 8, because the right-hand side of its assignment is expanded only when Y is *used*. Let us now consider a simply expanded variable.

```
X = 3
Y := $(X)
X = 8
```

In this case, the value of Y is 3 because the right-hand side of its assignment is expanded immediately, when the assignment is performed.

As can be seen from the previous examples, delayed expansion of recursively expanded variables has unusual, but often useful, side effects. Let us just briefly consider the two main benefits of delayed expansion:

- Forward variable references in assignments, even to variables that are still undefined, are not an issue.
- When a variable is eventually expanded, it makes use of the "latest" value of the variables it depends upon.

3.7.3 PATTERN RULES

Often, all files belonging to the same *group* or *category* (for example, object files) follow the same generation rules. In this case, rather than providing an explicit rule for each of them and lose generality in the Makefile, it is more appropriate and convenient to define a *pattern rule*.

As shown in the following short code example, a pattern rule applies to all files that match a certain *pattern*, which is specified within the rule *in place of the target*.

```
%.o :   %.c
        cc -c $<

kbd.o :   defs.h command.h
```

In particular:

Table 3.3
GNU make Automatic Variables

Var.	Description	Example value
$@	Target of the rule	kbd.o
$<	*First* prerequisite of the rule	kbd.c
$^	List of *all* prerequisites of the rule, delimited by blanks	kbd.c defs.h command.h
$?	List of prerequisites that are *more recent* than the target	defs.h
$*	Stem of the rule	kbd

- Informally speaking, in the pattern the character % represents any non-empty character string.
- The same character can be used in the prerequisites, too, to specify how they are related to the target.
- The command lines associated with a pattern rule can be customized, based on the specific target the rule is being applied for, by means of *automatic variables* like $<.
- It is possible to augment the prerequisites of a pattern rule by means of explicit rules without command lines, as shown at the end of the example.

More precisely, a target pattern is composed of three parts: a *prefix*, a % character, and a *suffix*. The prefix and/or suffix may be empty. A target name (which often is a file name) matches the pattern if it starts with the pattern prefix and ends with the pattern suffix. The non-empty sequence of characters between the prefix and the suffix is called the *stem*.

Since, as said previously, rule targets are often file names, directory specifications in a pattern are handled specially, to make it easier to write compact and general rules. In particular:

- If a target pattern does not contain any *slash*—which is the character that separates directory names in a file path specification—all directory names are removed from target file names before comparing them with the pattern.
- Upon a successful match, directory names are restored at the beginning of the stem. This operation is carried out before generating prerequisites.
- Prerequisites are generated by substituting the stem of the rule in the right-hand part of the rule, that is, the part that follows the colon (:).
- For example, file src/p.o satisfies the pattern rule %.o : %.c. In this case, the prefix is *empty*, the stem is src/p and the prerequisite is src/p.c because the src/ directory is removed from the file name before comparing it with the pattern and then restored.

When a pattern rule is applied, GNU make automatically defines several *automatic variables*, which become available in the corresponding command lines. Table 3.3 contains a short list of these variables and describes their contents.

As an example, the rightmost column of the table also shows the value that automatic variables would get if the rules above were applied to regenerate kbd.o, mentioned before, because defs.h has been modified.

To continue the example, let us assume that the Makefile we are considering contains the following additional rule. The rule updates library lib.a, by means of the ar tool, whenever any of the object files it contains (main.o kbd.o disk.o) is updated.

```
lib.a :   main.o kbd.o disk.o
      ar rs $@ $?
```

After applying the previous rule, kbd.o becomes more recent than lib.a, because it has just been updated. In turn, this triggers the application of the second rule shown above. While the second rule is being applied, the automatic variable corresponding to the target of the rule ($@) is set to lib.a and the list of prerequisites more recent than the target ($?) is set to kbd.o.

To further illustrate the use of automatic variables, we can also remark that we could use $^ instead of $? in order to completely rebuild the library rather than update it. This is because, as mentioned in Table 3.3, $^ contains the list of *all* prerequisites of the rule.

It is also useful to remark that GNU make comes with a large set of predefined, *built-in* rules. Most of them are pattern rules, and hence, they generally apply to a wide range of targets and it is important to be aware of their existence. They can be printed by means of the command-line option --print-data-base, which can also be abbreviated as -p.

For instance, there is a built-in pattern rule to generate an object file given the corresponding C source file:

```
%.o: %.c
        $(COMPILE.c) $(OUTPUT_OPTION) $<
```

The variables cited in the command line have got a built-in definition as well, that is:

```
COMPILE.c = $(CC) $(CFLAGS) $(CPPFLAGS) $(TARGET_ARCH) -c
OUTPUT_OPTION = -o $@
CC = cc
```

As a consequence, by appropriately definining some additional variables, for example CFLAGS, it is often possible to customize the behavior of a built-in rule instead of defining a new one.

When a rule in the Makefile overlaps with a built-in rule because it has the same target and prerequisites, *the former* takes precedence. This priority scheme has been designed to avoid undue interference of built-in rules the programmer may be unaware of, with any rule he/she explicitly put into his/her Makefile.

3.7.4 DIRECTIVES AND FUNCTIONS

GNU make provides a rather extensive set of *directives* and built-in *functions*. In general, directives control how the input information needed by GNU make is built, by taking it from different input files, and which parts of those input files are considered. Provided here is a glance at two commonly-used directives, namely:

- The `include <file>` directive instructs GNU make to temporarily *stop reading* the current `Makefile` at the point where the directive appears, read the additional `file`(s), and then continue.

  The `file` specification may contain a single file name or a list of names, separated by spaces. In addition, it may also contain variable and function expansions, as well as any wildcards understood and used by the shell to match file names.

- The `ifeq (<exp1>, <exp2>)` directive evaluates the two expressions `exp1` and `exp2`. If they are textually identical, then GNU make uses the `Makefile` section between `ifeq` and the next `else` directive; otherwise it uses the section between `else` and `endif`.

  In other words, this directive is similar to conditional statements in other programming languages. Directives `ifneq`, `ifdef`, and `ifndef` also exist, and do have the expected, intuitive meaning.

Concerning functions, the general syntax of a function call is

$$\$(<function> <arguments>)$$

where

- `function` represents the function name and `arguments` is a list of one or more arguments. At least one blank space is required to separate the function name from the first argument. Arguments are separated by commas.
- By convention, variable names are written in all capitals, whereas function names are in lowercase, to help readers distinguish between the two.

Arguments may contain references to:

- *Variables*, for instance: `$(subst a,b,$(X))`. This statement calls the function `subst` with 3 arguments: `a`, `b`, and the result of the expansion of variable `X`.
- Other, nested *function calls*, for example: `$(subst a,b,$(subst c,d,$(X)))`. Here, the third argument of the outer `subst` is the result of the inner, nested `subst`.

As for directives, in the following we are about to informally discuss only a few GNU make functions that are commonly found in `Makefiles`. Interested readers should refer to the full documentation of GNU make, available online [65], for in-depth information.

- The function $(subst from,to,text) replaces from with to in text. Both from and to must be simple text strings. For example:

$$\$(\texttt{subst .c,.o,p.c q.c}) \longrightarrow \texttt{p.o q.o}$$

- The function $(patsubst from,to,text) is similar to subst, but it is more powerful because from and to are patterns instead of text strings. The meaning of the % character is the same as in pattern rules.

$$\$(\texttt{patsubst \%.c,\%.o,p.c q.c}) \longrightarrow \texttt{p.o q.o}$$

- The function $(wildcard pattern ...) returns a list of names of existing files that match one of the given patterns. For example, $(wildcard *.c) evaluates to the list of all C source files in the current directory. wildcard is commonly used to set a variable to a list of file names with common characteristics, like C source files. Then, it is possible to further work on the list with the help of other functions and use the results as targets, as shown in the following example.

```
SRC = $(wildcard *.c)
ELF = $(patsubst %.c,%.elf,$(SRC))

all: $(ELF)

%.elf: %.c
        $(CC) -o $@ $<
```

- The function $(shell command) executes a shell command and captures its output as return value. For example, when executed on a Linux system:

$$\$(\texttt{shell uname}) \longrightarrow \texttt{Linux}$$

In this way, it is possible to set a variable to an operating system-dependent value and have GNU make do different things depending on the operating system it is running on, without providing separate Makefiles for all of them, which would be harder to maintain.

3.8 SUMMARY

This chapter provided an overview of the most peculiar aspects of a GCC-based cross-compilation toolchain, probably the most commonly used toolchain for embedded software development nowadays.

After starting with an overview of the workflow of the whole toolchain, which was the subject of Section 3.1, the discussion went on by focusing on specific components, like the compiler driver (presented in Section 3.2), the preprocessor (described in Section 3.3), and the runtime libraries (Section 3.5).

Due to its complexity, the discussion of the inner workings of the compiler has been postponed to Chapter 11, where it will be more thoroughly analyzed in the context of performance and memory footprint optimization.

Instead, this chapter went deeper into describing two very important, but often neglected, toolchain components, namely the linker (which was the subject of Section 3.4) and GNU make (discussed in Section 3.7).

Last, but not least, this chapter also provided some practical information on how to configure and build the toolchain components, in Section 3.6. More specific information on how to choose the exact version numbers of those components will be provided in Chapter 12, in the context of a full-fledged case study.

4 Execution Models for Embedded Systems

CONTENTS

A key design point of any embedded system is the selection of an appropriate *execution model* that, generally speaking, can be defined as the set of rules and constraints that organize the execution of the embedded system's activities.

Traditionally, many embedded systems—especially the smallest and simplest ones—were designed around the *cyclic executive* execution model that is simple, efficient, and easy to understand. However, it lacks modularity and its application becomes more and more difficult as the size and complexity of the embedded system increase.

For this reason, *task-based scheduling* is nowadays becoming a strong competitor, even if it is more complex for what concerns both software development and system requirements. This chapter discusses and compares the two approaches to help the reader choose the best one for a given design problem and introduce the more general topic of choosing the right execution model for the application at hand.

4.1 THE CYCLIC EXECUTIVE

The cyclic executive execution model is fully described in Reference [16] and supports the execution of a number of *periodic tasks*. A periodic task, usually denoted as τ_i, is an activity or action to be performed repeatedly, at constant time intervals that represent the period of the process. The period of task τ_i is denoted as T_i. The same term, cyclic executive, is also used to denote the program responsible for controlling task execution.

It must be remarked that, even though periodic tasks are probably one of the simplest ways to model execution, they still represent the most important software com-

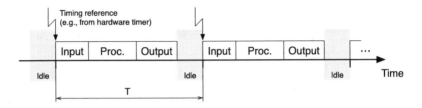

Figure 4.1 A prototypical cyclic executive.

ponents of an embedded, real-time system. For this reason, this book will mainly focus on this kind of task. Interested readers should refer to References [26, 111] for a more thorough and formal description of task scheduling theory, which includes more sophisticated and flexible task models. Reference [152] provides a comprehensive overview about the historical evolution of task scheduling theory and practice. To stay within the scope of the book, discussion will be limited to single-processor systems.

In its most basic form, a cyclic executive is very close to the typical intuitive structure of a simple embedded control system, whose time diagram is depicted in Figure 4.1. As shown in the figure, the activities carried out by the control system can be seen as three tasks, all with the same period T.

1. An *input* task interacts with input devices (e.g., sensors) to retrieve the input variables of the control algorithm.
2. A *processing* task implements the control algorithm and computes the output variables based on its inputs.
3. An *output* task delivers the values of the output variables to the appropriate output devices (e.g., actuators).

To guarantee that the execution period is fixed, regardless of unavoidable variations in task execution time, the cyclic executive makes use of a timing reference signal and synchronizes with it at the beginning of each cycle. Between the end of the output task belonging to the current cycle and the beginning of the input task belonging to the next cycle—that is, while the synchronization with the timing reference is being performed—the system is idle. Those areas are highlighted in gray in Figure 4.1.

In most embedded systems, hardware timers are available to provide accurate timing references. They can deliver timing signals by means of periodic interrupt requests or, in simple cases, the software can perform a polling loop on a counter register driven by the timer itself.

For what concerns the practical implementation, the cyclic executive corresponding to the time diagram shown in Figure 4.1 can be coded as an infinite loop containing several function calls.

```
while(1)
{
    WaitForReference();
    Input();
    Processing();
    Output();
}
```

In the listing above, the function `WaitForReference` is responsible for synchronizing loop execution with real time and ensuring that it is executed once every period T. The three functions `Input`, `Processing`, and `Output` correspond to the input, processing, and output tasks. The same code can also be seen as a *scheduling table*, a table of function calls in which each call represents a task.

Even though the example discussed so far is extremely simple, it already reveals several peculiar features of a typical cyclic executive. They must be remarked upon right from the beginning because they are very different from what task-based scheduling does, which is to be discussed in Section 4.4. Moreover, these differences have an important impact on programmers and affect the way they think about and implement their code.

In the cyclic executive implementation, tasks are mapped onto functions, which are called sequentially from the main loop. Task interleaving is decided offline, when the cyclic executive is designed, and then performed in a completely deterministic way. Since the schedule is laid out completely once and for all, it is a "proof by construction" that all tasks can be successfully executed and satisfy their timing constraints.

Moreover, task functions can freely access global variables to exchange data. Since the function execution sequence is fixed, no special care is needed to read and write these variables. For the same reason, enforcing producer–consumer constraints, that is, making sure that the value of a variable is not used before it is set, is relatively easy. Even complex data structures are straightforward to maintain because, by construction, a function will never be stopped "midway" by the execution of another function. As a consequence, a function will never find the data structure in an inconsistent state. The only exception to this general rule is represented by asynchronously executed code, such as signal and interrupt handlers.

4.2   MAJOR AND MINOR CYCLES

In theory, nothing prevents a programmer from crafting a cyclic executive completely by hand, according to his/her own design. From the practical point of view, it is often useful to make use of a well-understood structure, in the interest of software clarity and maintainability. For this reason, most cyclic executives are designed according to the following principles.

In order to accommodate tasks with different periods, an issue that was not considered in the simple example of Section 4.1, the scheduling table is divided into sections, or slices. Individual slices are called *minor cycles* and have all the same duration. The table as a whole is known as *major cycle*.

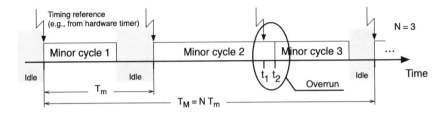

Figure 4.2   Major and minor cycles in a cyclic executive.

As shown in Figure 4.2, which depicts a cyclic executive in which the major cycle is composed of 3 minor cycles, minor cycle boundaries are synchronization points. At these points, the cyclic executive waits for a timing reference signal. As a consequence, the task activated at the very beginning of a minor cycle is synchronized with real time as accurately as possible. On the other hand, within a minor cycle, tasks are activated in sequence and suffer from any execution time jitter introduced by other tasks that precede them.

Minor cycle synchronization is also very useful to detect the most critical error that a cyclic executive may experience, that is, a cycle *overrun*. An overrun occurs when—contrary to what was assumed at design time—some task functions have an execution time that is longer than expected and the accuracy of the overall cyclic executive timing can no longer be guaranteed.

As an example, Figure 4.2 shows what happens if the task functions invoked in minor cycle 2 exceed the minor cycle length. If the cyclic executive does not handle this condition and simply keeps executing task functions according to the designed sequence, the whole minor cycle 3 is shifted to the right and its beginning is no longer properly aligned with the timing reference.

Overrun detection at a synchronization point occurs in two different ways depending on how synchronization is implemented:

1. If the timing reference consists of an interrupt request from a hardware timer, the cyclic executive will detect the overrun as soon as the interrupt occurs, that is, at time t_1 in Figure 4.2.
2. If the cyclic executive synchronizes with the timing reference by means of a polling loop, it will be able to detect the overrun only when minor cycle 2 eventually ends, that is, when the last task function belonging to that cycle returns. At that point, the cyclic executive should wait until the end of the cycle, but it will notice that the time it should wait for is already in the past. In the figure, this occurs at time t_2.

Generally speaking, the first method provides a more timely *detection* of overruns, as well as being an effective way to deal with indefinite non-termination of a task function. However, a proper *handling* of an overrun is often more challenging than detection itself. Referring back to Figure 4.2, it would be quite possible for the cyclic

Table 4.1
A Simple Task Set for a Cyclic Executive

$T_m = 20\,\text{ms}, T_M = 120\,\text{ms}$

Task τ_i	Period T_i (ms)	Execution time C_i (ms)	k_i
τ_1	20	6	1
τ_2	40	4	2
τ_3	60	2	3
τ_4	120	6	6

executive to abort minor cycle 2 at t_1 and immediately start minor cycle 3, but this may lead to significant issues at a later time. For instance, if a task function was manipulating a shared data structure when it was aborted, the data structure was likely left in an inconsistent state.

In addition, an overrun is most often indicative of a *design*, rather than an execution issue, at least when it occurs systematically. For this reason, overrun handling techniques are often limited to error reporting to the supervisory infrastructure (when it exists), followed by a system reset or shutdown.

Referring back to Figure 4.2, if we denote the minor cycle period as T_m and there are N minor cycles in a major cycle ($N = 3$ in the figure), the major cycle period is equal to $T_M = N T_m$. Periodic tasks with different periods can be placed within this framework by invoking their task function from one or more minor cycles. For instance, invoking a task function from all minor cycles leads to a task period $T = T_m$, while invoking a task function just in the first minor cycle gives a task period $T = T_M$.

Given a set of s periodic tasks $\tau_1, \ldots, \tau_s$, with their own periods $T_1, \ldots, T_s$, it is interesting to understand how to choose T_m and T_M appropriately, in order to accommodate them all. Additional requirements are to keep T_m as big as possible (to reduce synchronization overheads), and keep T_M as small as possible (to decrease the size of the scheduling table).

It is easy to show that an easy, albeit not optimal, way to satisfy these constraints is to set the minor cycle length to the *greatest common divisor* (GCD) of the task periods, and set the major cycle length to their *least common multiple* (LCM). Reference [16] contains information on more sophisticated methods. In formula:

$$T_m = \gcd(T_1, \ldots, T_s) \qquad (4.1)$$
$$T_M = \operatorname{lcm}(T_1, \ldots, T_s) \qquad (4.2)$$

When T_m and T_M have been chosen in this way, the task function corresponding to τ_i must be invoked every $k_i = T_i/T_m$ minor cycles. Due to Equation (4.1), it is guaranteed that every k_i is an integer and a sub-multiple of N, and hence, the schedule can actually be built in this way.

As an example, let us consider the task set listed in Table 4.1. Besides the task periods, the table also lists their execution time and the values k_i, computed as dis-

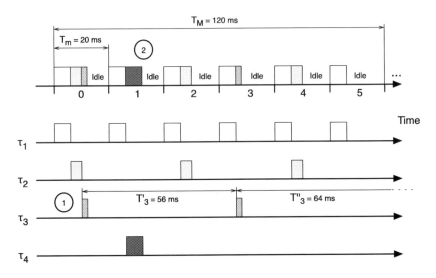

Figure 4.3   The cyclic executive schedule for the task set of Table 4.1.

cussed above. The execution time—usually denoted as C_i—is defined as the amount
of CPU time it takes to execute the task function corresponding to τ_i. It is especially
important to evaluate or, at least, estimate these values at design time, in order to
allocate task functions in the right minor cycles and avoid overruns.

In this example, from Equations (4.1) and (4.2), it is:

$$
\begin{aligned}
T_m &= \gcd(20, 40, 60, 120) = 20\,\text{ms} \\
T_M &= \operatorname{lcm}(20, 40, 60, 120) = 120\,\text{ms}
\end{aligned}
$$

and there are $N = 6$ minor cycles in every major cycle.

The corresponding cyclic executive schedule is shown at the top of Figure 4.3.
The bottom part of the figure shows the execution timeline of individual tasks and
demonstrates that all tasks are executed properly. This example is also useful to remark on two general properties of cyclic executives:

1. It is generally *impossible* to ensure that all tasks will be executed with their exact
   period for every instance, although this is true on average. In the schedule shown
   in Figure 4.3 this happens to τ_3, which is executed alternatively at time intervals
   of $T_3' = 56$ ms and $T_3'' = 64$ ms. In other words, on average the task period is still
   $T_3 = 60$ ms but every instance will suffer from an activation *jitter* of ± 4 ms around
   the average period.

   In this particular example, the jitter is due to the presence of τ_2 in minor cycle 0,
   whereas it is absent in minor cycle 3. In general, it may be possible to *choose*
   which tasks should suffer from jitter up to a certain extent, but it cannot be removed completely. In this case, swapping the activation of τ_2 and τ_3 in minor
   cycle 0 removed jitter from τ_3, but introduces some jitter on τ_2. Moving τ_2 so that

Table 4.2
A Task Set That Requires Task Splitting to Be Scheduled

$T_m = 20\,\text{ms}, T_M = 120\,\text{ms}$

Task τ_i	Period T_i (ms)	Execution time C_i (ms)	k_i
τ_1	20	6	1
τ_2	40	4	2
τ_3	60	2	3
τ_4	120	16	6

it is activated in the odd minor cycles instead of the even ones does not completely solve the problem, either.

2. There may be some *freedom* in where to place a certain task within the schedule. In the example, τ_4 may be placed in any minor cycle. In this case, the minor cycle is chosen according to secondary goals. For instance, choosing one of the "emptiest" minor cycles—as has been done in the example—is useful to reduce the likelihood of overrun, in case some of the C_i have been underestimated.

4.3 TASK SPLITTING AND SECONDARY SCHEDULES

In some cases, it may be impossible to successfully lay out a cyclic executive schedule by considering tasks as *indivisible* execution blocks, as has been done in the previous example. This is trivial in the case when the execution time of a task τ_i exceeds the minor cycle length, that is, $C_i > T_m$, but it may also happen in less extreme cases, when other tasks are present and must also be considered in the schedule.

Table 4.2 lists the characteristics of a task set very similar to the one used in the previous example (and outlined in Table 4.1), with one important difference. The execution time of τ_4 has been increased to $C_4 = 16\,\text{ms}$. Since the task periods have not been changed, T_m and T_M are still the same as before.

Even though it is $C_4 < T_m$, it is clearly impossible to fit τ_4 *as a whole* in any minor cycle, because none of them has enough idle time to accommodate it. It should also be remarked that the issue is not related to the availability of sufficient idle time in general, which would make the problem impossible to solve. Indeed, it is easy to calculate that in T_M, which is also the period of τ_4, the idle time left by the execution of τ_1, τ_2, and τ_3 amounts to 68 ms, which is much greater than C_4.

In cases like this, the only viable solution is to *split* the task (or tasks) with a significant C_i into two (or more) slices and allocate individual slices in different minor cycles.

As shown in Figure 4.4 it is possible, for example, to split τ_4 into two slices with equal execution time and allocate them in two adjacent minor cycles. Even though the overall execution timeline is still satisfactory, task splitting has two important, negative consequences:

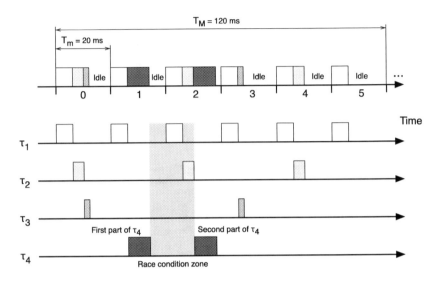

Figure 4.4 A possible task split for the task set of Table 4.2.

1. The criteria used to decide where a task must be split depend on task *timing* and not on its *behavior*. For this reason, when a task is split, it may be necessary to cut through its code in a way that has little to do with the internal, logic structure of the task itself. As a result, the code will likely be harder to understand and maintain.

2. Even more importantly, splitting a task as has been done with τ_4 in the example, introduces a *race condition zone*, or window, depicted as a gray area in Figure 4.4. Generally speaking, a race condition occurs whenever two tasks are allowed to access a shared data structure in an uncontrolled way. A race condition window is a time frame in which, due to the way tasks are scheduled, a race condition *may* occur.

   In this case, if τ_4 shares a data structure with τ_1 or τ_2—the tasks executed within the race condition zone—special care must be taken.

   - If τ_4 modifies the data structure in any way, the first part of τ_4 must be implemented so that it leaves the data structure in a consistent state, otherwise τ_1 and τ_2 may have trouble using it.
   - If τ_1 or τ_2 update the shared data structure, the first and second part of τ_4 must be prepared to find the data structure in two different states and handle this scenario appropriately.

   In both cases, it is clear that splitting a task does not merely require taking its code, dividing it into pieces and putting each piece into a separate function. Instead, some non-trivial modifications to the code are needed to deal with race conditions. These modifications add to code complexity and, in a certain sense, compromise one important feature of cyclic executives recalled in Section 4.1, that is, their ability to handle shared data structures in a straightforward, intuitive way.

Table 4.3
A Task Set Including a Task with a Large Period

$T_m = 20\,\text{ms}, T_M = 1200\,\text{ms}$

Task τ_i	Period T_i (ms)	Execution time C_i (ms)	k_i
τ_1	20	6	1
τ_2	40	4	2
τ_3	60	2	3
τ_4	**1200**	6	**60**

It is also useful to remark that what has been discussed so far is merely a simple introduction to the data sharing issues to be confronted when a more sophisticated and powerful execution model is adopted. This is the case of task-based scheduling, which will be presented in Section 4.4. To successfully solve those issues, a solid grasp of concurrent programming techniques is needed. This will be the topic of Chapter 5.

Besides tasks with a significant execution time, tasks with a large period, with respect to the others, may be problematic in cyclic executive design, too. Let us consider the task set listed in Table 4.3. The task set is very close to the one used in the first example (Table 4.1) but T_4, the period of τ_4, has been increased to 1200 ms, that is, ten times as before.

It should be remarked that tasks with a large period, like τ_4, are not at all uncommon in real-time embedded systems. For instance, such a task can be used for periodic status and data logging, in order to summarize system activities over time. For these tasks, a period on the order of 1 s is quite common, even though the other tasks in the same system, dealing with data acquisition and control algorithm, require much shorter periods.

According to the general rules to determine T_m and T_M given in Equations (4.1) and (4.2), it should be:

$$
\begin{aligned}
T_m &= \gcd(20, 40, 60, 1200) = 20\,\text{ms} \\
T_M &= \text{lcm}(20, 40, 60, 1200) = 1200\,\text{ms}
\end{aligned}
$$

In other words, the minor cycle length T_m would still be the same as before, but the major cycle length T_M would increase tenfold, exactly like T_4. As a consequence, a major cycle would now consist of $N = T_M / T_m = 60$ minor cycles instead of 6. What is more, as shown in Figure 4.5, 59 minor cycles out of 60 would still be exactly the same as before, and would take care of scheduling τ_1, τ_2, and τ_3, whereas just one (minor cycle 1) would contain the activation of τ_4.

The most important consequence of N being large is that the scheduling table becomes large, too. This not only increases the memory footprint of the system, but it also makes the schedule more difficult to visualize and understand. In many cases,

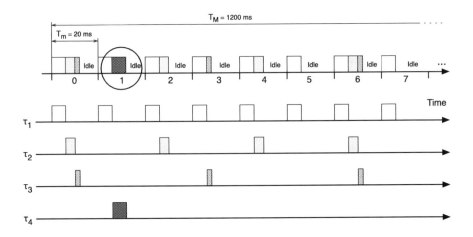

Figure 4.5   Schedule of the task set of Table 4.3, without secondary schedules.

T_M and N can be brought down to a more acceptable value by means of a *secondary schedule*.

In its simplest form, a secondary schedule can be designed by calculating T'_m and T'_M "as if" the task with a large period, τ_4 in our example, had a period equal to an integer sub-multiple of the real one, that is, a period $T'_4 = T_4/l$ where l is a positive integer.

The value l must be chosen so that it is as small as possible—leading to a T'_4 that is as close as possible to T_4—but giving a major cycle length T'_M that is still equal to the length T_M calculated according to the other tasks to be scheduled. When no suitable l can be found in this way, minor adjustments to the design period T_4 are often sufficient to work around the issue.

In the current example, the value of T_M calculated according to τ_1, τ_2, and τ_3 is

$$T_M = \text{lcm}(20, 40, 60) = 120 \, \text{ms}$$

Being $T_4 = 1200 \, \text{ms}$, the suitable value of l is $l = 10$ (leading to $T'_4 = 120 \, \text{ms}$ and $T'_M = 120 \, \text{ms}$) because any value lower than this would make $T'_M > T_M$.

Task τ_4 is then placed in the schedule according to the modified period T'_4 and its task function is surrounded by a *wrapper*. The wrapper invokes the actual task function once every l invocations and returns immediately in all the other cases. The resulting schedule is shown in Figure 4.6. As can be seen by looking at the figure, the schedule is still compact, even though τ_4 has been successfully accommodated.

On the downside, it should be noted that even though the wrapper returns immediately to the caller without invoking the corresponding task function most of the time (9 times out of 10 in the example), the execution time that must be considered during the cyclic executive design to accommodate the secondary schedule is still equal to C_4, that is, the execution time of τ_4.

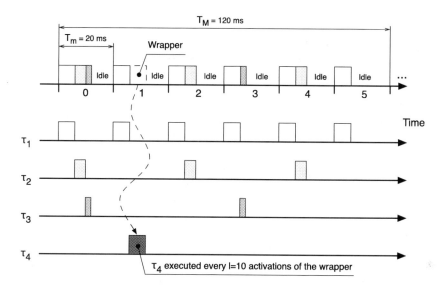

Figure 4.6   Schedule of the task set of Table 4.3, using a secondary schedule.

This is an extremely conservative approach because τ_4 is actually invoked only once every l executions of the wrapper (once out of 10 times in our example), and may make the cyclic executive design more demanding than necessary, or even infeasible.

Moreover, secondary schedules are unable to solve the issue of large T_M in general terms. For instance, when T_M is calculated according to Equation 4.2 and task periods are mutually prime, it will unavoidably be equal to the product of all task periods unless they are adjusted to make them more favorable.

4.4 TASK-BASED SCHEDULING

As seen in Section 4.1, in a cyclic executive, all tasks (or parts of them) are executed in a predefined and fixed order, based on the contents of the scheduling table. The concept of task is somewhat lost at runtime, because there is a single thread of execution across the main program, which implements the schedule, and the various task functions that are called one at a time.

In this section we will discuss instead a different, more sophisticated approach for task execution, namely *task-based scheduling*, in which the abstract concept of task is still present at runtime. In this execution model, tasks become the primary unit of scheduling under the control of an independent software component, namely the real-time *operating system*.

The main differences with respect to a cyclic executive are:

- The operating system itself (or, more precisely, one of its main components known as *scheduler*) is responsible for switching from one task to another. The switching points from one task to another are no longer hard-coded within the code and are chosen autonomously by the scheduler. As a consequence task switch may occur *anywhere* in the tasks, with very few exceptions that will be better discussed in Chapter 5.

- In order to determine the scheduling sequence, the scheduler must follow some criteria, formally specified by means of a *scheduling algorithm*. In general terms, any scheduling algorithm bases its work on certain task characteristics or attributes. For instance, quite intuitively, a scheduling algorithm may base its decision of whether or not to switch from one task to another on their relative importance, or *priority*. For this reason, the concept of task at runtime can no longer be reduced to a mere function containing the task code, as is done in a cyclic executive, but it must include these attributes, too.

The concept of task (also called *sequential process* in more theoretical descriptions) was first introduced in Reference [48] and plays a central role in a task-based system. It provides both an abstraction and a conceptual model of a code module that is being executed. In order to represent a task at runtime, operating systems store some relevant information about it in a data structure, known as *task control block* (TCB). It must contain all the information needed to represent the execution of a sequential program as it evolves over time.

As depicted in Figure 4.7, there are four main components directly or indirectly linked to a TCB:

1. The TCB contains a full copy of the *processor* state. The operating system makes use of this piece of information to switch the processor from one task to another. This is accomplished by saving the processor state of the previous task into its TCB and then restoring the processor state of the next task, an operation known as *context switch*.

   At the same time, the processor state relates the TCB to two other very important elements of the overall task state. Namely, the *program counter* points to the next instruction that the processor will execute, within the task's program code. The *stack pointer* locates the boundary between full and empty elements in the task stack.

   As can be inferred from the above description, the processor state is an essential part of the TCB and is always present, regardless of which kind of operating system is in use. Operating systems may instead differ on the details of *where* the processor state is stored.

   Conceptually, as shown in Figure 4.7, the processor state is *directly* stored in the TCB. Some operating systems follow this approach literally, whereas others store part or all of the processor state elsewhere, for instance in the task stack, and then make it accessible from the TCB through a pointer.

   The second choice is especially convenient when the underlying processor architecture provides hardware assistance to save and restore the processor state

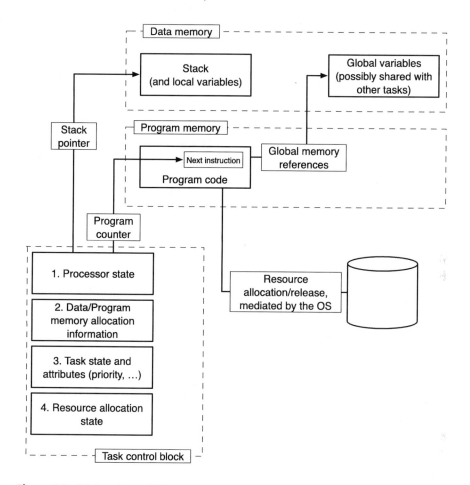

Figure 4.7 Main task control block components.

to/from a predefined, architecture-dependent location, which usually cannot be changed at will in software.

2. The *data and program memory allocation* information held in a TCB keep a record of the memory areas currently assigned to the task. The structure and content of this information heavily depends on the sophistication of the operating system.

For instance, very simple operating systems that only support a fixed number or statically created tasks may not need to keep this information at all, because data and program memory areas are assigned to the tasks at link time and the assignment never changes over time.

On the other hand, when the operating system supports the dynamic creation of tasks at runtime, it must at a minimum allocate the task stack dynamically and keep a record of the allocation. Moreover, if the operating system is also capable of loading executable images from a mass storage device on-demand, it should also keep a record of where the program code and global variable areas have been placed in memory.

3. The task *state* and *attributes* are used by the operating system itself to schedule tasks in an orderly way and support inter-task synchronization and communication. An informal introduction to these very important and complex topics will be given in the following, while more detailed information can be found in Chapter 5.

4. When resources allocation and release are mediated by the operating system, the task control block also contains the *resource allocation* state pertaining to the task. The word resource is used here in a very broad sense. It certainly includes all hardware devices connected to the system, but it may also refer to *software* resources.

Having the operating system work as a mediator between tasks and resources regarding allocation and release is a universal and well-known feature of virtually all general-purpose operating systems. After all, the goal of this kind of operating system is to support the coexistence of multiple application tasks, developed by a multitude of programmers.

Tasks can be started, stopped, and reconfigured at will by an interactive user, and hence, their characteristics and resource requirements are largely unknown to the operating system (and sometimes even to the user) beforehand. It is therefore not surprising that an important goal of the operating system is to keep resource allocation under tight control.

In a real-time embedded system, especially small ones, the scenario is very different because the task set to be executed is often fixed and well known in advance. Moreover, the user's behavior usually has little influence on its characteristics because, for instance, users are rarely allowed to start or stop real-time tasks in an uncontrolled way.

Even the relationship between tasks and resources may be different, leading to a reduced amount of *contention* for resource use among tasks. For instance, in a general-purpose operating system it is very common for application tasks to compete among each other to use a graphics coprocessor, and sharing this resource in an appropriate way is essential.

On the contrary, in a real-time system devices are often dedicated to a single purpose and can be used only by a single task. For example, an analog to digital converter is usually managed by a cyclic data acquisition task and is of no interest to any other tasks in the system.

For this reason, in simple real-time operating systems resources are often permanently and implicitly allocated to the corresponding task, so that the operating system itself is not involved in their management.

It is important to note that a correct definition of the information included in a TCB is important not only to thoroughly understand what a task *is*, but also how tasks are *managed* by the operating system. In fact, TCB contents represent the information that the operating system must save and restore when it wants to switch the CPU from one task to another in a transparent way.

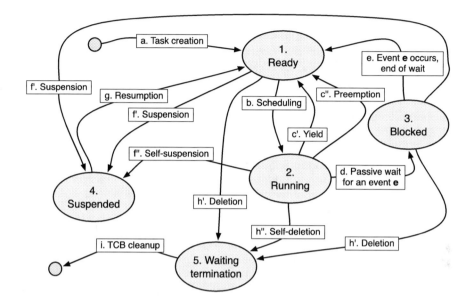

Figure 4.8   Task state diagram in the FREERTOS operating system.

4.5   TASK STATE DIAGRAM

In the previous section, we introduced the concept of task control block (TCB), a data structure managed and maintained by the operating system, which holds all the information needed to represent the execution of a sequential program. A particularly important piece of information held in the TCB is the notion of task *state*.

A commonly used way to describe in a formal way all the possible states a task may be in during its lifespan is to define a directed graph, called *task state diagram* (TSD) or, in the context of general-purpose operating systems, process state diagram (PSD). Although the details of the TSD layout vary from one operating system to another, the most important concepts are common to all of them. In particular:

- TSD *nodes* represent the possible task states.
- Its *arcs* represent transitions from one task state to another.

As a practical example, Figure 4.8 depicts the TSD defined by the FREERTOS operating system [17, 18], used as a case study throughout this book. Referring to the TSD, at any instant a process may be in one of the following states:

1. A task is in the *ready* state when it is eligible for execution but no processors are currently available to execute it, because all of them are busy with other activities. This is a common occurrence because the number of ready tasks usually exceeds the total number of processors (often just one) available in the system. A task does not make any progress when it is ready.

2. A task is *running* when it is actively being executed by a processor, and hence, makes progress. The number of tasks in the *running* state cannot exceed the total number of processors available in the system.

3. Tasks often have to wait for an external event to occur. For example:

   • A periodic task, after completing its activity in the current period, has to wait until the next period begins. In this case, the event is generated by the operating system's timing facility.

   • A task that issues an input–output operation to a device has to wait until the operation is complete. The completion event is usually conveyed from the device to the processor by means of an interrupt request.

   In these cases, tasks move to the *blocked* state. Once there, they no longer compete with other tasks for execution.

4. Tasks in the *suspended* state, like the ones in the *blocked* state, are not eligible for execution. The difference between the two states is that suspended tasks are not waiting for an event. Rather, they went into the *suspended* state either voluntarily or due to the initiative of another task. They can return to the *ready* state only when another task explicitly resumes them.

   Another difference is that there is usually an upper limit on the amount of time tasks are willing to spend in the *blocked* state. When the time limit (called time-out) set by a task expires before the event that the task was waiting for occurs, that task returns to the *ready* state with an error indication. On the contrary, there is no limit on the amount of time a task can stay in the *suspended* state.

5. When a task deletes itself or another task deletes it, it immediately ceases execution but its TCB is not immediately removed from the system. Instead, the task goes into the *waiting termination* state. It stays in that state until the operating system completes the cleanup operation associated with task termination. In FREERTOS, this is done by a system task, called the *idle task*, which is executed when the system is otherwise idle.

There are two kinds of state transition in a TSD.

   • A *voluntary* transition is performed under the control of the task that undergoes it, as a consequence of one explicit action it took.

   • An *involuntary* transition is not under the control of the task affected by it. Instead, it is the consequence of an action taken by another task, the operating system, or the occurrence of an external event.

Referring back to Figure 4.8, the following transitions are allowed:

   a. The *task creation* transition instantiates a new TCB, which describes the task being created. After creation, the new task is not necessarily executed immediately. However, it is eligible for execution and resides in the *ready* state.

   b. The operating system is responsible for picking up tasks in the *ready* state for execution and moving them into the *running* state, according to the outcome of its scheduling algorithm, whenever a processor is available for use. This action is usually called task *scheduling*.

c. A running task may voluntarily signal its willingness to relinquish the processor it is being executed on by asking the operating system to reconsider the scheduling decision it previously made. This is done by means of an operating system request known as *yield*, which corresponds to transition c' in the figure and moves the invoking task from the *running* state to the *ready* state.

The transition makes the processor previously assigned to the task available for use. This leads the operating system to run its scheduling algorithm and choose a task to run among the ones in the *ready* state. Depending on the scheduling algorithm and the characteristics of the other tasks in the *ready* state, the choice may or may not fall on the task that just yielded.

Another possibility is that the operating system itself decides to run the scheduling algorithm. Depending on the operating system, this may occur periodically or whenever a task transitions into the *ready* state from some other states for any reason. The second kind of behavior is more common with real-time operating systems because, by intuition, when a task becomes ready for execution, it may be "more important" than one of the running tasks from the point of view of the scheduling algorithm.

When this is the case, the operating system forcibly moves one of the tasks in the *running* state back into the *ready* state, with an action called *preemption* and depicted as transition c" in Figure 4.8. Then, it will choose one of the tasks in the *ready* state and move it into the *running* state by means of transition b.

One main difference between transitions c' and c" is therefore that the first one is voluntary, whereas the second one is involuntary.

d. The transition from the *running* to the *blocked* state is always under the control of the affected task. In particular, it is performed when the task invokes one of the operating system synchronization primitives to be discussed in Chapter 5, in order to wait for an event **e**.

It must be remarked that this kind of wait is very different from what can be obtained by using a polling loop because, in this case, no processor cycles are wasted during the wait.

e. When event **e** eventually occurs, the waiting task is returned to the *ready* state and starts competing again for execution against the other tasks. The task is not returned directly to the *running* state because it may or may not be the most important activity to be performed at the current time. As discussed for yield and preemption, this is a responsibility of the scheduling algorithm and not of the synchronization mechanism.

Depending on the nature of **e**, the component responsible for waking up the waiting task may be another task (when the wait is due to inter-task synchronization), the operating system timing facility (when the task is waiting for a time-related event), or an interrupt handler (when the task is waiting for an external event, such as an input–output operation).

f. The *suspension* and *self-suspension* transitions, denoted as f' and f" in the figure, respectively, bring a task into the *suspended* state. The only difference between the two transitions is that the first one is involuntary, whereas the second one is voluntary.

Table 4.4

A Task Set to Be Scheduled by the Rate Monotonic Algorithm

Task τ_i	Period T_i (ms)	Execution time C_i (ms)
τ_1	20	5
τ_2	40	10
τ_3	60	20

g. When a task is resumed, it unconditionally goes from the *suspended* state into the *ready* state. This happens regardless of which state it was in before being suspended.

An interesting side effect of this behavior is that, if the task was waiting for an event when the suspend/resume sequence took place, it may resume execution even though the event did not actually take place. In this case, the task being resumed will receive an error indication from the blocking operating system primitive.

h. Tasks permanently cease execution by means of a *deletion* or *self-deletion* transition. These two transitions have identical effects on the affected task. The only difference is that, in the first case, deletion is initiated by another task, whereas in the second case the task voluntarily deletes itself. As discussed above, the TCB of a deleted task is not immediately removed from the system. A side effect of a self-deletion transition is that the operating system's scheduling algorithm will choose a new task to be executed.

i. The *TCB cleanup* transition removes tasks from the *waiting termination* state. After this transition is completed, the affected tasks completely disappear from the system and their TCB may be reused for a new task.

4.6   RACE CONDITIONS IN TASK-BASED SCHEDULING

In Section 4.3, a *race condition* was informally defined as an issue that hinders program correctness when two or more tasks are allowed uncontrolled access to some shared variables or, more generally, a *shared object*. As was remarked there, in a cyclic executive the issue can easily be kept under control because race condition zones appear only as a consequence of task splitting and, even in that case, their location in the schedule is well known in advance.

On the contrary, the adoption of task-based scheduling makes the extent and location of race condition zones hard to predict. This is because the task switching points are now chosen autonomously by the operating system scheduler instead of being hard-coded in the code. Due to this fact, the switching points may even change from one task activation to another.

For example, Figure 4.9 shows how the task set specified in Table 4.4 is scheduled by the Rate Monotonic scheduling algorithm [110]. According to this algorithm,

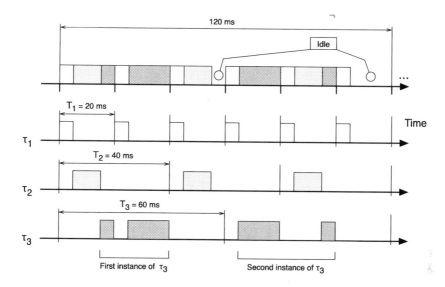

Figure 4.9 Rate Monotonic scheduling of the task set specified in Table 4.4.

which is very commonly used in real-time systems, tasks are assigned a fixed priority that is inversely proportional to their periods. As a consequence, Table 4.4 lists them in decreasing priority order. At any instant, the scheduler grants the processor to the highest-priority task ready for execution.

As can be seen in the figure, even though the task set being considered is extremely simple, the lowest-priority task τ_3 is not only preempted in different places from one instance to another, but by different tasks, too. If, for instance, tasks τ_2 and τ_3 share some variables, no race condition occurs while the first instance of τ_3 is being executed, because the first instance of τ_3 is preempted only by τ_1. On the other hand, the second instance of τ_3 is preempted by both τ_1 and τ_2 and race condition may occur in this case.

As the schedule becomes more complex due to additional tasks, it rapidly becomes infeasible to foresee all possible scenarios. For this reason, a more thorough discussion about race conditions and how to address them in a way that is *independent* from the particular scheduling algorithm in use is given here. Indeed, dealing with race conditions is perhaps the most important application of the task synchronization techniques to be outlined in Chapter 5.

For the sake of simplicity, in this book race conditions will be discussed in rather informal terms, looking mainly at their implication from the concurrent programming point of view. Interested readers should refer, for instance, to the works of Lamport [105, 106] for a more formal description. From the simple example presented above, it is possible to identify two general, *necessary* conditions for a race condition to occur:

1. Two or more tasks must be executed concurrently, leaving open the possibility of a context switch occurring among them. In other words, they must be within a race condition zone, as defined above.
2. These tasks must be actively working on the same shared object when the context switch occurs.

It should also be noted that the conditions outlined above are not yet *sufficient* to cause a race condition because, in any case, the occurrence of a race condition is also a *time-dependent* issue. In fact, even though both necessary conditions are satisfied, the context switch must typically occur at very specific locations in the code to cause trouble. In turn, this usually makes the race condition probability very low and makes it hard to reproduce, analyze, and fix.

The second condition leads to the definition of *critical region* related to a given shared object. This definition is of great importance not only from the theoretical point of view, but also for the practical design of concurrent programs. The fact that, for a race condition to occur, two or more tasks must be actively working on the same shared object leads to classifying the code belonging to a task into two categories.

1. A usually large part of a task's code implements operations that are *internal* to the task itself, and hence, do not make access to any shared data. By definition, all these operations cannot lead to any race condition because the second necessary condition described above is not satisfied.

   From the software development point of view, an important consequence of this observation is that these pieces of code can be safely disregarded when the code is analyzed to reason about and avoid race conditions.
2. Other parts of the task's code indeed make access to shared data. Therefore, those regions of code must be looked at more carefully because they may be responsible for a race condition if the other necessary conditions are met, too.

   For this reason, they are called *critical regions* or *critical sections* with respect to the shared object(s) they are associated with.

Keeping in mind the definition of critical region just given, we can imagine that race conditions on a certain shared object can be avoided by allowing only one task to be within a critical region, pertaining to that object, within a race condition zone. Since, especially in a task-based system, race condition zones may be large and it may be hard to determine their locations in advance, a more general solution consists of enforcing the *mutual exclusion* among all critical regions pertaining to the same shared object *at any time*, without considering race condition zones at all.

Traditionally, the implementation of mutual exclusion among critical regions is based on a *lock-based* synchronization protocol, in which a task that wants to access a shared object, by means of a certain critical region, must first of all *acquire* some sort of *lock* associated with the shared object and possibly wait, if it is not immediately available.

Afterwards, the task is allowed to use the shared object freely. Even though a context switch occurs at this time, it will not cause a race condition because any

other task trying to enter a critical region pertaining to the same shared object will
be blocked.

When the task has completed its operation on the shared object and the shared
object is again in a consistent state, it must *release* the lock. In this way, any other
task can acquire it and be able to access the shared object in the future.

In other words, in order to use a lock-based synchronization protocol, critical regions must be "surrounded" by two auxiliary pieces of code, usually called the critical region *entry* and *exit* code, which take care of acquiring and releasing the lock,
respectively. For some kinds of task synchronization techniques, better described in
Chapter 5, the entry and exit code must be invoked explicitly by the task itself, and
hence, the overall structure of the code strongly resembles the one outlined above. In
other cases, for instance when using message passing primitives among tasks, critical regions as well as their entry/exit code may be "hidden" within the inter-task
communication primitives and be invisible to the programmer, but the concept is still
the same.

For the sake of completeness, it must also be noted that mutual exclusion implemented by means of lock-based synchronization is by far the most common, but it is
not the only way to solve the race condition problem. It is indeed possible to avoid
race conditions *without* any lock, by using *lock-free* or *wait-free* inter-task communication techniques.

Albeit a complete description of lock-free and wait-free communication is outside the scope of this book—interested readers may refer to References [2, 3, 4,
70, 72, 104] for a more detailed introduction to this subject and its application to
real-time embedded systems—some of their advantages against lock-based synchronization are worth mentioning anyway. Moreover, a simple, hands-on example of
how wait-free communication works—in the context of communication between a
device driver and a hardware controller—will be given in Section 8.5.

It has already been mentioned that, during the lock acquisition phase, a task τ_a
blocks if another task τ_b is currently within a critical region associated with the same
lock. The block takes place regardless of the relative priorities of the tasks. It lasts at
least until τ_b leaves the critical region and possibly more, if other tasks are waiting
to enter their critical region, too.

As shown in Figure 4.10 if, for any reason, τ_b is delayed (or, even worse, halted
due to a malfunction) while it is within its critical region, τ_a and any other tasks
willing to enter a critical region associated with the same lock will be blocked and
possibly be unable to make any further progress.

Even though τ_b proceeds normally, if the priority of τ_a is higher than the priority
of τ_b, the way mutual exclusion is implemented goes against the concept of task
priority, because a higher-priority task is forced to wait until a lower-priority task
has completed part of its activities. Even though, as will be shown in Chapter 6, it is
possible to calculate the worst-case blocking time suffered by higher-priority tasks
for this reason and place an upper bound on it, certain classes of applications may
not tolerate it in any case.

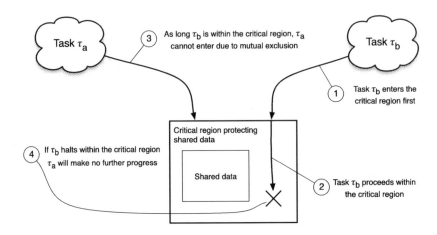

Figure 4.10   Shortcoming of lock-based synchronization when a task halts.

Lock-free and wait-free communication do not require mutual exclusion, and hence, they completely solve the blocking time issue. With lock-free communication it is indeed possible that an operation on a shared object fails, due to a concurrent object access from another task. However, the probability of such a failure is usually very low and it is possible to make the overall failure probability negligible by retrying the operation a small number of times.

The main drawback of lock-free and wait-free communication is that their inner workings are considerably more complex than any traditional inter-task communication and synchronization mechanisms. However, the recent development and widespread availability of open-source libraries containing a collection of lock-free data structures, for example, the Concurrent Data Structures library (libcds) [100] will likely bring what is today considered an advanced topic in concurrent programming into the mainstream.

4.7 SUMMARY

This chapter provided an overview of two main execution models generally adopted in embedded system design and development. The first one, described in Sections 4.1 and 4.2, has the clear advantages of being intuitive for programmers and efficient for what concerns execution.

On the other hand, it also has several shortcomings, outlined in Section 4.3, which hinder its applicability as system size and complexity grow. Historically, this consideration led to the introduction of more sophisticated execution models, in which the concept of task is preserved and plays a central role at runtime.

The basic concepts of task-based scheduling were illustrated in Sections 4.4, while Section 4.5 contains more information about the task state diagram, a key data structure that represents the evolution of a task through its lifetime in the system.

However, task-based scheduling also brings out a whole new class of programming complications, which arose only in rare cases with the cyclic executive model related to race conditions in shared data access.

In this chapter, Section 4.6 introduced the reader to the general problem of race conditions in task-based scheduling and how to address it by correctly identifying critical regions in the code and managing them appropriately by means of mutual exclusion. The same section also provided some pointers to alternate ways of solving the problem by means of lock- and wait-free inter-task communication techniques.

Staying with traditional techniques, a proper implementation of mutual exclusion requires an adequate knowledge of a programming technique known as concurrent programming, which will be the subject of the next chapter.

5 Concurrent Programming Techniques

CONTENTS

A side effect of the adoption of task-based scheduling is the need for task communication and synchronization. All operating systems invariably provide a rich selection of communication and synchronization primitives but, without a solid theoretical background, it is easy to misuse them and introduce subtle, time-dependent errors in the software.

This chapter provides the necessary background and bridges it into software development practice by presenting the main communication and synchronization primitives provided by the FREERTOS open-source operating system [18]. On the one hand, this RTOS is small/simple enough to be discussed in a small amount of space. On the other hand, it is comprehensive enough to give readers a thorough introduction to the topic.

Interested readers could refer to the FREERTOS reference manual [19], for more detailed information. The next chapter contains an introduction to real-time task scheduling algorithms and the related scheduling analysis theory for a few simple cases of practical relevance.

5.1 TASK MANAGEMENT

In all real-time operating systems, the concept of task as unit of scheduling plays a central role. Therefore, they all provide a set of primitives to instantiate new tasks and manage them, and FREERTOS is no exception. Table 5.1 summarizes the main task management and scheduling primitives provided by FREERTOS. To use them, it is necessary to include the main FREERTOS header file, `FreeRTOS.h`, followed by `task.h`.

In the rightmost column of the table, some primitives are marked as *optional*. This is because FREERTOS, like other operating systems designed for small em-

Table 5.1

Task Management Primitives of FREERTOS

Function	Purpose	Optional
vTaskStartScheduler	Start the scheduler	-
vTaskEndScheduler	Stop the scheduler	-
xTaskCreate	Create a new task	-
vTaskDelete	Delete a task given its handle	*
uxTaskPriorityGet	Get the priority of a task	*
vTaskPrioritySet	Set the priority of a task	*
vTaskSuspend	Suspend a specific task	*
vTaskResume	Resume a specific task	*
xTaskResumeFromISR	Resume a specific task from an ISR	*
xTaskIsTaskSuspended	Check whether a task is suspended	*
vTaskSuspendAll	Suspend all tasks but the running one	-
xTaskResumeAll	Resume all tasks	-
uxTaskGetNumberOfTasks	Return current number of tasks	-

bedded systems, can be configured to achieve different trade-offs between footprint and the richness of the operating system interface. Namely, a small set of operating system primitives is always provided, whereas others can be included or excluded from the configuration by defining appropriate macros in the system configuration file FreeRTOSConfig.h. Readers are referred to References [18, 19] for further information about FREERTOS configuration.

When using FREERTOS, all tasks share the same address space among them and with the operating system itself. This approach keeps the operating system as simple as possible, and makes it easy and efficient to share variables, because all tasks automatically and implicitly share memory. On the other hand, it becomes harder to protect tasks from each other with respect to illegal memory accesses, but it should be noted that many microcontrollers intended for embedded applications lack any hardware support for this protection anyway.

For some processor architectures, FREERTOS is able to use a memory protection unit (MPU), when available, to implement a limited form of memory access protection between tasks and the operating system itself. In this case, the operating system interface becomes slightly more complex because it becomes necessary to configure memory protection and specify how it applies to individual tasks. Even though a full discussion of this feature is outside the scope of this book, Chapter 15 contains more information about the inner workings of typical MPU hardware. Interested readers will find more information about software-related aspects in Reference [19].

Practically, the operating system is implemented as a library of object modules. Hence, as described in Chapter 3, the application program—comprising a set of tasks—is then linked against it when the application's executable image is built, exactly as any other library. The application and the operating system modules are therefore bundled together in the resulting executable image.

For reasons more thoroughly discussed in Chapter 2, the executable image is usually stored in a nonvolatile memory within the target system (flash memories are in widespread use nowadays). When the system is turned on, the image entry point is invoked either directly, through the processor reset vector, or indirectly, by means of a minimal boot loader. When no boot loader is present, the executable image must also include an appropriate startup code, which takes care of initializing the target hardware before executing the `main()` C-language application entry point.

Another important aspect to remark on is that, when `main()` gets executed, the operating system scheduler is *not yet active*. It can be explicitly started and stopped by means of the following function calls:

```
void vTaskStartScheduler(void);
void vTaskEndScheduler(void);
```

It should be noted that the function `vTaskStartScheduler` reports errors back to the caller in an unusual way:

- When it is able to start the scheduler successfully, `vTaskStartScheduler` *does not return* to the caller at all because, in that case, execution proceeds with the tasks that have been created before starting the scheduler.
- On the other hand, `vTaskStartScheduler` may return to the caller for two distinct reasons, one normal and one abnormal:
  1. The scheduler was successfully started, but one of the tasks then stopped it by invoking `vTaskEndScheduler`.
  2. The scheduler was not started at all because an error occurred.

The two scenarios in which `vTaskStartScheduler` returns to the caller are clearly very different because, in the first case, the return is delayed and it usually occurs when the application is shut down in an orderly manner, for instance, at the user's request. On the contrary, in the second case the return is almost immediate and the application tasks are never actually executed.

However, since `vTaskStartScheduler` has no return value, there is no trivial way to distinguish between them. If the distinction is important for the application being developed, then the programmer must make the necessary information available on his or her own, for example, by setting a shared flag after a full and successful application startup so that it can be checked by the code that follows `vTaskStartScheduler`.

It is possible to create a new FreeRTOS task either before or after starting the scheduler, by calling the `xTaskCreate` function:

```
BaseType_t xTaskCreate(
    TaskFunction_t pvTaskCode,
    const char * const pcName,
    unsigned short usStackDepth,
    void *pvParameters,
    UBaseType_t uxPriority,
    TaskHandle_t *pvCreatedTask);
```

The only difference between the two cases is that, if a task is created when the scheduler is already running, it is eligible for execution immediately and may even preempt its creator. On the other hand, tasks created before starting the scheduler are not eligible for execution until the scheduler is started.

The arguments of xTaskCreate are defined as follows:

- pvTaskCode is a pointer to a function returning void and with one void * argument. It represents the entry point of the new task, that is, the function that the task will start executing from. This function must be designed to *never* return to the caller because this operation has undefined results in FREERTOS.

- pcName, a constant string of characters, represents the human-readable name of the task being created. The operating system simply stores this name along with the other task information it keeps track of, without interpretation, but it is useful when inspecting the operating system data structures, for example, as happens during debugging. The maximum length of the name actually stored by the operating system is limited by a configuration parameter; longer names are silently truncated.

- usStackDepth indicates how much memory must be allocated for the task stack. It is not expressed in bytes, but in machine *words*, whose size depends on the underlying hardware architecture. If necessary, the size of a stack word can be calculated by looking at the StackType_t data type, defined in the architecture-dependent part of FREERTOS itself.

- pvParameters is a void * pointer that will be passed to the task entry point, indicated by pvTaskCode, when task execution starts. The operating system does not interpret this pointer in any way. It is most commonly used to point to a shared memory structure that holds the task parameters and, possibly, its return values.

- uxPriority represents the initial, or baseline priority of the new task, expressed as a non-negative integer. The symbolic constant tskIDLE_PRIORITY, defined in the operating system's header files, gives the priority of the idle task, that is, the lowest priority in the system, and higher values correspond to higher priorities.

- pvCreatedTask points to the task *handle*, which will be filled if the task has been created successfully. The handle must be used to refer to the new task in the future by means of other operating system primitives.

The return value of xTaskCreate is a status code. If its value is pdPASS, the function was successful in creating the new task and filling the memory area pointed by pvCreatedTask with a valid handle, whereas any other value means that an error occurred and may convey more information about the nature of the error. The FREERTOS header projdefs.h, automatically included by FreeRTOS.h, contains the full list of error codes that may be returned by any operating system function.

For what concerns task execution, FREERTOS implements a *fixed-priority* scheduler in which, informally speaking, the highest-priority task ready for execution gets executed. As described in Sections 6.1 and 6.2, this scheduler is not only very intuitive and easy to understand, but also has several very convenient properties for real-time applications.

The total number of priority levels available for use is set in the operating system configuration. It must be determined as a trade-off between application requirements and overhead, because the size of several operating system data structures, and hence, the operating system memory requirements, depend on it. The currently configured value is available in the symbolic constant `configMAX_PRIORITIES`. Hence, the legal range of priorities in the system goes from `tskIDLE_PRIORITY` to `tskIDLE_PRIORITY+configMAX_PRIORITIES−1`, extremes included.

After creation, a task can be deleted by means of the function

```
void vTaskDelete(TaskHandle_t xTaskToDelete);
```

Its only argument, `xTaskToDelete`, is the handle of the task to be deleted. A task may also delete itself and, in this case, `vTaskDelete` will not return to the caller.

It is important to remark that, in FREERTOS, task deletion is *immediate* and *unconditional*. Therefore, the target task may be deleted and cease execution at any time and location in the code. This may lead to unforeseen side effects when, generally speaking, the task is holding some *resources* at the time of deletion. As a consequence, those resources will never be released and made available to other tasks again. See Section 5.3 for an example of this undesirable scenario.

Moreover, for technical reasons, the memory dynamically allocated to the task by the operating system (for instance, to store its stack) cannot be freed immediately during the execution of `vTaskDelete` itself; this duty is instead delegated to the idle task. If the application makes use of `vTaskDelete`, it is important to ensure that a portion of the processor time is available to the idle task. Otherwise, the system may run out of memory not because there is not enough, but because the idle task was unable to free it fast enough before it needed to be reused.

After creation, it is possible to change the priority of a task or retrieve it by means of the functions

```
void vTaskPrioritySet(TaskHandle_t xTask,
    UBaseType_t uxNewPriority);
```

```
UBaseType_t uxTaskPriorityGet(TaskHandle_t xTask);
```

Both functions take a task handle, `xTask`, as their first argument. The special value NULL can be used as a shortcut to refer to the calling task.

The function `vTaskPrioritySet` modifies the priority of a task after it has been created, and `uxTaskPriorityGet` returns the current priority of the task. It should, however, be noted that both the priority given at task creation and the priority set by `vTaskPrioritySet` represent the *baseline* priority of the task.

Instead, `uxTaskPriorityGet` returns the *active* priority of a task, which may differ from its baseline priority when the *priority inheritance* protocol for semaphore-based mutual exclusion, to be discussed in Sections 5.3 and 6.2, is in use.

The functions `vTaskSuspend` and `vTaskResume` take an argument of type `xTaskHandle` according to the following prototypes:

```
void vTaskSuspend(TaskHandle_t xTaskToSuspend);
void vTaskResume(TaskHandle_t xTaskToResume);
```

They are used to suspend and resume the execution of the task identified by their argument, respectively. In other words, these functions are used to move a task into and from state 4 of the task state diagram shown in Figure 4.8. For `vTaskSuspend`, the special value `NULL` can be used to suspend the invoking task, whereas, obviously, it makes no sense for a task to attempt to resume itself.

It should be noted that `vTaskSuspend` may suspend the execution of a task at an arbitrary point. Like `vTaskDelete`, it must therefore be used with care because any resources that the task is holding are not implicitly released while the task is suspended. As a consequence, any other tasks willing to access the same resources may have to wait until the suspended task is eventually resumed.

FREERTOS, like most other monolithic operating systems, does not hold a full task control block for interrupt handlers, and hence, they are not full-fledged tasks. One of the consequences of this design choice is that interrupt handlers cannot invoke many operating system primitives, in particular the ones that may block the caller. In other cases a specific variant of some primitives is provided, to be used specifically by interrupt handlers, instead of the normal one.

For this reason, for instance, calling `vTaskSuspend(NULL)` from an interrupt handler is forbidden. For related operating system design constraints, interrupt handlers are also not allowed to suspend regular tasks by invoking `vTaskSuspend` with a non-`NULL` `xTaskHandle` as argument. The function

```
UBaseType_t xTaskResumeFromISR(TaskHandle_t xTaskToResume);
```

is the variant of `vTaskResume` that must be used to resume a task from an interrupt handler, also known as interrupt service routine (ISR) in the FREERTOS nomenclature.

Since, as said above, interrupt handlers do not have a full-fledged, dedicated task context in FREERTOS, `xTaskResumeFromISR` cannot immediately and automatically perform a full context switch to a new task when needed, as its regular counterpart would do. An explicit context switch would be necessary, for example, when a low-priority task is interrupted and the interrupt handler resumes a higher-priority task.

On the contrary, `xTaskResumeFromISR` merely returns a nonzero value in this case, in order to make the invoking interrupt handler aware of the situation. In response to this indication, the interrupt handler must invoke the FREERTOS scheduling algorithm, as better discussed in Chapter 8, so that the higher-priority task just resumed will be considered for execution immediately after interrupt handling ends.

The *optional* function

```
UBaseType_t xTaskIsTaskSuspended(TaskHandle_t xTask);
```

can be used to tell whether or not a certain task, identified by its handle `xTask`, is currently suspended. Its return value is nonzero if the task is suspended, and zero otherwise.

The function

```
void vTaskSuspendAll(void);
```

suspends all tasks *but* the calling one. Symmetrically, the function

```
UBaseType_t xTaskResumeAll(void);
```

resumes all tasks suspended by `vTaskSuspendAll` and performs a context switch when necessary, for instance, when the priority of one of the resumed tasks is higher than the priority of the invoking task. In this case, the invoking task is later notified that it lost the processor for this reason because it will get a nonzero return value from `xTaskResumeAll`. Since the context switch is performed immediately within `xTaskResumeAll` itself, this function cannot be called from an interrupt handler.

Contrary to what could be expected, both `vTaskSuspendAll` and `xTaskResumeAll` are extremely efficient on single-processor systems, such as those targeted by FREERTOS. In fact, these functions are not implemented by suspending and resuming all tasks one by one—as their names would instead suggest—but by temporarily disabling the operating system scheduler. This operation requires little more work than updating a shared counter.

The last function related to task management simply returns the number of tasks currently present in the system, regardless of their states:

```
UBaseType_t uxTaskGetNumberOfTasks(void);
```

The count includes the calling task itself, as well as blocked and suspended tasks. Moreover, it may also include some tasks that have been deleted by `vTaskDelete`. This is a side effect of the delayed dismissal of the operating system's data structures associated with a task when the task is deleted.

5.2 TIME AND DELAYS

Most of the processing performed by a real-time system ought be strongly correlated, quite intuitively, with the notion of *time*. Accordingly, FREERTOS provides a small set of primitives, listed in Table 5.2, which help tasks to *keep track* of the elapsed time and *synchronize* their activities with time, by delaying their execution. In order to use them, it is necessary to include `FreeRTOS.h`, followed by `task.h`.

More sophisticated timing facilities are optionally available in recent versions of the operating system, whose discussion is outside the scope of this book. For those, authoritative reference information can be found in Reference [19]. It should however be noted that the additional sophistication comes at a cost. First of all, they

Table 5.2
Time-Related Primitives of FREERTOS

Function	Purpose	Optional
xTaskGetTickCount	Return current time, in ticks	-
xTaskGetTickCountFromISR	Return current time, in ticks, to an ISR	-
vTaskDelay	Relative time delay	*
vTaskDelayUntil	Absolute time delay	*

consume more system resources, because their implementation is based on an additional system task, or *daemon*. Secondly, their interface is more complex and less intuitive.

In addition to the time-related primitives just introduced, all FREERTOS primitives that may potentially block the caller, such as those discussed in Sections 5.3 and 5.4, allow the calling task to specify an upper bound to the blocking time, and hence, avoid unpredictable disruptions to its execution schedule.

In FREERTOS the data type `TickType_t` is used to represent both the current time and a time interval. The current time is simply represented by the integer number of clock *ticks* elapsed from an arbitrary absolute time reference, that is, when the operating system scheduler was first started. As a consequence, the reference changes whenever the system is booted and the FREERTOS representation of time, by itself, is not enough to provide a correct notion of real-world, wall-clock time.

For this, applications must rely on dedicated hardware components like real-time clocks (RTC), which are fortunately widely available nowadays, even on very low-cost microcontrollers. As an added benefit, those components are usually able to keep track of wall-clock time even when the system is shut down, provided they are given an adequate backup power supply.

The length of a tick depends on the operating system configuration and, to some extent, on hardware capabilities. The configuration macro `configTICK_RATE_HZ` represents the configured tick frequency in Hertz and may be useful to convert back and forth between the usual time measurement units and clock ticks. For instance, the quantity `1000000/configTICK_RATE_HZ` is the tick length, approximated as an integer number of microsecond.

The function

```
TickType_t xTaskGetTickCount(void);
```

returns the current time, expressed as the integral number of ticks elapsed since the operating system scheduler was started. Barring low-level implementation details, FREERTOS maintains its notion of current time by incrementing a *tick counter* at the frequency specified by `configTICK_RATE_HZ`. This function simply returns the current value of the tick counter to the caller.

In most microcontrollers, the tick counter data type is either a 16- or 32-bit unsigned integer, depending on the configuration. Therefore, it is important to consider

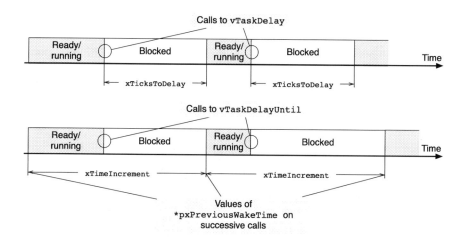

Figure 5.1   Relative versus absolute delays.

that the FREERTOS time counter will sooner or later wrap around and restart counting from zero. For instance, an unsigned, 32-bit counter incremented 1000 times per second—a common configuration choice for FREERTOS—will wrap around after about 1193 hours, that is, slightly more than 49 days.

It is therefore crucial that any application that must keep functioning correctly for a longer time without rebooting—as many real-time applications must do—is aware of the wraparound and handles it appropriately if it manipulates time values directly. If this is not the case, the application will be confronted with time values that suddenly "jump into the past" when a wraparound occurs, with imaginable consequences. The delay functions, to be discussed next, already handle time counter wraparound automatically, and hence, no special care is needed to use them.

Two distinct delay functions are available, depending on whether the delay should be *relative*, that is, measured with respect to the instant at which the delay function is invoked, or *absolute*, that is, until a certain instant in the future.

```
void vTaskDelay(const TickType_t xTicksToDelay);

void vTaskDelayUntil(TickType_t *pxPreviousWakeTime,
    const TickType_t xTimeIncrement);
```

- The function vTaskDelay implements a relative time delay. It blocks the calling task for xTicksToDelay ticks, then returns. As shown at the top of Figure 5.1, the time interval is relative to the time of the call, which is implicitly taken as a time reference point. The amount of delay from that reference point is fixed and is equal to xTicksToDelay.

- On the other hand, the function `vTaskDelayUntil` implements an absolute time delay. Referring to the bottom of Figure 5.1, the wake-up time for a certain invocation of the function is calculated as the wake-up time of the previous invocation, pointed by argument `pxPreviousWakeTime`, plus the time interval `xTimeIncrement`.

  Hence, the wake-up time is unrelated to the function invocation time and the amount of delay introduced by `vTaskDelayUntil`, which varies from an invocation to another, so that the delay between wake-up events is kept constant and equal to `xTimeIncrement`, regardless of what the task did in between.

  In extreme cases, the function might not block at all if the prescribed wake-up time is already in the past. Before returning, the function also increments the value pointed by `pxPreviousWakeTime` by `xTimeIncrement`, so that it is ready for the next call.

The right function to be used—and, as a consequence, the kind of delay introduced in task execution—depends on the purpose of the delay. A relative delay may be useful, for instance, if an I/O device must be allowed a certain amount of time to respond to a command. In this case, the delay must be measured from when the command has actually been sent to the device, and a relative delay makes sense.

On the other hand, an absolute delay is better when a task has to carry out an operation periodically because it guarantees that the period will stay constant even if the amount of time the task spends in the ready and running states—represented by the gray rectangles in Figure 5.1—varies from one instance to another.

5.3   SEMAPHORES

The first definition of a *semaphore* as a general intertask synchronization mechanism is due to Dijkstra [48]. The original proposal was based on busy wait, but most implementations found within contemporary operating systems use passive wait instead, without changing its semantics.

Over the years, semaphores have successfully been used to address many problems of practical significance in diverse concurrent programming domains even though, strictly speaking, they are not powerful enough to solve *every* concurrent programming problem that can be conceived [102].

Another reason of their popularity is that they are easy to implement in an efficient way. For this reason, virtually all operating systems offer semaphores as an intertask synchronization method.

According to its abstract definition, a semaphore is an object that contains two items of information, as shown in the right part of Figure 5.2:

- a *value* v, represented as a nonnegative integer, and
- a *queue* of tasks q, which are waiting on the semaphore.

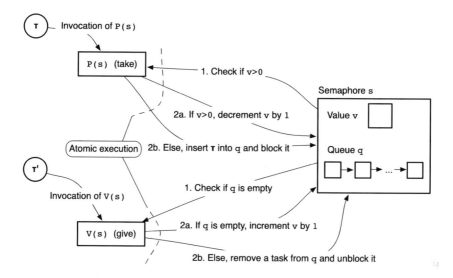

Figure 5.2   Abstract structure of a semaphore and behavior of its primitives.

The initial value of the semaphore is chosen (explicitly or implicitly) by the programmer upon semaphore creation, while the queue of waiting tasks is always initially empty.

Neither the value, nor the queue associated with a semaphore can be accessed or manipulated *directly* after initialization, albeit some implementation may make the current value of a semaphore available to programmers.

The only way to interact with a semaphore, and possibly modify its value and queue as a consequence, is to invoke the following abstract primitives, whose behavior is also depicted in Figure 5.2. An important assumption about those primitives is that their implementation ensures that they are executed *atomically*, that is, as indivisible units in which no context switches occur.

1. The primitive P(s)—called `take` in the FREERTOS nomenclature—when invoked on a semaphore s checks whether or not the current value of s is strictly greater than zero.
   - If the value of s is strictly greater than zero, P(s) decrements the value by one and returns to the caller. The calling task is not blocked in this case.
   - Otherwise, a reference to the calling task is inserted into the queue associated with s and the task is moved into the *Blocked* state of the task state diagram discussed in Section 4.5.
2. The primitive V(s)—called `give` by FREERTOS—when invoked on a semaphore s checks whether or not the queue associated with s is currently empty.
   - If the queue is empty, V(s) simply increments the value of s by one and returns to the caller. After the increment, the value of s will be strictly positive.

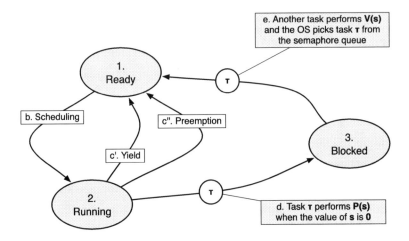

Figure 5.3 Task state diagram states and transitions involved in semaphore operations.

- Otherwise, the value of s is certainly zero. In this case, V(s) picks one of the tasks referenced by the queue associated with s, removes the reference from the queue, and unblocks it. In other words, referring to the task state diagram, the selected task is moved back to the *ready* state, so that it is eligible for execution again.

It must be noted that semaphore primitives are tied to the task state diagram because their execution may induce the transition of a task from one state to another. Figure 5.3 is a simplified version of the full FREERTOS task state diagram (Figure 4.8) that highlights the nodes and arcs involved in semaphore operations.

As shown in the figure, the transition of a certain task τ from the *running* to the *blocked* state caused by P(s) is voluntary because it depends on, and is caused by the invocation of the semaphore primitive by τ itself. On the contrary, the transition of τ back into the *ready* state is involuntary because it depends on an action performed by another task, namely, the task that invokes the primitive V(s). It cannot be otherwise because, as long as τ is blocked, it cannot proceed with execution and cannot perform any action on its own.

Another aspect worth mentioning is that, when a task (τ in this case) is unblocked as a consequence of a V(s), it goes into the *ready* state rather than *running*. Recalling how these states have been defined in Section 4.5, this means that the task is now *eligible* for execution again, but it does *not* imply that it shall *resume execution* immediately. This is useful to keep a proper separation of duties between the task synchronization primitives being described here and the task scheduling algorithms that will be presented in depth in Section 6.1. In this way:

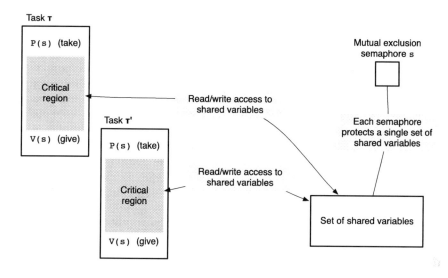

Figure 5.4   Usage of a semaphore and its primitives for mutual exclusion.

- The correct use of synchronization primitives stipulates whether or not a task may proceed in its execution, for instance, to avoid race conditions.
- The scheduling algorithms decide in which order tasks eligible to proceed are actually executed based on other criteria, for instance, their relative priority.

For similar reasons, the way semaphore queues are managed and, most importantly, *which* task is unblocked by V(s) among those found in the semaphore queue, is left unspecified in the abstract definition of semaphore. This is because the queue management strategy must often be chosen in concert with the scheduling algorithm, in order to ensure that task synchronization and scheduling work well together. More information about this topic will be given in Section 6.2. The same considerations are also true for the message passing primitives, which embed a synchronization mechanism and will be the topic of Section 5.4.

A widespread application of semaphores is to ensure mutual exclusion among a number of critical regions that make access to the same set of shared variables, a key technique to avoid race conditions as described in Section 4.6. Namely, the critical regions are associated with a semaphore s, whose initial value is 1. The critical region's entry and exit code are the primitives P(s) and V(s). In other words, as shown in Figure 5.4, those primitives are placed like "brackets" around the critical regions they must protect.

A full formal proof of correctness of the mutual exclusion technique just described is beyond the scope of this book. However, its workings will be described in a simple case, in order to give readers at least a reasonable confidence that it is indeed correct. As shown in Figure 5.4, the example involves two concurrent tasks, τ and τ', both willing to enter their critical regions. Both critical regions are associated with the same set of shared variables and are protected by the same semaphore s.

- For the sake of the example, let us imagine that τ executes its critical region entry code (that is, the primitive P(s)) first. It will find that the value of s is 1 (the initial value of the semaphore), it will decrement the value to 0, and it will be allowed to proceed into its critical region immediately, without blocking.
- If any other tasks, like τ' in the figure, execute the primitive P(s) they will be blocked because the current value of semaphore s is now 0. This is correct and necessary to enforce mutual exclusion, because task τ has been allowed to enter its critical region and it has not exited from it yet.
- It must also be noted that two or more executions of P(s) cannot "overlap" in any way because they are executed atomically, one after another. Therefore, even though multiple tasks may invoke P(s) concurrently, the operating system enforces an ordering among them to ensure that P(s) indeed works as intended.
- Task τ' will be blocked at least until task τ exits from the critical region and executes the critical region exit code, that is, V(s). At this point one of the tasks blocked on the semaphore and referenced by the semaphore queue, if any, will be unblocked.
- Assuming the choice fell on τ', it will proceed when the operating system scheduling algorithm picks it up for execution, then τ' will enter its critical region. Mutual exclusion is still ensured because any tasks previously blocked on the semaphore are still blocked at the present time. Moreover, any additional tasks trying to enter their critical region and invoking a P(s) will be blocked, too, because the semaphore value it still 0.
- When τ' exits from the critical region and executes V(s) two different outcomes are possible:
  1. If some other tasks are still blocked on the semaphore, the process described above repeats, allowing another task into its critical region. In this case, the semaphore value stays at 0 so that any "new" tasks trying to enter their critical region will be blocked.
  2. If the queue associated with s is empty, that is, no tasks are blocked on the semaphore, V(s) increments the value of s by one. As a result, the value of s becomes 1 and this brings the semaphore back to its initial state. Any task trying to enter its critical region in the future will be allowed to do so without blocking, exactly as happened to task τ in this example.

Another very important purpose of semaphores, besides mutual exclusion, is *condition synchronization*. With this term, we denote a form of synchronization in which a task must be blocked until a certain event happens or a condition is fulfilled.

For instance, let us consider an application in which a task τ_1 must transfer some information to another task, τ_2. As shown in Figure 5.5, an intuitive way to

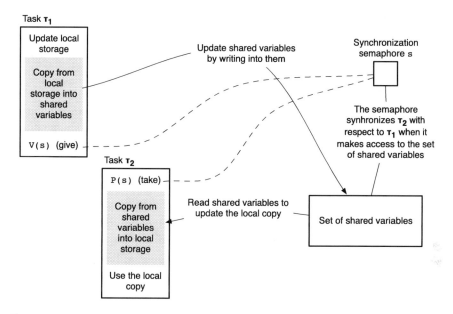

Figure 5.5 Usage of a semaphore and its primitives for condition synchronization.

implement the transfer is to allocate a set of shared variables large enough to hold the information to be transferred. Then, τ_1 writes the information into the shared variables after preparing it and τ_2 reads from the same variables, in order to create its local copy of the same information and use it.

Of course, τ_2 must synchronize with τ_1 so that it does not start reading from the shared variables before τ_1 has completely written into them the information to be transferred. If this is not the case, τ_2 may retrieve and use corrupted or inconsistent information. Namely, the goal is to *block* τ_2 until a certain condition is fulfilled. In this case, the condition is that the shared variables have been completely filled by τ_1.

We neglect for the time being that τ_1 may be "too fast" for τ_2, that is, we assume that τ_1 will never provide new information to be shared before τ_2 is done with the old one. Under this assumption the problem can be solved by using a *synchronization semaphore* s initialized to 0. Referring again to Figure 5.5, the semaphore is used by τ_1 and τ_2 in the following way:

- Before reading from the shared variables, τ_2 performs a P(s). Since the initial value of the semaphore is 0 this primitive blocks τ_2, unless τ_1 already filled the shared variables beforehand, as will be described in the following.
- After writing into the shared variables, τ_1 performs a V(s). This primitive has two distinct outcomes depending on whether or not τ_2 already executed its P(s) in the past:
  1. If τ_2 already executed its P(s) and is currently blocked, it will be unblocked. Then, τ_2 is free to proceed and read from the shared variables.

2. If, informally speaking, τ_2 "is late" and did not execute its P(s) yet, the value of s becomes 1. This signifies that the shared variables have been updated by τ_1 and are ready to be read by τ_2 in the future without further delay. In fact, when τ_2 will eventually execute P(s), it will simply bring the value of s back to 0 without blocking.

As shown in the previous example, synchronization semaphores are usually more difficult to use correctly with respect to mutual exclusion semaphores, because there is no fixed template to follow. Namely, both the initial value of synchronization semaphores and the position of the primitives that operate on them within the code may vary widely, depending on the purpose of the semaphores.

Even though there is only one theoretical definition of semaphore, the FREERTOS operating system actually provides four different semaphore implementations, or "flavors." This is a choice made by most operating systems, because each kind of semaphore represents a different trade-off between features and efficiency. It is the responsibility of the programmer to choose the best kind of semaphore case by case, bearing in mind which purpose a certain semaphore has in the system. The following paragraphs briefly describe the peculiarities of each kind of semaphore.

1. *Counting semaphores* are the most general kind of semaphore provided by FREERTOS. They are equivalent in most respects to the abstract semaphores discussed previously, but they are also the slowest. The main difference with respect to abstract semaphores is that, in the FREERTOS implementation, the maximum value that the semaphore value may legitimately assume must be declared in advance, when the semaphore is created.
2. *Binary semaphores* are less powerful than counting semaphores, because they have a maximum value as well as an initial value of one. As a consequence, their value can only be either one or zero, but they can still be used for either mutual exclusion or task synchronization. In return for this restriction, their implementation is somewhat faster than the one of counting semaphores.
3. *Mutex semaphores* are even more limited than binary semaphores, the additional restriction being that they must *only* be used as mutual exclusion semaphores. In other words, the P(s) and V(s) primitives on a mutex semaphore s must always appear in pairs and must be placed as brackets around critical regions. As a consequence, mutex semaphores cannot be used for task synchronization.
   In exchange for this, mutex semaphores implement *priority inheritance*. As discussed in Section 6.2, this is a key feature to guarantee a well-defined and acceptable interaction between task synchronization and scheduling from the point of view of real-time execution.
4. *Recursive mutex semaphores* have the same features as ordinary mutex semaphores. In addition, they also support the so-called *recursive* locks and unlocks, in which tasks are allowed to contain two (or more) nested critical regions, controlled by the same semaphore s and delimited by their own P(s) and V(s) brackets.

Table 5.3
Semaphore Creation/Deletion Primitives of FREERTOS

Function	Purpose	Optional
xSemaphoreCreateCounting	Create a counting semaphore	*
xSemaphoreCreateBinary	Create a binary semaphore	-
xSemaphoreCreateMutex	Create a mutex semaphore	*
xSemaphoreCreateRecursiveMutex	Create a recursive mutex	*
vSemaphoreDelete	Delete a semaphore (of any kind)	-

When s is an ordinary mutex semaphore, this arrangement leads to an indefinite wait because, when a task τ attempts to enter the inner critical region, it finds that the value of s is 0 and blocks. Clearly, this is an unwelcome side effect of the fact that τ itself previously entered the outer critical regions. In fact, as a general rule, τ blocks until the outer critical region is abandoned by the task that is currently executing within it. However, in this scenario that task is τ itself, and hence, τ will never be unblocked.

On the other hand, when s is a recursive mutex semaphore, the system prevents the issue by automatically taking and giving the semaphore only at the outermost critical region boundaries, as it should be.

The four kinds of semaphore just described are created by means of distinct functions, listed in Table 5.3. After creation, most kinds of semaphore (except recursive mutual exclusion semaphores) are used in a uniform way, by means of the same set of functions. In order to call any semaphore-related function, it is necessary to include two operating system headers, namely FreeRTOS.h and semphr.h. The first one contains general macros, data type definitions, and function prototypes pertaining to FREERTOS as a whole. The second one holds specific definitions related to semaphores.

The function

```
SemaphoreHandle_t xSemaphoreCreateCounting(
    UBaseType_t uxMaxCount,
    UBaseType_t uxInitialCount);
```

creates a counting semaphore with a given maximum (uxMaxCount) and initial (uxInitialCount) value. When successful, it returns to the caller a valid semaphore handle, a pointer to the internal operating system data structure representing the semaphore, of type SemaphoreHandle_t. If semaphore creation fails, the function returns a NULL pointer.

To create a binary semaphore, the function xSemaphoreCreateBinary should be used instead as:

```
SemaphoreHandle_t xSemaphoreCreateBinary(void);
```

Table 5.4

Semaphore Manipulation Primitives of FREERTOS

Function	Purpose	Optional
xSemaphoreTake	Perform a P() on a semaphore	-
xSemaphoreGive	Perform a V() on a semaphore	-
xSemaphoreTakeFromISR	P() from an interrupt handler	-
xSemaphoreGiveFromISR	V() from an interrupt handler	-
xSemaphoreTakeRecursive	P() on a recursive mutex	*
xSemaphoreGiveRecursive	V() on a recursive mutex	*

Like for `xSemaphoreCreateCounting`, the return value is either a valid semaphore handle upon successful completion, or a NULL pointer upon failure. Instead, since both the maximum and initial value of a binary semaphore are constrained to be 1, they are not explicitly indicated and the argument list is therefore empty. It should also be noted that binary semaphores are the only kind of semaphore that is always available for use in FREERTOS, regardless of how it has been configured. All the others are optional.

Mutual exclusion semaphores are created by means of two different functions, depending on whether the recursive lock and unlock feature is desired or not:

```
SemaphoreHandle_t xSemaphoreCreateMutex(void);
SemaphoreHandle_t xSemaphoreCreateRecursiveMutex(void);
```

Like the others, these creation functions also return either a valid semaphore handle upon successful completion, or a NULL pointer upon failure. All mutual exclusion semaphores are unlocked when they are first created—that is, their initial value is 1—and priority inheritance is always enabled for them.

A semaphore can be deleted by invoking the function:

```
void vSemaphoreDelete(SemaphoreHandle_t xSemaphore);
```

Its only argument is the handle of the semaphore to be deleted. This function must be used with care because the semaphore is destroyed immediately, even though there are some tasks waiting on it. Moreover, it is the programmer's responsibility to ensure that a semaphore handle will never be used by any task after the corresponding semaphore has been destroyed, a constraint that may not be trivial to enforce in a concurrent programming environment.

After being created, semaphores are acted upon by means of the functions listed in Table 5.4. Most kinds of semaphore except recursive, mutual exclusion semaphores are acted upon by means of the functions `xSemaphoreTake` and `xSemaphoreGive`, the FREERTOS counterpart of P() and V(), respectively. As shown in the following prototypes, both take a semaphore handle `xSemaphore` as their first argument:

```
BaseType_t xSemaphoreTake(SemaphoreHandle_t xSemaphore,
    TickType_t xBlockTime);
```

```
BaseType_t xSemaphoreGive(SemaphoreHandle_t xSemaphore);
```

Comparing the prototypes with the abstract definition of the same primitives, two differences are evident:

1. The abstract primitive P() may block the caller for an unlimited amount of time. Since, for obvious reasons, this may be inconvenient in a real-time system, the function xSemaphoreTake has a second argument, xBlockTime, that specifies the maximum blocking time. In particular:
   - If the value is portMAX_DELAY (a symbolic constant defined when the main FREERTOS header file is included), the function blocks the caller until the semaphore operation is complete. In other words, when this value of xBlockTime is used, the concrete function behaves in the same way as its abstract counterpart.
     For this option to be available, the operating system must be configured to support task suspend and resume, as described in Section 5.1.
   - If the value is 0 (zero), the function returns an error indication to the caller when the operation cannot be performed immediately.
   - Any other value is interpreted as the maximum amount of time the function will possibly block the caller, expressed as an integral number of clock *ticks*. See Section 5.2 for more information about time measurement and representation in FREERTOS.
2. According to their abstract definition, neither P() nor V() can ever fail. On the contrary, their real-world implementation may encounter an error for a variety of reasons. For example, as just described, xSemaphoreTake fails when a finite timeout is specified and the operation cannot be completed before the timeout expires. For this reason, the return value of xSemaphoreTake and xSemaphoreGive is a *status code*, which is pdTRUE if the operation was successful. Otherwise, they return pdFALSE.

Another important difference between the abstract definition of P() and V() with respect to their actual implementation is that, for reasons that will be better described in Chapter 8, neither xSemaphoreTake nor xSemaphoreGive can be invoked from an interrupt handler. In their place, the following two functions must be used:

```
BaseType_t xSemaphoreTakeFromISR(SemaphoreHandle_t xSemaphore,
    BaseType_t *pxHigherPriorityTaskWoken);
```

```
BaseType_t xSemaphoreGiveFromISR(SemaphoreHandle_t xSemaphore,
    BaseType_t *pxHigherPriorityTaskWoken);
```

Like all other FREERTOS primitives that can be invoked from an interrupt handler, these functions *never* block the caller. In addition, they return to the caller—in the variable pointed by pxHigherPriorityTaskWoken—an indication of whether or not they unblocked a task with a priority higher than the interrupted task.

The interrupt handler should use this information, as discussed in Chapter 8, to determine whether or not it should request the execution of the FREERTOS task scheduling algorithm before exiting. As before, the return value of these functions can be either `pdTRUE` or `pdFALSE`, depending on whether they were successful or not.

The last pair of special functions to be presented is the counterpart of `xSemaphoreTake` and `xSemaphoreGive` which must be used with recursive mutual exclusion semaphores:

```
BaseType_t xSemaphoreTakeRecursive(SemaphoreHandle_t xMutex,
    TickType_t xBlockTime);
```

```
BaseType_t xSemaphoreGiveRecursive(SemaphoreHandle_t xMutex);
```

Both their arguments and return values are the same as `xSemaphoreTake` and `xSemaphoreGive`, respectively.

One last aspect worth mentioning is the interaction between semaphore use and task deletion that, as discussed in Section 5.1 is immediate and unconditional. The high-level effect of deleting a task while it is *within* a critical region is therefore the same as the terminated task never exited from the critical region for some reason. Namely, no other tasks will ever be allowed to enter a critical region controlled by the same semaphore in the future.

Since this usually corresponds to a complete breakdown of any concurrent program, the direct invocation of `vTaskDelete` should usually be avoided, and it should be replaced by a more sophisticated deletion mechanism. One simple solution is to send a deletion *request* to the target task by some other means—for instance, one of the intertask communication mechanisms described in this section and the next one. The target task must be designed so that it responds to the request by terminating itself at a well-known location in its code after any required cleanup operation has been carried out and any mutual exclusion semaphore it held has been released.

Besides using semaphores, an alternative way to implement mutual exclusion is by means of the task suspend and resume primitives discussed in Section 5.1. In fact, in a single-processor system, `vTaskSuspendAll` opens a mutual exclusion region because the first task that successfully executes it will effectively prevent all other tasks from being executed until it invokes `xTaskResumeAll`.

As a side effect, any task executing between `vTaskSuspendAll` and `xTaskResumeAll` implicitly gets the highest possible priority in the system, except interrupt handlers. This consideration indicates that the method just described has two main shortcomings:

1. Any FREERTOS primitive that might block the caller for any reason even temporarily, or might require a context switch, must not be used within this kind of critical region. This is because blocking the only task allowed to run would completely lock up the system, and it is impossible to perform a context switch with the scheduler disabled.

2. Protecting critical regions with a sizable execution time in this way would probably be unacceptable in many applications because it leads to a large amount of unnecessary blocking. This is especially true for high-priority tasks, because if one of them becomes ready for execution while a low-priority task is engaged in a critical region of this kind, it will not run immediately, but only at the end of the critical region itself.

5.4 MESSAGE PASSING

The semaphore-based synchronization and communication methods discussed in Section 5.3 rely on two distinct and mostly independent mechanisms to implement synchronization and communication among tasks. In particular:

- Semaphores are essentially able to pass a *synchronization* signal from one task to another. For example, a semaphore can be used to block a task until another task has accomplished a certain action and make sure they operate together in a correct way.
- Data transfer takes place by means of *shared variables* because semaphores, by themselves, are unable to do so. Even though semaphores do have a value, it would be impractical to use it for data transfer, due to the way semaphores were conceived and defined.

Seen in a different way, the role of semaphores is to coordinate and enforce mutual exclusion and precedence constraints on task actions, in order to ensure that their access to shared variables takes place in the right sequence and at the appropriate time. On the other hand, semaphores are not actively involved in data transfer.

Message passing takes a radically different approach to task synchronization and communication by providing a *single* mechanism that is able, by itself, to provide both synchronization and data transfer at the same time and with the same set of primitives.

In this way, the mechanism not only works at a higher level of abstraction and becomes easier to use, but also it can be adopted with minimal updates when shared memory is not necessarily available. This happens, for example, in distributed systems where the communicating tasks may be executed by distinct computers linked by a communication network.

In its abstract form, a message passing mechanism is based upon two basic primitives, defined as follows:

- A *send* primitive, which transfers a certain amount of information, called a *message*, from one task to another. The invocation of the primitive may imply a synchronization action that blocks the caller if the data transfer cannot take place immediately.
- A *receive* primitive, which allows the calling task to retrieve the contents of a message sent to it by another task. Also in this case, the calling task blocks if the message it is seeking is not immediately available, thus synchronizing the receiver with the sender.

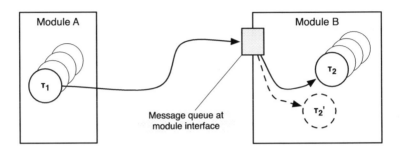

Figure 5.6   Indirect naming scheme in message passing.

Even if this definition still lacks many important lower-level details that will be discussed later, it is already clear that the most apparent effect of message passing primitives is to transfer a certain amount of information from the sending task to the receiving one. At the same time, message passing primitives also incorporate synchronization because they may block the caller when needed.

It should also be noted that, with message passing, mutual exclusion is not a concern because any given message is *never shared* among tasks. In fact, the message passing mechanism works as if the message were instantaneously copied from the sender to the receiver, so that message ownership is implicitly passed from the sender to the receiver when the message is transferred.

In this way, even if the sender modifies its local copy of a message after sending it, this will not affect the message already sent in the past. Symmetrically, the receiver is allowed to modify a message it received. This action is local to the receiver and does not affect the sender's local copy of the message in any way.

A key design aspect of a message passing scheme is how the sender identifies the intended recipient of a message and, symmetrically, how a receiver indicates which senders it is willing to receive messages from. In FREERTOS, the design is based upon an *indirect naming scheme*, in which the *send* and *receive* primitives are associated by means of a third, intermediate entity, known as *message queue*. Other popular names for this intermediate entity, used by other operating systems and in other contexts, are *mailbox* or *channel*.

As shown in Figure 5.6, even though adopting an indirect naming scheme looks more complex than, for instance, directly naming the recipient task when sending, it is indeed advantageous to software modularity and integration.

Let us assume that two software modules, *A* and *B*, each composed of multiple tasks and with a possibly elaborate internal structure, have to synchronize and communicate by means of message passing. In particular, a task τ_1 within module *A* has to send a message to a recipient within module *B*.

If we chose to directly name the recipient task when sending a message, then task τ_1 needs to know the internal structure of module *B* accurately enough to determine that the intended recipient is, for instance, τ_2 as is shown in the figure.

If the internal architecture of module B is later changed, so that the intended recipient becomes τ_2' instead of τ_2, module A must be updated accordingly. Otherwise, communication will no longer be possible or, even worse, messages may reach the wrong task.

On the contrary, if communication is carried out with an indirect naming scheme, module A and its task τ_1 must only know the name of the *message queue* that module B is using for incoming messages. Since the name of the mailbox is part of the interface of module B to the external world, it will likely stay the same even if the implementation or the internal design of the module itself changes over time.

An additional effect of communicating through a message queue rather that directly between tasks is that the relationship among communicating tasks becomes very flexible, albeit more complex. In fact, four scenarios are possible:

1. The simplest one is a *one-to-one* relationship, in which one task sends messages to another through the message queue.
2. A *many-to-one* relationship is also possible, in which multiple tasks send messages to a message queue, and a single task receives and handles them.
3. In a *one-to-many* relationship, a single task feeds messages into a message queue and multiple tasks receive from the queue. Unlike the previous two scenarios, this one and the following one cannot easily be implemented with direct inter-task communication.
4. The *many-to-many* relationship is the most complex one, comprising multiple sending and receiving tasks all operating on the same message queue.

Establishing a one-to-many or a many-to-many relationship does not allow the sending task(s) to determine exactly which task—among the receiving ones—will actually receive and handle its message. However, this may still be useful, for instance, to conveniently handle concurrent processing in software modules acting as servers.

In this case, those software modules will contain a number of equivalent "worker" tasks, all able to handle a single request at a time. All of them will be waiting for requests using the same message queue located at the module's boundary. When a request arrives, one of the workers will be allowed to proceed and get it. Then, the worker will process the request and provide an appropriate reply to the requesting task. Meanwhile, the other workers will still be waiting for additional requests and may start working on them concurrently.

As mentioned earlier, message passing primitives incorporate both data transfer *and* synchronization aspects. Albeit the data transfer mechanism by itself is straightforward, in order to design a working concurrent application, it is important to look deeper into how message queue synchronization works.

As illustrated in Figure 5.7, message queues enforce two synchronization constraints, one for each side of the communication:

1. The receive primitive blocks the caller when invoked on an *empty* message queue. The invoking task will be unblocked when a message is sent to the queue. If multiple tasks are blocked trying to receive from a message queue when a message

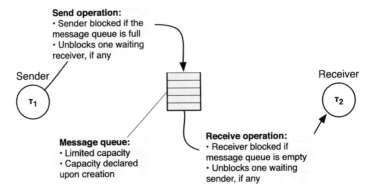

Figure 5.7 Message queue synchronization model.

arrives, the operating system selects and unblocks exactly one of them. That task receives the incoming message, whereas the others wait until further messages are sent to the queue.

2. The send primitive blocks the caller when the message queue it operates on is *full*, that is, the number of messages it currently contains is equal to its maximum capacity, declared upon creation. When a task receives a message from the queue, the operating system selects one of the tasks waiting to send, unblocks it, and puts its message into the message queue. The other tasks keep waiting until more space becomes available in the message queue.

It should also be noted that many operating systems (including FREERTOS) offer a *nonblocking* variant of the send and receive primitives. Even though these variants may sometimes be useful from the software development point of view, they will not be further discussed here because, in that case, synchronization simply does not occur.

Another popular variant is a *timed* version of send and receive, in which it is possible to specify the maximum amount of time the primitives are allowed to block the caller. If the operation cannot be completed within the allotted time, the caller is unblocked anyway and the primitives return an error indication.

Under this synchronization model, considering the case in which the message queue is neither empty nor full, messages flow *asynchronously* from the sender to the receiver. No actual synchronization between those tasks actually occurs, because neither of them is blocked by the message passing primitives it invokes.

When necessary, stricter forms of synchronization can be implemented starting from these basic primitives. In particular, as shown in Figure 5.8:

- In a *synchronous* message transfer, often called *rendezvous*, the sending task τ_1 is blocked until the receiving task τ_2 is ready to receive the message. Moreover, as before, the receiving task is blocked by the receive primitive until a message is available.

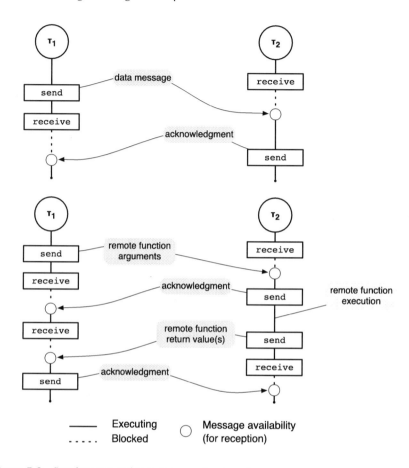

Figure 5.8 Synchronous and remote invocation transfers.

As shown on the top half of Figure 5.8, a synchronous transfer from τ_1 to τ_2 can be realized by means of two asynchronous message transfers going in opposite directions. The first transfer (from τ_1 to τ_2) carries the actual message to be transferred, and the second one (from τ_2 to τ_1) holds an acknowledgment.

A peculiarity of the second message transfer is that it is not used to actually move data between tasks—the transferred message may, in fact, be empty—but only for its synchronization semantics. Its purpose is to block τ_1 until τ_2 has successfully received the data message and subsequently sent the reply.

- A *remote invocation* message transfer, also known as *extended rendezvous*, enforces an even stricter synchronization between communicating tasks. Namely, when task τ_1 sends a remote invocation message to task τ_2, it is blocked until a reply message is sent back from τ_2 to τ_1. Symmetrically—

and here is the difference with respect to a synchronous transfer—τ_2 is blocked until that reply has been successfully received by τ_1.

As the name suggests, this synchronization model is often used to imitate a function call, or invocation, using message passing. As in a regular function call, the requesting task τ_1 prepares the arguments of the function it wants task τ_2 to execute. Then, it puts them into a request message and sends the message to τ_2, which will be responsible to execute the requested function. At the same time, τ_1 blocks and will not proceed further, until the function results become available.

Meanwhile, τ_2 has received the request and, through a local computation, executes the requested function and eventually generates the reply message. When the reply is ready, τ_2 sends it to τ_1 and unblocks it. It should be noted that at this stage τ_2 blocks until the reply has been received by τ_1. In this way, τ_1 can make sure that the reply has reached its intended destination, or at least be notified if an error occurred.

As shown in the bottom half of Figure 5.8, a remote invocation from τ_1 to τ_2 can be implemented in terms of two synchronous message transfers going in opposite directions, that is, a total of four asynchronous message transfers.

Although, as has been shown previously, asynchronous message transfers can be used as a "building block" for constructing the most sophisticated ones—and are therefore very flexible—they have some drawbacks as well. As has been remarked, for instance, in Reference [25] the most important concern is probably that asynchronous message transfers give "too much freedom" to the programmer, somewhat like the "goto" statement of unstructured sequential programming.

The resulting programs are therefore more complex to understand and check for correctness, also due to the proliferation of explicit message passing primitives in the code. This observation leads us to remark on the importance of good programming practice to avoid using those message passing primitives directly if at all possible, but encapsulate them within higher-level communication functions, which implement more structured semantics.

The FREERTOS implementation of message queues follows very closely the abstract definition just given, with some added features tailored for real-time applications. Incidentally, FREERTOS makes use of message queues to implement semaphores, too. For this reason, all the primitives that operate on message queues, summarized in Table 5.5, are always available and cannot be removed from the operating system configuration. To use them, it is necessary to include the operating system headers FreeRTOS.h and queue.h.

The function xQueueCreate creates a new message queue and has the following prototype:

```
QueueHandle_t xQueueCreate(
    UBaseType_t uxQueueLength,
    UBaseType_t uxItemSize);
```

Table 5.5
Message Passing Primitives of FREERTOS

Function	Purpose	Optional
xQueueCreate	Create a message queue	-
vQueueDelete	Delete a message queue	-
xQueueSendToBack	Send a message	-
xQueueSend	Same as xQueueSendToBack	-
xQueueSendToFront	Send a high-priority message	-
xQueueOverwrite	Forcefully send	-
xQueueSendToBackFromISR	Send a message from interrupt handler	-
xQueueSendFromISR	Same as xQueueSendToBackFromISR	-
xQueueSendToFrontFromISR	Send a h.-p. message from interrupt handler	-
xQueueOverwriteFromISR	Forcefully send from interrupt handler	-
xQueueReceive	Receive a message	-
xQueueReceiveFromISR	Receive a message from interrupt handler	-
xQueuePeek	Nondestructive receive	-
xQueuePeekFromISR	Nondestructive receive from interrupt handler	-
uxQueueMessagesWaiting	Query queue length	-
uxQueueMessagesWaitingFromISR	Query queue length from interrupt handler	-
xQueueIsQueueEmptyFromISR	Check if a queue is empty from interrupt handler	-
xQueueIsQueueFullFromISR	Check if a queue is full from interrupt handler	-

Its two arguments specify the maximum number of elements the newly created message queue can contain, uxQueueLength, and the size of each element, uxItemSize, expressed in bytes. Upon successful completion, the function returns a valid message queue handle to the caller, which represents the message queue just created and must be used for any subsequent operation on it. If an error occurs, the function returns a NULL pointer instead.

When a message queue is no longer in use, it is useful to delete it, in order to reclaim the memory allocated to it for future use. This is done by means of the function:

```
void vQueueDelete(QueueHandle_t xQueue);
```

It should be noted that the deletion of a FREERTOS message queue takes place *immediately* and is never delayed, even though some tasks are currently blocked because they are engaged in a send or receive primitive on the message queue itself. The effect of the deletion on the waiting tasks depends on whether or not they specified a time limit for the execution of the primitive:

- if they did so, they will receive an error indication when a timeout occurs;
- otherwise, they will be blocked forever.

Since this kind of behavior—especially the second one—is often undesirable in a real-time system, it is the responsibility of the programmer to ensure that message queues are deleted only when no tasks are blocked on them, when those tasks will no longer be needed in the future, or when they have other ways to recover (for instance, by means of a timeout mechanism).

After a message queue has been successfully created and its xQueue handle is available for use, it is possible to send a message to it by means of the functions

```
BaseType_t xQueueSendToBack(
    QueueHandle_t xQueue,
    const void *pvItemToQueue,
    TickType_t xTicksToWait);

BaseType_t xQueueSendToFront(
    QueueHandle_t xQueue,
    const void *pvItemToQueue,
    TickType_t xTicksToWait);
```

The first function, xQueueSendToBack, sends a message to the back of a message queue. The function xQueueSend is totally equivalent to xQueueSendToBack and has the same prototype. The message to be sent is pointed by the pvItemToQueue argument, whereas its size is implicitly assumed to be equal to the size of a message queue item, as declared when the queue was created.

The last argument, xTicksToWait, specifies the maximum amount of time allotted to the operation. The values it can assume are the same as for semaphore primitives. Namely:

- If the value is portMAX_DELAY, the behavior of the primitive is the same as its abstract counterpart. When the message queue is full, the function blocks the caller until the space it needs becomes available. For this option to be available, the operating system must be configured to support task suspend and resume, as described in Section 5.1.
- If the value is 0 (zero), the function returns an error indication to the caller when the operation cannot be performed immediately because the message queue is full at the moment.
- Any other value is interpreted as the maximum amount of time the function will possibly block the caller, expressed as an integral number of clock *ticks*. See Section 5.2 for more information about ticks and time measurement in FREERTOS.

The return value of xQueueSendToBack is pdPASS if the function was successful, whereas any other value means than an error occurred. In particular, the error code errQUEUE_FULL means that the function was unable to send the message within the maximum amount of time specified by xTicksToWait because the queue was full.

FREERTOS message queues are normally managed in first-in, first-out (FIFO) order. However, a high-priority message can be sent using the xQueueSendToFront function instead of xQueueSendToBack. The only difference between those two functions is that xQueueSendToFront sends the message to the *front* of the message queue, so that it passes over the other messages stored in the queue and will be received before them.

Another possibility is to forcefully send a message to a message queue even though it is full by *overwriting* one message already stored into it in the past. This is generally done on message queues with a capacity of one, with the help of the following function:

```
BaseType_t xQueueOverwrite(
    QueueHandle_t xQueue,
    const void * pvItemToQueue);
```

Unlike for the previous functions, the argument xTicksToWait is not present because this function always completes its work immediately. As before, the return value indicates whether the function was successful or not.

Neither   xQueueSendToBack   nor   xQueueSendToFront   nor xQueueOverwrite can be invoked from an interrupt handler. Instead, either xQueueSendToBackFromISR or xQueueSendToFrontFromISR or xQueueOverwriteFromISR must be called in this case, according to the following prototypes:

```
BaseType_t xQueueSendToBackFromISR(
    QueueHandle_t xQueue,
    const void *pvItemToQueue,
    BaseType_t *pxHigherPriorityTaskWoken);

BaseType_t xQueueSendToFrontFromISR(
    QueueHandle_t xQueue,
    const void *pvItemToQueue,
    BaseType_t *pxHigherPriorityTaskWoken);

BaseType_t xQueueOverwriteFromISR(
    QueueHandle_t xQueue,
    const void *pvItemToQueue
    BaseType_t *pxHigherPriorityTaskWoken);
```

As before, the function xQueueSendFromISR is totally equivalent to xQueueSendToBackFromISR.

The differences of these functions with respect to their regular counterparts are:

- They never block the caller, and hence, they do not have a xTicksToWait argument. In other words, they always behave as if the timeout were 0, so that they return an error indication to the caller if the operation cannot be concluded immediately.
- The argument pxHigherPriorityTaskWoken points to a BaseType_t variable. The function will set the referenced variable to either pdTRUE or pdFALSE, depending on whether or not it awakened a task with a priority higher than the task which was running when the interrupt handler started.

The interrupt handler should use this information, as discussed in Chapter 8, to determine if it should invoke the FREERTOS scheduling algorithm before exiting.

Messages are always received from the front of a message queue by means of the following functions:

```
BaseType_t xQueueReceive(
    QueueHandle_t xQueue,
    void *pvBuffer,
    TickType_t xTicksToWait);

BaseType_t xQueuePeek(
    QueueHandle_t xQueue,
    void *pvBuffer,
    TickType_t xTicksToWait);
```

The first argument of these functions is a message queue handle xQueue, of type QueueHandle_t, which indicates the message queue they will work upon.

The second argument, pvBuffer, is a pointer to a memory buffer into which the function will store the message just received. The memory buffer must be large enough to hold the message, that is, at least as large as a message queue item.

The last argument, xTicksToWait, specifies how much time the function should wait for a message to become available if the message queue was completely empty when the function was invoked. The valid values of xTicksToWait are the same as already mentioned when discussing xQueueSendToBack.

The function xQueueReceive, when successful, *removes* the message it just received from the message queue, so that each message sent to the queue is received exactly once. On the contrary, the function xQueuePeek simply *copies* the message into the memory buffer indicated by the caller without removing it for the queue.

The return value of xQueueReceive and xQueuePeek is pdPASS if the function was successful, whereas any other value means that an error occurred. In particular, the error code errQUEUE_EMPTY means that the function was unable to receive a message within the maximum amount of time specified by xTicksToWait because the queue was empty. In this case, the buffer pointed by pvBuffer will not contain any valid message after these functions return to the caller and its contents shall not be used.

The functions xQueueReceiveFromISR and xQueuePeekFromISR are the variants of xQueueReceive and xQueuePeek, respectively, which must be used within an interrupt handler. Neither of them block the caller. Moreover, xQueueReceiveFromISR returns to the caller—in the variable pointed by pxHigherPriorityTaskWoken—an indication on whether or not it awakened a task with a higher priority than the interrupted one.

There is no need to do the same for xQueuePeekFromISR because it never frees any space in the message queue, and hence, it never wakes up any tasks. The two functions have the following prototype:

```
BaseType_t xQueueReceiveFromISR(
    QueueHandle_t xQueue,
    void *pvBuffer,
    BaseType_t *pxHigherPriorityTaskWoken);

BaseType_t xQueuePeekFromISR(
    QueueHandle_t xQueue,
    void *pvBuffer);
```

The functions belonging to the last group:

```
UBaseType_t
    uxQueueMessagesWaiting(const QueueHandle_t xQueue);

UBaseType_t
    uxQueueMessagesWaitingFromISR(const QueueHandle_t xQueue);

BaseType_t
    xQueueIsQueueEmptyFromISR(const QueueHandle_t xQueue);

BaseType_t
    xQueueIsQueueFullFromISR(const QueueHandle_t xQueue);
```

query various aspects of a message queue status. In particular,

- `uxQueueMessagesWaiting` and `uxQueueMessagesWaitingFromISR` return the number of items currently stored in the message queue `xQueue`. As usual, the latter variant must be used when the invoker is an interrupt handler.
- `xQueueIsQueueEmptyFromISR` and `xQueueIsQueueFullFromISR` return the Boolean value `pdTRUE` if the message queue `xQueue` is empty (or full, respectively) and `pdFALSE` otherwise. Both can be invoked safely from an interrupt handler.

These functions should be used with caution because, although the information they return is certainly accurate at the time of the call, the scope of its validity is somewhat limited. It is worth mentioning, for example, that the information may *no longer* be valid and should not be relied upon when any subsequent message queue operation is attempted because other tasks may have changed the queue status in the meantime.

For example, the preventive execution of `uxQueueMessageWaiting` by a task, with a result greater than zero, is not enough to guarantee that the same task will be able to immediately conclude a `xQueueReceive` with a non-zero `xTicksToWait` in the immediate future.

This is because other tasks, or interrupt handlers, may have received messages from the queue and emptied it completely in the meantime. On the contrary, `xQueueReceive` primitive, with `xTicksToWait` equal to 0, has been specifically designed to work as intended in these cases.

5.5   SUMMARY

This chapter contains an introduction to *concurrent programming* from the practical point of view, within the context of a real-world real-time operating system for embedded applications. Starting with the all-important concept of *task*, the basic unit of scheduling in real-time execution that has been discussed in Section 5.1, it was then possible to introduce readers to the main inter-task communication and synchronization mechanisms. This was the subject of Sections 5.3 and 5.4, which described the basic concept as well as the practical aspects of semaphore-based synchronization and message-passing, respectively.

Because this book focused on *real-time embedded systems*, rather than general-purpose computing, Section 5.2 also presented in detail how real-time operating systems manage time and timed delays and which primitives they make available to users for this purpose. The discussion precedes inter-process communication and synchronization because, in a real-time operating environment, the latter are obviously subject to timing constraints, too. These constraints are most often expressed by means of a timeout mechanism, which has many analogies with a timed delay.

6 Scheduling Algorithms and Analysis

CONTENTS

In the previous chapter, we informally introduced the notion of *scheduling algorithm*, that is, the set of rules that leads a task-based operating system to choose and enforce a specific task execution order. We also introduced the concept of task *priority* as a simple example of an attribute attached to each task, which guides and affects the scheduling algorithm's choices.

It is now time to discuss the topic in more detail and slightly more formally. Even though most of the theoretical derivations will still be left out for conciseness, the main purpose of this section and the following one is to provide an introduction to scheduling and scheduling analysis theory.

On the one hand, this will provide interested readers with enough background information to further pursue the matter by means of more advanced books, like [25, 26, 39, 111]. On the other hand, the same background is also helpful to justify, for instance, why the Rate Monotonic scheduling algorithm is currently very popular for real-time execution. In fact, this is not only because it is simple to implement and its behavior is easy to understand by intuition, as shown in the example of Section 4.6, but is also due to its significant theoretical properties.

6.1 SCHEDULING ALGORITHMS FOR REAL-TIME EXECUTION

First of all, let us clarify what is the *role* of the scheduling algorithm in real-time task execution and, in particular, what is its relationship with the task synchronization and communication primitives described in Sections 5.3 and 5.4.

In any application comprising multiple, concurrently-executed tasks, the exact order in which tasks execute is not completely specified and constrained by the application itself. In fact, the communication and synchronization primitives described in Sections 5.3 and 5.4 enforce only as many task execution ordering constraints as necessary to ensure that the *results* produced by the application are correct in all cases.

For example, a mutual exclusion semaphore (Section 5.3) ensures that only one task at a time is allowed to operate on shared data. Similarly, a message sent from one task to another (Section 5.4) can force the receiving task to wait until the sending task has completed a computation and, at the same time, transfers the results from one task to the other.

Despite these correctness-related constraints, the application will still exhibit a significant amount of nondeterminism because, in many cases, the execution of its tasks may *interleave* in different ways without violating any of those constraints. For example, as discussed in Section 5.4, a message queue puts in effect a synchronization between senders and receivers only when it is either empty or full.

When this is not the case, senders and receivers are free to proceed and interleave their execution arbitrarily. The application results will of course be the same in all cases and they will be correct, provided the application was designed properly. However, its *timings* may vary considerably from one execution to another due to different interleavings.

Therefore, if some tasks have a constraint on how much time it takes to complete them, a constraint also known as *response time deadline*—as is common in a real-time system—only *some* of the interleavings that are acceptable from the point of view of correctness will also be adequate to satisfy those additional constraints.

As a consequence, in a real-time system it is necessary to *further restrict* the nondeterminism, beyond what is done by the communication and synchronization primitives, to ensure that the task execution sequence will not only produce correct results in all cases, but will also lead tasks to meet their deadlines. This is exactly what is done by real-time scheduling algorithms.

Under these conditions, scheduling analysis techniques—briefly presented in Section 6.2—are able to establish whether or not all tasks in the system will be able to meet their deadlines and, using more complex techniques, calculate the *worst-case* response time of each task, too. For the time being, interrupts will be left out of the discussion for simplicity. Interrupt handling and its effects on real-time performance will be further discussed in Chapter 8.

Even neglecting interrupts, it turns out that assessing the worst-case timing behavior of an arbitrarily complex concurrent application is very difficult. For this reason, it is necessary to introduce a simplified *task model*, which imposes some restrictions on the structure of the application to be considered for analysis and its tasks.

The simplest model, also known as the *basic* task model, will be the starting point for the discussion. It has the following characteristics:

1. The application consists of a fixed number of tasks, and that number is known in advance. All tasks are created when the application starts executing.
2. Tasks are *periodic*, with fixed and known periods, so that each task can be seen as an infinite sequence of *instances* or *jobs*. Each task instance becomes ready for execution at regular time intervals, that is, at the beginning of each task period.
3. Tasks are completely *independent* of each other. They neither synchronize nor communicate in any way.

4. As outlined above, timing constraints are expressed by means of *deadlines*. For a given task, a deadline represents an upper bound on the response time of its instances that must always be satisfied. In the basic task model the deadline of each task is equal to its period. In other words, the previous instance of a task must always be completed before the next one becomes ready for execution.
5. The worst-case execution time of each task—that is, the maximum amount of processor time it may possibly need to complete any of its instances when the task is executed in isolation—is fixed and can be computed offline.
6. All system's overheads, for example, context switch times, are negligible.

Although the basic model just introduced is very convenient and leads to interesting results concerning theoretical scheduling analysis, it also has some shortcomings that limit its direct application to real-world scenarios.

In particular, the requirement about task independence rules out, for instance, mutual exclusion and synchronization semaphores, as well as message passing. Therefore, it is somewhat contrary to the way concurrent systems are usually designed—in which tasks necessarily interact with one another—and must be relaxed to make scheduling analysis useful in practice.

Other improvements to the model, which are possible but will not be further discussed in this book, due to lack of space, are:

- The deadline of a task is not always the same as its period. For instance, a deadline *shorter* than the period is of particular interest to model tasks that are executed infrequently but, when they are, must be completed with tight timing constraints.
- Some tasks are *sporadic* rather than periodic. This happens, for instance, when the execution of a task is triggered by an event external to the system.
- In a modern hardware architecture, it may be difficult to determine an upper bound on a task execution time which is at the same time *accurate* and *tight*. This is because those architectures include hardware components (like caches, for example), in which the average time needed to complete an operation may differ from the worst-case time by several orders of magnitude.

Table 6.1 summarizes the notation that will be used throughout this chapter to discuss scheduling algorithms and their analysis. Even though it is not completely standardized, the notation proposed in the table is the one adopted by most textbooks and publications on the subject. In particular:

- The symbol τ_i, already introduced in the previous sections, represents the i-th task in the system. Unless otherwise specified, tasks are enumerated by decreasing priority, so that the priority of τ_j is greater than the priority of τ_i if and only if $j < i$.
- Periodic tasks consist of an infinite number of repetitions, or *instances*. When it is necessary to distinguish an instance of τ_i from another, we use

Table 6.1

Notation for Real-Time Scheduling Algorithms and Analysis

Symbol	Meaning
τ_i	The i-th task
$\tau_{i,j}$	The j-th instance of the i-th task
T_i	The period of task τ_i
D_i	The relative deadline of task τ_i
C_i	The worst-case execution time of task τ_i
R_i	The worst-case response time of task τ_i
$r_{i,j}$	The release time of $\tau_{i,j}$
$d_{i,j}$	The absolute deadline of $\tau_{i,j}$
$f_{i,j}$	The response time of $\tau_{i,j}$

the notation $\tau_{i,j}$, which indicates the j-th instance of τ_i. Instances are enumerated according to their temporal order, so that $\tau_{i,j}$ precedes $\tau_{i,k}$ in time if and only if $j < k$, and the first instance of τ_i is usually written as $\tau_{i,0}$.

- In a periodic task τ_i individual instances are *released*, that is, they become ready for execution, at regular time intervals. The distance between two adjacent releases is the *period* of the task, denoted by T_i.
- The symbol D_i represents the *deadline* of τ_i expressed in *relative* terms, that is, with respect to the release time of each instance. In the model, the relative deadline is therefore the same for all instances of a given task. Moreover, in the following it will be assumed that $D_i = T_i$ $\forall i$ for simplicity.
- The worst-case execution time of τ_i is denoted as C_i. As outlined above, the worst-case execution time of a task is the maximum amount of processor time needed to complete any of its instances when the task is executed *in isolation*, that is, without the presence of any other tasks in the system. It is important to note that a task *execution* time shall not be confused with its *response* time, to be described next.
- The worst-case response time of τ_i, denoted as R_i, represents the maximum amount of time needed to complete any of its instances when the task is executed *together with* all the other tasks in the system. It is therefore $R_i \geq C_i$ $\forall i$ because, by intuition, the presence of other tasks can only worsen the completion time of τ_i. For instance, the presence of a higher-priority task τ_j may lead the scheduler to temporarily stop executing τ_i in favor of τ_j when the latter becomes ready for execution.
- Besides considering timing parameters pertaining to a task τ_i as a whole— like its period T_i, deadline D_i, and response time R_i—it is sometimes important to do the same at the instance level, too. In this case, $r_{i,j}$ is used to denote the release time of the j-th instance of τ_i, that is:

$$r_{i,j} = \phi_i + jT_i, \quad j = 0, 1, \ldots \tag{6.1}$$

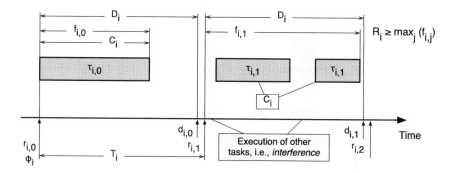

Figure 6.1 Notation for real-time scheduling algorithms and analysis.

where ϕ_i represents the initial *phase* of τ_i, that is, the absolute time at which its first instance $\tau_{i,0}$ is released.

- Similarly, $d_{i,j}$ represents the absolute deadline of the j-th instance of τ_i, which is:

$$d_{i,j} = r_{i,j} + D_i \ . \tag{6.2}$$

An important difference between D_i and $d_{i,j}$ is that the former is a *relative* quantity, which is measured with respect to the release time of each instance of τ_i and is the same for all instances. On the contrary, the latter is an *absolute* quantity that represents the instant in time at which task instance $\tau_{i,j}$ must necessarily already be completed in order to satisfy its timing constraints. As a consequence, $d_{i,j}$ is different for each instance of τ_i.

- Last, $f_{i,j}$ denotes the response time of task instance $\tau_{i,j}$. It is again a relative quantity, measured with respect to the release time of the corresponding instance, that is, $r_{i,j}$. The difference between $f_{i,j}$ and R_i is that the former represents the *actual* response time of $\tau_{i,j}$, a specific instance of τ_i, whereas the latter is the *worst-case* response time among all (infinite) instances of τ_i. Hence, it is $R_i \geq \max_j (f_{i,j})$. We stipulate that R_i may be greater than the maximum instance-by-instance response time $f_{i,j}$ to take into account the fact that, due to the way R_i is calculated, it may be a *conservative* bound.

One of the useful consequences of introducing a formal task model is that it is now possible to be more precise in defining what, so far, has been described as "satisfying timing constraints," in a generic way. According to the above definitions, all tasks in a set meet their deadline—and hence, they all satisfy their timing constraints—if and only if $R_i \leq D_i \ \forall i$.

Figure 6.1 depicts the notation just introduced in graphical form and further highlights the difference between C_i and R_i when other tasks are executed concurrently with τ_i. The left part of the figure, related to instance $\tau_{i,0}$, shows how that instance is executed when there are no other tasks in the system. Namely, when the instance is released at $r_{i,0}$ and becomes ready for execution, it immediately transitions to the running state of the TSD and stays in that state until completion. As a consequence, its response time $f_{i,0}$ will be the same as its execution time C_i.

On the contrary, the right part of Figure 6.1 shows what may happen if instance $\tau_{i,1}$ is executed concurrently with other higher-priority tasks. In this case:

- If a higher-priority task is being executed when $\tau_{i,1}$ is released, $\tau_{i,1}$ does not immediately transition to the running state. Instead, it stays in the ready state until the higher-priority task has been completed.
- If a higher-priority task is released while $\tau_{i,1}$ is being executed, the operating system may temporarily stop executing $\tau_{i,1}$ in favor of the higher-priority task, with an action known as *preemption*.

For both these reasons, the response time of $\tau_{i,1}$, denoted as $f_{i,1}$ in the figure, may become significantly longer that its execution time C_i because the task instance endures a certain amount of *interference* from higher-priority tasks. A very important goal of defining a satisfactory real-time scheduling algorithm, along with an appropriate way of analyzing its behavior, is to ensure that $f_{i,j}$ is bounded for any instance j of task i. Moreover, it must also be guaranteed that, for all tasks, the resulting worst-case response time R_i is acceptable with respect to the task deadline, that is, $R_i \leq D_i \ \forall i$.

The *Rate Monotonic* (RM) scheduling algorithm for single-processor systems, introduced by Liu and Leyland [110] assigns to each task in the system a *fixed* priority, which is inversely proportional to its period T_i. Tasks are then selected for execution according to their priority, that is, at each instant the operating system scheduler chooses for execution the ready task with the highest priority. Preemption of lower-priority tasks in favor of higher-priority ones is performed, too, as soon as a higher-priority task becomes ready for execution. An example of how Rate Monotonic schedules a simple set of tasks, listed in Table 4.4, was shown in Figure 4.9.

It should be noted that the Rate Monotonic priority assignment takes into account only the task period T_i, and not its execution time C_i. In this way, tasks with a shorter period are expected to be executed before the others. Intuitively, this makes sense because we are assuming $D_i = T_i$, and hence, tasks with a shorter period have less time available to complete their work. On the contrary, tasks with a longer period can afford giving precedence to more urgent tasks and still be able to finish their execution in time.

This informal reasoning can be confirmed with a mathematical proof of optimality, that is, it has been proved that Rate Monotonic is the best scheduling policy among all the *fixed* priority scheduling policies when the *basic task model* is considered. In particular, under the following assumptions:

1. Every task τ_i is periodic with period T_i.
2. The relative deadline D_i for every task τ_i is equal to its period T_i.
3. Tasks are scheduled preemptively and according to their priority.
4. There is only one processor.

It has been proved [110] that, if a given set of periodic tasks with fixed priorities can be scheduled so that all tasks meet their deadlines by means of a certain scheduling algorithm A, then the Rate Monotonic algorithm is able to do the same, too.

Another interesting mathematical proof about the Rate Monotonic answers a question of significant, practical relevance. From the previous discussion, it is already clear that the response time of a task instance depends on *when* that instance is released with respect to the other tasks in the system, most notably the higher-priority ones. This is because the relative position of task instance release times affects the amount of interference the task instance being considered endures and, as a consequence, the difference between its response time $f_{i,j}$ with respect to the execution time C_i.

It is therefore interesting to know what is the relative position of task instance release times that leads to the worst possible response time R_i. The *critical instant theorem* [110] provides a simple answer to this question for Rate Monotonic. Namely, it states that a *critical instant* for a task instance occurs when it is released together with an instance of all higher-priority tasks. Moreover, releasing a task instance at a critical instant leads that instance to have the worst possible response time R_i among all instances of the same task.

Therefore, in order to determine R_i for the Rate Monotonic algorithm, it is unnecessary to analyze, simulate, or experimentally evaluate the system behavior for *any* possible relationship among task instance release times, which may be infeasible or very demanding. Instead, it is enough to look at the system behavior in a single scenario, in which task instances are released at a critical instant.

Given that the Rate Monotonic algorithm has been proved to be optimal among all fixed-priority scheduling algorithms, it is still interesting to know if it is possible to "do better" than Rate Monotonic, by relaxing some constraints on the structure of the scheduler and add some complexity to it. In particular, it is interesting to investigate the scenario in which task priorities are no longer constrained to be fixed, but may change over time instead. The answer to this question was given by Liu and Layland in [110], by defining a dynamic-priority scheduling algorithm called earliest deadline first (EDF) and proving it is optimal among all possible scheduling algorithms, under some constraints.

The EDF algorithm selects tasks according to their absolute deadlines. That is, at each instant, tasks with earlier deadlines receive higher priorities. According to (6.1) and (6.2), the absolute deadline $d_{i,j}$ of the j-th instance (job) of task τ_i is given by

$$d_{i,j} = \phi_i + jT_i + D_i \ . \tag{6.3}$$

From this equation, it is clear that the priority of a given task τ_i as a whole changes dynamically, because it depends on the current deadline of its active instance. On the other hand, the priority of a given task instance $\tau_{i,j}$ is still fixed, because its deadline is computed once and for all by means of (6.3) and it does not change afterward.

This property also gives a significant clue on how to simplify the practical implementation of EDF. In fact, EDF implementation does not require that the scheduler continuously monitors the current situation and rearranges task priorities when needed. Instead, task priorities shall be updated only when a new task instance is released. Afterwards, when time passes, the priority order among active task instances does not change, because their absolute deadlines do not move.

As happened for RM, the EDF algorithm works well according to intuition, because it makes sense to increase the priority of more "urgent" task instances, that is, instances that are getting closer to their deadlines without being completed yet. The same reasoning has also been confirmed in [110] by a mathematical proof. In particular, considering the basic task model complemented by the following assumptions:

1. Tasks are scheduled preemptively;
2. There is only one processor.

It has been proved that EDF is the optimal scheduling algorithm. The definition of optimality used in the proof is the same one adopted for Rate Monotonic. Namely, the proof shows that, if *any* task set is schedulable by *any* scheduling algorithm under the hypotheses of the theorem, then it is also schedulable by EDF.

6.2   SCHEDULING ANALYSIS

In the previous section, it has been shown that the Rate Monotonic (RM) and Earliest Deadline First (EDF) are optimal scheduling algorithms within the scope of the basic task model and in their own class, that is, fixed-priority and dynamic-priority algorithms.

Even though those results are of great theoretical value, they still do not answer a rather fundamental question that arises during the design of a real-time software application. In fact, in practice, software designers are interested to know whether or not a certain set of tasks they are working with is schedulable by means of the scheduling algorithm (most often, Rate Monotonic) available on the real-time operating system of their choice.

Moreover, in some cases, a simple "yes or no" answer may not give designers all the information they need to be confident that their system will work correctly with reasonable margins. For instance, a designer may not be satisfied to just know that all tasks in the set will meet their deadline (that is, $R_i \leq D_i$ $\forall i$) but he/she may also want to know the actual value of R_i, in order to judge how far or how close his/her tasks are from missing their deadlines.

In this book, the scope of the analysis will be limited to the Rate Monotonic algorithm only. Similar analysis methods also exist for EDF and other scheduling algorithms, but are considerably more complex than for RM. Readers are referred to other publications, for instance [25, 26, 39, 111], for a complete description of those methods.

The first scheduling analysis method to be discussed here is probably the simplest one. It can be applied to *single-processor* systems and is based on a quantity known as *processor utilization factor*, usually denoted as U. Formally, U is defined as:

$$U = \sum_{i=1}^{N} \frac{C_i}{T_i} \tag{6.4}$$

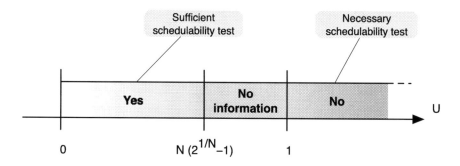

Figure 6.2 U-based schedulability tests for Rate Monotonic.

where, according to the notation presented in Table 6.1, the fraction C_i/T_i represents the fraction of processor time spent executing task τ_i. The processor utilization factor is therefore a measure of the computational load imposed on the processor by a given task set. Accordingly, the computational load associated with a task increases when its execution time C_i increases and/or its period T_i decreases.

Although U can be calculated in a very simple way, it does provide useful insights about the schedulability of the task set it refers to. First of all, an important theoretical result identifies task sets that are certainly *not schedulable*. Namely, if $U > 1$ for a given task set, then the task set is not schedulable, regardless of the scheduling algorithm.

Besides the formal proof—which can be found in [110]—this result is quite intuitive. Basically, it states that it is impossible to allocate to the tasks a fraction of processor time U that exceeds the total processor time available, that is, 1. It should also be noted that this result merely represents a *necessary* schedulability condition and, by itself, it does not provide any information when $U \leq 1$.

Further information is instead provided by a *sufficient* schedulability test for Rate Monotonic. Informally speaking, it is possible to determine a threshold value for U so that, if U is below that threshold, the task set can certainly be scheduled by Rate Monotonic, independently of all the other characteristics of the task set itself. More formally, it has been proved that if

$$U = \sum_{i=1}^{N} \frac{C_i}{T_i} \leq N(2^{1/N} - 1) \, , \tag{6.5}$$

where N is the number of tasks in the task set, then the task set is certainly schedulable by Rate Monotonic. Interested readers will find the complete proof in [110] and, in a more refined form, in [47].

Combined together, the two conditions just discussed can be summarized as shown in Figure 6.2. Given that, in a single-processor system, valid values of U vary between 0 and 1, the two conditions together identify three ranges of values of U with different schedulability properties:

Table 6.2

Task Set with $U \simeq 0.9$ Schedulable by Rate Monotonic

Task τ_i	Period T_i (ms)	Execution time C_i (ms)
τ_1	50	30
τ_2	100	20
τ_3	200	20

1. if $0 \leq U \leq N(2^{1/N} - 1)$, the task set is certainly *schedulable*, because it passes the sufficient test;
2. if $U > 1$, the task set is certainly *not schedulable*, because it fails the necessary test;
3. if $N(2^{1/N} - 1) < U \leq 1$ the tests give *no information* about schedulability and the task set may or may not be schedulable.

When the processor utilization factor U of a task set falls in the third range, further analysis—to be performed with more complex and sophisticated techniques—is necessary to determine whether or not the task set is schedulable. The following two examples highlight that, within this "uncertainty area," two task sets with the same value of U may behave very differently for what concerns schedulability. According to the critical instant theorem, the system will be analyzed at a critical instant for all tasks—that is, when the first instance of each task is released at $t = 0$—because we are looking for the worst possible behavior in terms of timings.

The characteristics of the first task set, called A in the following, are listed in Table 6.2. The task set has a processor utilization factor $U_A = 0.9$. Since, for $N = 3$ tasks, it is:

$$N(2^{1/N} - 1)\Big|_{N=3} \simeq 0.78 \, , \tag{6.6}$$

neither the necessary nor the sufficient schedulability test provide any information.

The corresponding scheduling diagram, drawn starting at a critical instant according to the Rate Monotonic algorithm is shown in Figure 6.3. As is commonly done in this kind of diagram, it consists of three main parts:

1. The top of the diagram summarizes the execution time, period, and deadline of the tasks in the task set. The execution time of each task is represented by a gray block, while the period (that coincides with the deadline in the task model we are using) is depicted with a dashed horizontal line ending with a vertical bar. Different shades of gray are used to distinguish one task from another.
2. The bottom of the diagram shows when task instances are released. There is a horizontal time line for each task, with arrows highlighting the release time of that task's instances along it.
3. The mid part of the diagram is a representation of how the Rate Monotonic scheduler divides the processor time among tasks. Along this time line, a gray block

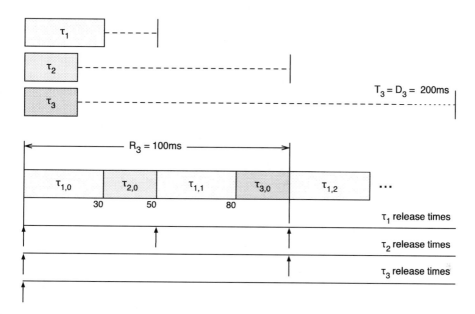

Figure 6.3 Scheduling diagram for the task set of Table 6.2.

signifies that the processor is executing a certain task instance, written in the block itself, and empty spaces mean that the processor is idle because no tasks are ready for execution at that time. The numbers near the bottom right corner of task instances represent the absolute time when the instance ends.

For this particular task set, the Rate Monotonic scheduler took the following scheduling decisions:

- At $t = 0$, the first instance of all three tasks has just been released and all of them are ready for execution. The scheduler assigns the processor to the highest-priority instance, that is, $\tau_{1,0}$.
- At $t = 30$ ms, instance $\tau_{1,0}$ has completed its execution. As a consequence, the scheduler assigns the processor to the highest-priority task instance still ready for execution, that is, $\tau_{2,0}$.
- At $t = 50$ ms, instance $\tau_{2,0}$ completes its execution, too, just in time for the release of the next instance of τ_1. At this point, both $\tau_{3,0}$ and $\tau_{1,1}$ are ready for execution. As always, the scheduler picks the highest-priority task instance for execution, $\tau_{1,1}$ in this case.
- At $t = 80$ ms, when instance $\tau_{1,1}$ is completed, the only remaining task instance ready for execution is $\tau_{3,0}$ and its execution eventually begins.
- The execution of $\tau_{3,0}$ concludes at $t = 100$ ms. At the same time, new instances of both τ_1 and τ_2 are released. The processor is assigned to $\tau_{1,2}$ immediately, as before.

Table 6.3

Task Set with $U \simeq 0.9$ Not Schedulable by Rate Monotonic

Task τ_i	Period T_i (ms)	Execution time C_i (ms)
τ_1	50	30
τ_2	75	20
τ_3	100	3

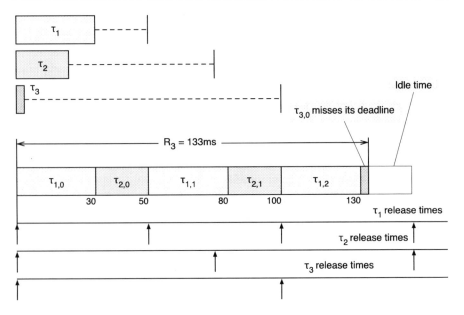

Figure 6.4   Scheduling diagram for the task set of Table 6.3.

By examining the diagram, it is evident that all task instances released at $t = 0$ met their deadlines. For example, the response time of $\tau_{3,0}$ is $f_{3,0} = 100$ ms. Since $\tau_{3,0}$ was released at a critical instant, we can also conclude that the worst-case response time of τ_3 is $R_3 = f_{3,0} = 100$ ms. Being $D_3 = 100$ ms, the condition $R_3 \leq D_3$ is therefore satisfied.

Overall, it turns out that Rate Monotonic was able to successfully schedule this particular task set, including the lowest-priority task τ_3, even though no further idle time remained between the release of its first instance, $\tau_{3,0}$ and the corresponding deadline $d_{3,0} = 100$ ms.

By contrast, the outcome is very different for another task set, called B in the following, even though it is very similar to the previous one and has the same processor utilization factor $U_B \simeq 0.9$. (actually, the value of U is even slightly lower in the second case with respect to the first). Table 6.3 lists the tasks parameters of task set B and Figure 6.4 shows its scheduling diagram, drawn in the same way as the previous one.

In this case, Rate Monotonic was unable to schedule the task set because instance $\tau_{3,0}$, released at a critical instant, concluded its execution at $f_{3,0} = 133$ ms. Due to the critical instant theorem, this corresponds to a worst-case response time $R_3 = 133$ ms well beyond the deadline $D_3 = 100$ ms.

It should be noted that the failure of Rate Monotonic is not due to a lack of processor time. In fact, saying $U \simeq 0.9$ means that, *on average*, the task set only requires 90% of the available processor time to be executed. This is also clear from the scheduling diagram, which shows that a significant amount of idle time—where the processor is not executing any task—remains between $t = 133$ ms (where $\tau_{3,0}$ concludes its execution) and $t = 150$ ms (where a new instance of both τ_1 and τ_2 is released).

On the contrary, the different outcome of Rate Monotonic scheduling when it is applied to task sets A and B depends on the relationship among the periods and execution times of the tasks in the task sets, which is less favorable in task set B, even though U remains the same in both cases.

At the same time, the scheduling diagram also shows that *no other tasks* can be executed at all in the interval $[0, 130]$ ms when $\tau_{1,0}$ and $\tau_{2,0}$ are released together at $t = 0$. As a consequence, as long as the deadline of task τ_3 is $D_3 \leq 130$ ms, it will miss the deadline regardless of how small C_3 is.

Overall, these simple examples lead us to observe that, at least in some cases, the value of U does not provide enough information about the schedulability of a set of tasks. Hence, researchers developed more sophisticated tests, which are more complex than the U-based tests discussed so far, but are able to provide a definite answer about the schedulability of a set of tasks, without any uncertainty areas.

Among them, we will focus on a method known as response time analysis (RTA) [13, 14]. With respect to the U-based tests, it is slightly more complex, but it is an *exact* (both necessary and sufficient) schedulability test that can be applied to any fixed-priority assignment scheme on single-processor systems.

Moreover, it does not just give a "yes or no" answer to the schedulability question, but calculates the worst-case response times R_i individually for each task. It is therefore possible to compare them with the corresponding deadlines D_i to assess whether all tasks meet their deadlines or not and judge how far (or how close) they are from missing their deadlines as well.

According to RTA, the response time R_i of task τ_i can be calculated by considering the following recurrence relationship:

$$w_i^{(k+1)} = C_i + \sum_{j \in hp(i)} \left\lceil \frac{w_i^{(k)}}{T_j} \right\rceil C_j , \qquad (6.7)$$

in which:

- $w_i^{(k+1)}$ and $w_i^{(k)}$ are the $(k+1)$-th and the k-th estimate of R_i, respectively. Informally speaking, Equation (6.7) provides a way to calculate the next estimate of R_i starting from the previous one.

- The first approximation $w_i^{(0)}$ of R_i is chosen by letting $w_i^{(0)} = C_i$, which is the smallest possible value of R_i.
- $hp(i)$ denotes the set of indices of the tasks with a priority higher than τ_i. For Rate Monotonic, the set contains the indices j of all tasks τ_j with a period $T_j < T_i$.

It has been proved that the succession $w_i^{(0)}, w_i^{(1)}, \ldots, w_i^{(k)}, \ldots$ defined by (6.7) is monotonic and nondecreasing. Two cases are then possible:

1. If the succession does not converge, there exists at least one scheduling scenario in which τ_i does not meet its deadline D_i, regardless of the specific value of D_i.
2. If the succession converges, it converges to R_i, and hence, it will be $w_i^{(k+1)} = w_i^{(k)} = R_i$ for some k. In this case, τ_i meets its deadline in every possible scheduling scenario if and only if the worst-case response time provided by RTA is $R_i \leq D_i$.

As an example, let us apply RTA to the task sets listed in Tables 6.2 and 6.3. For what concerns the first task set, considering task τ_1 we can write:

$$w_1^{(0)} \quad = \quad C_1 = 30\,\text{ms} \tag{6.8}$$

$$w_1^{(1)} \quad = \quad C_1 + \sum_{j \in hp(1)} \left\lceil \frac{w_i^{(k)}}{T_j} \right\rceil C_j = C_1 = 30\,\text{ms} \ . \tag{6.9}$$

In fact, the set $hp(1)$ is empty because τ_1 is the highest-priority (shortest-period) task in the set. The succession converges and, as a consequence, it is:

$$R_1 = 30\,\text{ms} \ . \tag{6.10}$$

For what concerns task τ_2, it is:

$$w_2^{(0)} \quad = \quad C_2 = 20\,\text{ms} \tag{6.11}$$

$$w_2^{(1)} \quad = \quad C_2 + \left\lceil \frac{w_2^{(0)}}{T_1} \right\rceil C_1 = 20 + \left\lceil \frac{20}{50} \right\rceil 30 = 20 + 30 = 50\,\text{ms} \tag{6.12}$$

$$w_2^{(2)} \quad = \quad 20 + \left\lceil \frac{50}{50} \right\rceil 30 = 20 + 30 = 50\,\text{ms} \ . \tag{6.13}$$

In this case, $hp(2) = \{1\}$ because τ_1 has a higher priority than τ_2. The succession converges, and hence, it is:

$$R_2 = 50\,\text{ms} \ . \tag{6.14}$$

The analysis of the lowest-priority task τ_3, for which $hp(3) = \{1,2\}$ proceeds in the same way:

$$w_3^{(0)} = C_3 = 20\,\text{ms} \tag{6.15}$$

$$w_3^{(1)} = C_3 + \left\lceil \frac{w_2^{(0)}}{T_1} \right\rceil C_1 + \left\lceil \frac{w_2^{(0)}}{T_2} \right\rceil C_2$$

$$= 20 + \left\lceil \frac{20}{50} \right\rceil 30 + \left\lceil \frac{20}{100} \right\rceil 20 = 20 + 30 + 20 = 70\,\text{ms} \tag{6.16}$$

$$w_3^{(2)} = 20 + \left\lceil \frac{70}{50} \right\rceil 30 + \left\lceil \frac{70}{100} \right\rceil 20 = 20 + 60 + 20 = 100\,\text{ms} \tag{6.17}$$

$$w_4^{(3)} = 20 + \left\lceil \frac{100}{50} \right\rceil 30 + \left\lceil \frac{100}{100} \right\rceil 20 = 20 + 60 + 20 = 100\,\text{ms} \tag{6.18}$$

Also in this case, the succession converges (albeit convergence requires more iterations than before) and we can conclude that $R_3 = 100\,\text{ms}$. Quite unsurprisingly, the worst-case response times just obtained from RTA coincide perfectly with the ones determined from the scheduling diagram in Figure 6.3, with the help of the critical instant theorem. The main advantage of RTA with respect to the scheduling diagram is that the former can be automated more easily and is less error-prone when performed by hand.

Let us consider now the task set listed in Table 6.3. For τ_1 and τ_2, the RTA results are the same as before, that is:

$$w_1^{(0)} = C_1 = 30\,\text{ms} \tag{6.19}$$

$$w_1^{(1)} = C_1 = 30\,\text{ms} \tag{6.20}$$

and

$$w_2^{(0)} = C_2 = 20\,\text{ms} \tag{6.21}$$

$$w_2^{(1)} = 20 + \left\lceil \frac{20}{50} \right\rceil 30 = 20 + 30 = 50\,\text{ms} \tag{6.22}$$

$$w_2^{(1)} = 20 + \left\lceil \frac{50}{50} \right\rceil 30 = 20 + 30 = 50\,\text{ms} , \tag{6.23}$$

from which we conclude that $R_1 = 30\,\text{ms}$ and $R_2 = 50\,\text{ms}$. For what concerns τ_3, the lowest-priority task, it is:

$$w_3^{(0)} = C_3 = 3\,\text{ms} \tag{6.24}$$

$$w_3^{(1)} = 3 + \left\lceil \frac{3}{50} \right\rceil 30 + \left\lceil \frac{3}{75} \right\rceil 20 = 3 + 30 + 20 = 53\,\text{ms} \tag{6.25}$$

$$w_3^{(2)} = 3 + \left\lceil \frac{53}{50} \right\rceil 30 + \left\lceil \frac{53}{75} \right\rceil 20 = 3 + 60 + 20 = 83\,\text{ms} \tag{6.26}$$

$$w_3^{(3)} = 3 + \left\lceil \frac{83}{50} \right\rceil 30 + \left\lceil \frac{83}{75} \right\rceil 20 = 3 + 60 + 40 = 103\,\text{ms} \tag{6.27}$$

$$w_3^{(4)} = 3 + \left\lceil \frac{103}{50} \right\rceil 30 + \left\lceil \frac{103}{75} \right\rceil 20 = 3 + 90 + 40 = 133\,\text{ms} \tag{6.28}$$

$$w_3^{(5)} = 3 + \left\lceil \frac{133}{50} \right\rceil 30 + \left\lceil \frac{133}{75} \right\rceil 20 = 3 + 90 + 40 = 133\,\text{ms}\ . \tag{6.29}$$

Therefore, $R_3 = 133\,\text{ms}$, as also confirmed by the scheduling diagram shown in Figure 6.4.

In summary, the RTA method proceeds as follows:

1. The worst-case response time R_i is individually calculated for each task τ_i in the task set, by means of the recurrence relationship (6.7) and the associated succession.
2. If, at any point, either a diverging succession is encountered or $R_i > D_i$ for some i, then the task set is not schedulable because at least task τ_i misses its deadline in some scheduling scenarios.
3. Otherwise, the task set is schedulable, and the worst-case response time is known for all tasks.

It is worth noting that, unlike the U-based scheduling tests discussed previously, this method no longer assumes that the relative deadline D_i is equal to the task period T_i. On the contrary, it is also able to handle the more general case in which $D_i \le T_i$, even though this aspect will not be further discussed in the following. Moreover, the method works with any fixed-priority ordering, and not just with the Rate Monotonic priority assignment, as long as hp(i) is defined appropriately for all i and a preemptive scheduler is in use.

Another useful property of RTA is that it is more flexible than U-based tests and is easily amenable to further extensions, for instance, to consider the effect of task interaction on schedulability. These extensions aim at removing one important limitation of the basic task model used so far and bring it closer to how real-world tasks behave.

For simplicity, in this book, the discussion will only address the following two main kinds of interaction. Readers are referred, for instance, to [26, 39, 111] for more detailed and comprehensive information about the topic.

1. Task interactions due to *mutual exclusion*, a ubiquitous necessity when dealing with shared data, as shown in Section 5.3.

2. Task *self suspension*, which takes place when a task waits for any kind of external event.

The second kind of interaction includes, for instance, the case in which a task interacts with a hardware device by invoking an Input–Output operation and then waits for the results. Other examples, involving only tasks, include semaphore-based task synchronization, outlined in Section 5.3 and message passing, discussed in Section 5.4.

In a real-time system any kind of task interaction, mutual exclusion in particular, must be designed with care, above all when the tasks involved have different priorities. In fact, a high-priority task may be *blocked* when it attempts to enter its critical region if a lower-priority task is currently within a critical region controlled by the same semaphore. From this point of view, the mutual exclusion mechanism is hampering the task priority scheme. This is because, if the mutual exclusion mechanism were not in effect, the high-priority task would always be preferred over the lower-priority one for execution.

This phenomenon is called *priority inversion* and, if not adequately addressed, can adversely affect the schedulability of the system, to the point of making the response time of some tasks completely unpredictable, because the priority inversion region may last for an *unbounded* amount of time and lead to an *unbounded priority inversion*.

Even though proper software design techniques may alleviate the issue—for instance, by avoiding useless or redundant critical regions—it is also clear that the problem cannot be completely solved in this way unless all forms of mutual exclusion, as well as all critical regions, are banned from the system. This is indeed possible, by means of lock-free and wait-free communication, but those techniques could imply a significant drawback in software design and implementation complexity.

On the other hand, it is possible to *improve* the mutual exclusion mechanism in order to guarantee that the worst-case blocking time endured by each individual task in the system is bounded. The worst-case blocking time can then be calculated and used to refine the response time analysis (RTA) method discussed previously, in order to determine their worst-case response times.

The example shown in Figure 6.5 illustrates how an unbounded priority inversion condition may arise, even in very simple cases. In the example, the task set is composed of three tasks, τ_1, τ_2, and τ_3, listed in decreasing priority order, scheduled by a preemptive, fixed-priority scheduler like the one specified by Rate Monotonic. As shown in the figure, τ_1 and τ_3 share some data and protect their data access with two critical regions controlled by the same semaphore s. The third task τ_2 does not share any data with the others, and hence, does not contain any critical region.

- The example starts when neither τ_1 nor τ_2 is ready for execution and the lowest-priority task τ_3 is running.
- After a while τ_3 executes a P(s) to enter its critical region. It does not block, because the value of s is currently 1. Instead, τ_3 proceeds beyond the

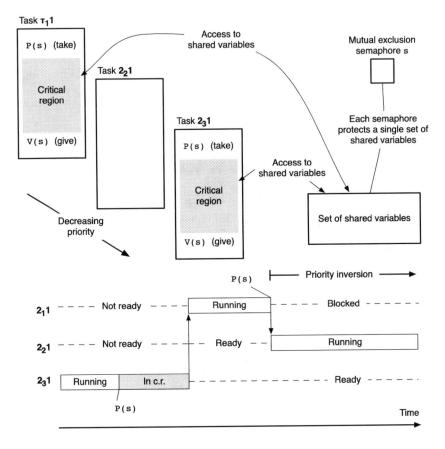

Figure 6.5 Unbounded priority inversion.

critical region boundary and keeps running. At the same time, the value of s becomes 0.

- While τ_3 is running, τ_1—the highest-priority task in the system—becomes ready for execution. Since the scheduler is preemptive, this event causes an immediate preemption of τ_3 (which is still within its critical region) in favor of τ_1.
- While τ_1 is running, τ_2 becomes ready for execution, too. Unlike in the previous case, this event does not have any immediate effect on scheduling, because the priority of τ_2 is lower than the one of the running task τ_1.
- After some time, τ_1 tries to enter its critical region by executing the entry code P(s), like τ_3 did earlier. However, the semaphore primitive *blocks* τ_1 because the value of s is currently 0.

- When τ_1 blocks, the scheduler picks τ_2 for execution, because it is the highest-priority task which is ready at the moment.

This point in time is the beginning of the priority inversion region because τ_1, the highest-priority task in the system is blocked due to a lower-priority task, τ_3, and the system is executing the mid-priority task τ_2.

It is important to remark that, so far, nothing "went wrong" in the system. In fact, τ_1 has been blocked for a sensible reason (it must be prevented from accessing shared data while τ_3 is working on them) and the execution of τ_2 is in perfect adherence to how the scheduling algorithm has been specified. However, a crucial question is *for how long* will the priority inversion region last.

Referring back to Figure 6.5 it is easy to observe that:

- The amount of time τ_1 will be forced to wait does not depend on τ_1 itself. Since it is blocked, there is no way it can directly affect its own future computation, and it will stay in this state until τ_3 exits from its critical region.
- The blocking time of τ_1 does not completely depend on τ_3, either. In fact, τ_3 is ready for execution, but it will not proceed (and will not leave its critical region) as long as there is any higher-priority task ready for execution, like τ_2 in the example.

As a consequence, the duration of the priority inversion region does not depend completely on the tasks that are actually sharing data, τ_1 and τ_3 in the example. Instead, it also depends on the behavior of *other tasks*, like τ_2, which have nothing to do with τ_1 and τ_3. Indeed, in a complex software system built by integrating multiple components, the programmers who wrote τ_1 and τ_3 may even be unaware that τ_2 exists.

The presence of multiple, mid-priority tasks makes the scenario even worse. In fact, it is possible that they take turns entering the ready state so that at least one of them is in the ready state at any given time. In this case, even though none of them monopolizes the processor by executing for an excessive amount of time, when they are taken as a whole they may prevent τ_3 (and hence, also τ_1) from being executed at all.

In other words, it is possible that a group of *mid-priority* tasks like τ_2, in combination with a *low-priority* task τ_3, prevents the execution of the *high-priority* task τ_1 for an unbounded amount of time, which is against the priority assignment principle and obviously puts schedulability at risk. It is also useful to remark that, as happens for many other concurrent programming issues, this is not a systematic error. Rather, it is a *time-dependent* issue that may go undetected when the system is bench tested.

Considering again the example shown in Figure 6.5, it is easy to notice that the underlying reason for the unbounded priority inversion is the preemption of τ_3 by τ_1 while it was within its critical region. If the preemption were somewhat delayed after τ_3 exited from the critical region, the issue would not occur, because there would be no way for mid-priority tasks like τ_2 to make the priority inversion region unbounded.

This informal reasoning can indeed be formally proved and forms the basis of a family of methods—called *priority ceiling protocols* and fully described in [153]—to avoid unbounded priority inversion. In a single-processor system, a crude implementation of those methods consists of completely forbidding preemption during the execution of critical regions.

As described in Sections 5.1 and 5.3, this also implements mutual exclusion and it may be obtained by disabling the operating system scheduler or, even more drastically, turning interrupts off. In this way, any task that successfully enters a critical region also gains the highest possible priority in the system so that no other task can preempt it. The task goes back to its regular priority as soon as it exits from the critical region.

Even though this method has the clear advantage of being extremely simple to implement, it also introduces by itself a significant amount of a different kind of blocking. Namely, any higher-priority task like τ_2 that becomes ready while a low-priority task τ_3 is within a critical region will not get executed—and we therefore consider it to be blocked by τ_3—until τ_3 exits from the critical region and returns to its regular priority.

The problem has been solved anyway because the amount of blocking endured by tasks like τ_2 is indeed bounded by the maximum amount of time τ_3 may spend within its critical region, which is finite if τ_3 has been designed in a proper way. Nevertheless, we are now potentially blocking many tasks which were not blocked before, and it turns out that most of this extra blocking is actually unessential to solve the unbounded priority inversion problem. For this reason, this way of proceeding is only appropriate for very short critical regions. On the other hand, a more sophisticated approach—which does not introduce as much extra blocking—is needed in the general case.

Nevertheless, the underlying idea is useful, that is, a better cooperation between the *synchronization* mechanism used for mutual exclusion and the processor *scheduler* can indeed solve the unbounded priority inversion problem. Namely, the *priority inheritance* protocol, proposed by Sha, Rajkumar, and Lehoczky [153] and implemented on most real-time operating systems—including FREERTOS—is implemented by enabling the mutual exclusion mechanism to *temporarily boost* task priorities, and hence, affect scheduling decisions.

Informally speaking, the general idea behind the priority inheritance protocol is that, if a task τ is blocking a set of n higher-priority tasks $\tau_1, \ldots, \tau_n$ at a given instant, it will temporarily inherit the highest priority among them. This temporary priority boost lasts until the blocking is in effect and prevents any mid-priority task from preempting τ and unduly make the blocking experienced by $\tau_1, \ldots, \tau_n$ longer than necessary or unbounded.

More formally, the priority inheritance protocol assumes that the following assumptions hold:

- The tasks are under the control of a fixed-priority scheduler and are executed by a single processor.

- If there are two or more highest-priority tasks ready for execution, the scheduler picks them in first-come first-served (FCFS) order, that is, they are executed in the same order as they became ready.
- Semaphore wait queues are ordered by priority so that, when a task executes a V(s) on a semaphore s and there is at least one task waiting on s, the highest-priority waiting task is unblocked and becomes ready for execution.

The priority inheritance protocol itself consists of the following set of rules:

1. When a task τ_1 attempts to enter a critical region that is "busy"—because its controlling semaphore has been taken by another task τ_2—it blocks. At the same time, τ_2 *inherits* the priority of τ_1 if the current priority of τ_1 is higher than its own.
2. As a consequence, τ_2 executes the rest of its critical region with a priority at least equal to the priority it just inherited. In general, a task inherits the highest active priority among all tasks it is blocking.
3. When task τ_2 exits from the critical region and it is no longer blocking any other task, it goes back to its baseline priority.
4. Otherwise, if τ_2 is still blocking some other tasks—this may happen when critical regions are nested into each other—it inherits the highest active priority among them.

Under these assumptions, it has been proved that, if there are a total of K semaphores $S_1, \ldots, S_K$ in the system and critical regions are *not nested*, the worst-case blocking time experienced by each instance of task τ_i when using the priority inheritance protocol is finite and is bounded by a quantity B_i, calculated as:

$$B_i = \sum_{k=1}^{K} \text{usage}(k,i)C(k) \ . \tag{6.30}$$

In the equation above,

- $\text{usage}(k,i)$ is a function that returns 1 if semaphore S_k is used by (at least) one task with a priority less than the priority of τ_i, and also by (at least) one task with a priority higher than or equal to the priority of τ_i, *including τ_i itself*. Otherwise, $\text{usage}(k,i)$ returns 0.
- $C(k)$ is the worst-case execution time among all critical regions associated with, or guarded by, semaphore S_k.

It should be noted that the bound B_i given by (6.30) is often "pessimistic" when applied to real-world scenarios, because

- It assumes that if a certain semaphore *can possibly block* a task, it *will indeed block* it.
- For each semaphore, the blocking time suffered by τ_i is always assumed to be equal to the worst-case execution time of the longest critical region guarded by that semaphore, even though that critical region is never entered by τ_i itself.

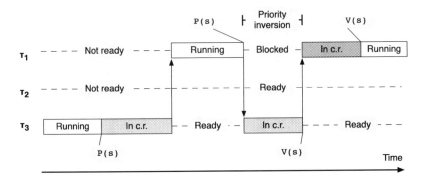

Figure 6.6 Priority inheritance protocol.

However, it is an acceptable compromise between the tightness of the bound it calculates and its computational complexity. Better algorithms exist and are able to provide a tighter bound of the worst-case blocking time, but their complexity is also higher.

As an example, let us apply the priority ceiling algorithm to the simple example shown in Figure 6.5, which led to unbounded priority inversion if not handled with care. The corresponding time diagram in depicted in Figure 6.6.

- The priority inheritance protocol does nothing—and everything proceeds as in the previous case—until task τ_1 blocks on P(s).
- At that time, the priority inheritance protocol boosts the priority of τ_3 because τ_3 has a lower priority than τ_1 and is blocking it.
- As a consequence, τ_2 stays in the ready state even though it is ready for execution. The scheduler picks up τ_3 for execution instead, because it is the highest-priority ready task.
- Task τ_3 proceeds through its critical region. Within a finite amount of time, τ_3 exits from it and executes V(s).
- Since now τ_3 is no longer blocking τ_1 (τ_1 has just been unblocked by V(s)), and it is not blocking any other tasks as well, the priority inheritance protocol restores the priority of τ_3 to its baseline value, that is, the lowest priority in the system.
- Task τ_1 is now chosen for execution, because it is ready and has the highest priority in the system. It enters its critical region and proceeds.

The example shows that, by means of the priority inheritance protocol, the length of the priority inversion region is now bounded and limited to the region highlighted in the figure. Namely, the worst-case length of the priority inversion region is now equal to the maximum amount of time that τ_3 can possibly spend within its critical region, corresponding to the light gray boxes drawn on τ_3's time line.

However, it is also useful to remark that, within the priority inversion region, τ_3 now blocks *both* τ_1 and τ_2, whereas only τ_1 was blocked in the previous example.

This fact leads us to observe that the priority inheritance protocol implies a trade-off between enforcing an upper bound on the length of priority inversion regions and introducing additional blocking in the system, as any other algorithm dealing with unbounded priority inversion does.

Hence, for the priority inheritance protocol, we identify *two* distinct kinds of blocking:

1. *Direct blocking* occurs when a high-priority task tries to acquire a shared resource—for instance, get access to some shared data by taking a mutual exclusion semaphore—while the resource is held by a lower-priority task. This is the kind of blocking affecting τ_1 in this case. Direct blocking is an unavoidable consequence of mutual exclusion and ensures the *consistency* of the shared resources.
2. *Push-through blocking* is a consequence of the priority inheritance protocol and is the kind of blocking experienced by τ_2 in the example. It occurs when an intermediate-priority task (like τ_2) is not executed even though it is ready because a lower-priority task (like τ_3) has inherited a higher priority. This kind of blocking may affect a task even if it does not actually use any shared resource, but it is necessary to avoid *unbounded priority inversion*.

To conclude the example, let us apply (6.30) to the simple scenario being considered and calculate the worst-case blocking time for each task. Since there is only one semaphore in the system, that is, $K = 1$, the formula becomes:

$$B_i = \text{usage}(1, i)C(1) \; , \tag{6.31}$$

where $\text{usage}(1, i)$ is a function that returns 1 if semaphore s is used by (at least) one task with a priority less than the priority of τ_i, and also by (at least) one task with a priority higher than or equal to the priority of τ_i. Otherwise, $\text{usage}(1, i)$ returns 0. Therefore:

$$\text{usage}(1,1) \; = \; 1 \tag{6.32}$$
$$\text{usage}(1,2) \; = \; 1 \tag{6.33}$$
$$\text{usage}(1,3) \; = \; 0 \; . \tag{6.34}$$

Similarly, $C(1)$ is the worst-case execution time among all critical regions associated with, or guarded by, semaphore s. Referring back to Figure 6.6, $C(1)$ is the maximum between the length of the light gray boxes (critical region of τ_3) and the dark gray box (critical region of τ_1). As a result, we can write:

$$B_1 \; = \; C(1) \tag{6.35}$$
$$B_2 \; = \; C(1) \tag{6.36}$$
$$B_3 \; = \; 0 \; . \tag{6.37}$$

These results confirm that both τ_1 and τ_2 can be blocked (by τ_3)—in fact, $B_1 \neq 0$ and $B_2 \neq 0$—while the only task that does not suffer any blocking is the lowest-priority one, τ_3, because $B_3 = 0$. On the other hand, they also confirm the "pessimism" of (6.30) because the worst-case amount of blocking is calculated to be the *maximum* between the two critical regions present in the system. Instead, it is clear from the diagram that the length of the dark-gray critical region, the critical region of τ_1, cannot affect the amount of blocking endured by τ_1 itself.

Considering mutual exclusion as the *only* source of blocking in the system is still not completely representative of what happens in the real world, in which tasks also invoke external operations and wait for their completion. For instance, it is common for tasks to start an input–output (I/O) operation and wait until it completes or a timeout expires. Another example would be to send a message to another task and then wait for an answer.

In general, all scenarios in which a task voluntarily suspends itself for a variable amount of time, provided this time has a known and finite upper bound, are called *self-suspension* or *self-blocking*. The analysis presented here is based on Reference [144], which addresses schedulability analysis in the broader context of real-time synchronization for multiprocessor systems. Interested readers are referred to [144] for further information and the formal proof of the statements discussed in this book.

Contrary to what can be expected by intuition, the effects of self-suspension are not necessarily *local* to the task that is experiencing it. On the contrary, the self-suspension of a high-priority task may hinder the schedulability of lower-priority tasks and, possibly, make them no longer schedulable. This is because, after self-suspension ends, the high-priority task may become ready for execution (and hence, preempt the lower-priority task) at the "wrong time" and have a greater impact on their worst-case response time with respect to the case in which the high-priority task runs continuously until completion.

Moreover, when considering self-suspension, even the critical instant theorem that, as shown previously, plays a central role in the schedulability analysis theory we leveraged so far, is no longer directly applicable to compute the worst-case interference that a task may be subject to.

In any case, the worst-case extra blocking endured by task τ_i due to its own self-suspension, as well as the self-suspension of higher-priority tasks, denoted B_i^{SS}, can still be calculated efficiently and can be written as

$$B_i^{SS} = S_i + \sum_{j \in hp(i)} \min(C_j, S_j) \ . \tag{6.38}$$

In the above formula:

- S_i is the worst-case self-suspension time of task τ_i.
- $hp(i)$ denotes the set of task indexes with a priority higher than τ_i.
- C_j is the execution time of task τ_j.

Informally speaking, according to (6.38) the worst-case blocking time B_i^{SS} due to self-suspension endured by task τ_i is given by the sum of its own worst-case self-suspension time S_i plus a contribution from each of the higher-priority tasks. The individual contribution of task τ_j to B_i^{SS} is given by its own worst-case self-suspension time S_j, but it never exceeds its execution time C_j.

In addition, since the self-suspension of a task has an impact on how it interacts with other tasks, as well as on the properties of the interaction, it becomes necessary to refine the analysis of worst-case blocking due to mutual exclusion.

There are several different ways to consider the combined effect of self-suspension and mutual exclusion on worst-case blocking time calculation. Perhaps the most intuitive one, presented in References [111, 144], introduces the notion of task *segments*—that is, portions of task execution delimited by a self-suspension. Accordingly, if task τ_i performs Q_i self-suspensions during its execution, it is said to contain $Q_i + 1$ segments.

Task segments are considered to be completely independent from each other for what concerns blocking due to mutual exclusion. Stated in an informal way, the calculations consider that each task goes back to the worst possible blocking scenario after each self-suspension. Following this approach, the worst-case blocking time B_i of τ_i due to mutual exclusion, calculated as specified in (6.30), becomes the worst-case blocking time endured by *each individual task segment* and is denoted as B_i^{1S}.

Hence, the worst-case blocking time of task τ_i due to mutual exclusion, B_i^{TI}, is given by

$$B_i^{TI} = (Q_i + 1)B_i^{1S} \ . \tag{6.39}$$

Combining (6.38) and (6.39), the total worst-case blocking time B_i that affects τ_i, considering both self-suspension directly and its effect on mutual exclusion blocking, can be written as:

$$
\begin{aligned}
B_i &= B_i^{SS} + B_i^{TI} \\
&= S_i + \sum_{j \in hp(i)} \min(C_j, S_j) + (Q_i + 1) \sum_{k=1}^{K} usage(k, i)C(k)
\end{aligned}
\tag{6.40}
$$

It is useful to remark that, in the above formula, C_j and $C(k)$ have got two different meanings that should not be confused despite the likeness in notation, namely:

- C_j is the execution time of a *specific task*, τ_j in this case, while
- $C(k)$ is the worst-case execution time of *any task* within any critical region guarded by S_k.

The value of B_i calculated by means of (6.40) can then be used to extend RTA and consider the blocking time in worst-case response time calculations. Namely, the basic recurrence relationship (6.7) can be rewritten as:

$$w_i^{(k+1)} = C_i + B_i + \sum_{j \in hp(i)} \left\lceil \frac{w_i^{(k)}}{T_j} \right\rceil C_j \ , \tag{6.41}$$

where B_i represents the worst-case blocking time calculated according to (6.40).

It has been proved that the new recurrence relationship still has the same properties as the original one. In particular, if the succession $w_i^{(0)}, w_i^{(1)}, \ldots, w_i^{(k)}, \ldots$ converges, it still provides the worst-case response time R_i for an appropriate choice of $w_i^{(0)}$. On the other hand, if the succession does not converge, τ_i is surely not schedulable. As before, setting $w_i^{(0)} = C_i$ provides a sensible initial value for the succession.

The main difference is that the new formulation is *pessimistic*, instead of necessary and sufficient, because the bound B_i on the worst-case blocking time is not tight. Therefore it may be practically impossible for a task to ever incur in a blocking time equal to B_i, and hence, experience the worst-case response time calculated by (6.41).

A clear advantage of the approach just described is that it is very simple and requires very little knowledge about the internal structure of the tasks. For instance, it is unnecessary to know exactly *where* the self-suspension operations are. Instead, it is enough to know *how many* there are, which is much simpler to collect and maintain as software evolves with time. However, the disadvantage of using such a limited amount of information is that it makes the method extremely conservative. Thus, the B_i calculated in this way is definitely not a tight upper bound for the worst-case blocking time and may widely overestimate it in some cases.

More sophisticated and precise methods do exist, such as that described in Reference [103]. However, as we have seen in several other cases, the price to be paid for a tighter upper bound for the worst-case blocking time is that much more information is needed. For instance, in the case of [103], we need to know not only how many self suspensions each task has got, but also their exact location within the task. In other words, we need to know the execution time of each individual task segment, instead of the task execution time as a whole.

6.3 SUMMARY

This chapter contains the basics of task-based scheduling algorithms. After introducing some basic nomenclature, models, and concepts related to this kind of scheduling, it went along describing what are probably the most widespread real-time scheduling algorithms, known as rate monotonic (RM) and earliest deadline first (EDF).

The second part of the chapter presented a few basic scheduling analysis techniques for the RM algorithm, able to mathematically prove whether or not a set of periodic tasks satisfies their deadlines, by considering some of their basic characteristics, like their period and execution time.

The analysis started with a very simple task model and was later extended to include some additional aspects of task behavior of practical interest, for instance, their interaction when accessing shared data in mutual exclusion.

The analysis methods mentioned in this chapter can also be applied to the EDF algorithm, but they become considerably more complex than for RM, both from the theoretical point of view, and also for what concerns their practical implementation. For this reason, they have not been considered in this book.

7 Configuration and Usage of Open-Source Protocol Stacks

CONTENTS

Many microcontrollers are nowadays equipped with at least one integrated Ethernet controller. It is therefore relatively easy—by just adding an external Ethernet physical transceiver (PHY)—to add network connectivity to an embedded system.

In turn, this extra feature can be quite useful for a variety of purposes, for instance remote configuration, diagnostics, and system monitoring, which are becoming widespread requirements, even on relatively low-end equipment.

This chapter shows how to configure and use an open-source protocol stack, namely LWIP [51], which is capable of handling the ubiquitous TCP, UDP, and IP protocols, and how to interface it with the underlying real-time operating system.

Another topic addressed by this chapter is how to use the protocol stack effectively from the application tasks, by choosing the most suitable LWIP application programming interface depending on the applications at hand.

On the other hand, due to its greater complexity, the way of writing a device driver in general and, more specifically, interfacing the protocol stack with an Ethernet controller by means of its device driver, will be the subject of a chapter by itself, Chapter 8.

7.1 INTRODUCTION TO THE LWIP PROTOCOL STACK

The LWIP protocol stack [51, 52] is an open-source product that provides a lightweight implementation of TCP/IP and related protocols. The main LWIP design

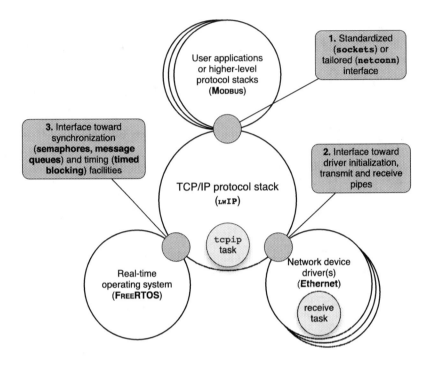

Figure 7.1   Interfaces and tasks of the LWIP protocol stack.

goals are portability, as well as small code and data size, making it especially suited for small-scale embedded systems.

It interacts with other system components by means of three main interfaces, depicted in Figure 7.1.

1. Toward user applications or higher-level protocol stacks, LWIP offers a choice between two different APIs, informally called *netconn* and *sockets*, to be described in Sections 7.4 and 7.6, respectively. As an example, Chapter 12 describes how an open-source MODBUS TCP protocol stack [119, 120] can be layered on top of LWIP by means of one of these APIs.
2. In order to support network communication, LWIP requires appropriate *device drivers* for the network devices in use. An example of how to design and implement a device driver for the embedded Ethernet controller of the NXP LPC2468 [125, 126] and LPC1768 [127, 128] microcontrollers.
3. Last, but not least, LWIP needs operating system support to work. Operating system services are accessed by means of a third interface, which is discussed in Section 7.2, taking the FREERTOS operating system as an example.

Going into greater detail, Figure 7.2 depicts the non-portable code modules of LWIP (gray rectangles) and illustrates how they relate to the various APIs described previously (light gray rectangles) and to hardware devices (dark gray rectangles).

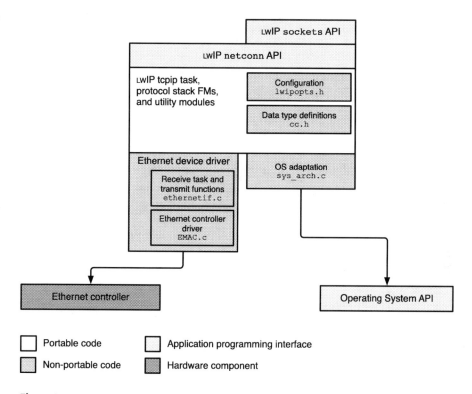

Figure 7.2 Nonportable modules of LWIP.

Portable code modules, which indeed constitute the vast majority of LWIP code, are shown as one single white rectangle.

As can be seen from the figure, the main non-portable modules, that is, the modules that should possibly be modified or rewritten when LWIP is ported to a new processor architecture, toolchain, or software project, are:

- The operating system adaptation layer, `sys_arch.c`, to be outlined in Section 7.2.
- The network device driver that, in the case of an Ethernet controller, is typically implemented in the modules `ethernetif.c` and `EMAC.c` and will be thoroughly described in Chapter 8.
- The header `lwipopts.h`, which contains all the LWIP configuration information. Its contents will be described in Section 7.3.
- Moreover, the header `cc.h` contains architecture or compiler-dependent definitions for some data types widely used by the protocol stack. They are quite straightforward to implement and will not be further discussed in the following for conciseness.

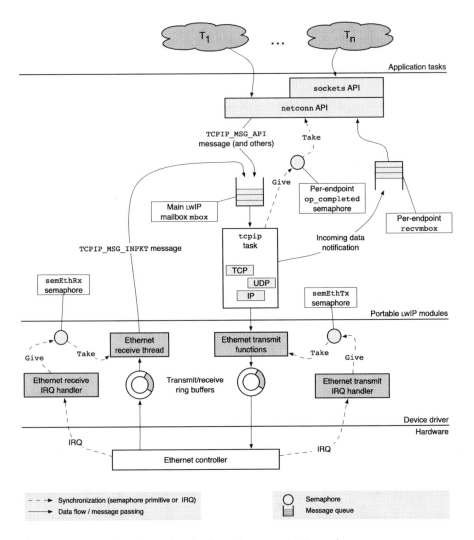

Figure 7.3   Internal LWIP synchronization and communication paths.

Going into even more detail, Figure 7.3 shows how LWIP is structured internally and how its different layers and components communicate and synchronize with each other along the network transmit and receive paths.

In its default configuration, LWIP carries out all protocol stack operations within a single, sequential task, shown in the middle of the figure. Although it is called *tcpip* task (or thread) in the LWIP documentation, this task actually takes care of all protocols supported by LWIP (not only TCP and IP) and processes all frames received from the network.

The main interactions of this thread with the other LWIP components take place by means of several inter-task synchronization devices, as described in the following.

- The mailbox `mbox` is the main way other components interact with the tcpip thread. It is used by network interface drivers to push received frames into the protocol stack, and by application threads to send network requests, for instance to send a UDP datagram. In Figure 7.3, a solid arrow represents either a data transfer or a message passing operation, comprising both data transfer and synchronization.
- At the application level, there are zero or more active network communication endpoints at any instant. A per-endpoint semaphore, `op-completed`, is used to block an application task when it requests a synchronous operation on that endpoint, and hence, it must wait until it is completed.

  In the figure, a synchronization semaphore is represented by a light gray circle, and a dashed arrow stands for a synchronization primitive, either `Take` or `Give`. The same kind of arrow is also used to denote the activation of an IRQ handler by hardware.
- Any application data received from the network is made available to application tasks by means of a per-endpoint mailbox, `recvmbox`. The mailbox interaction also blocks the application task when no data are available to be received.
- Along the transmit path, the main LWIP thread calls the transmit functions of the network interface driver directly, whenever it wants to transmit a frame. The functions are asynchronous, that is, they simply enqueue the frame for transmission and return to the caller.

The per-endpoint mailbox `acceptmbox` (used to enqueue information about incoming connections for listening TCP endpoints), as well as the data structures related to memory management, are not discussed here for simplicity. Some additional information about network buffer management will be provided in Section 7.5.

It is also worth noting that the lowest part of Figure 7.3 depicts the typical structure of a network device driver and its synchronization points with interrupt handlers and, eventually, with hardware. This structure is not further discussed here, but it will be recalled and expanded in Chapter 8.

7.2   OPERATING SYSTEM ADAPTATION LAYER

The LWIP operating system porting layer, also called *operating system emulation layer* in the LWIP documentation [52] resides in the source module `sys_arch.c` and makes the protocol stack portable with respect to the underlying operating system.

It provides a uniform interface to several operating system services, in the four main areas listed in Table 7.1. In addition, it also implements several support functions for tracing and debugging, not discussed here for conciseness.

As an example, the same table shows how the main functions and data types to be implemented by the porting layer have been mapped onto the facilities provided by the FREERTOS operating system. Most of the mapping is quite intuitive, taking into

Table 7.1

LWIP Operating System Interface and Corresponding FREERTOS Facilities

LWIP interface	FREERTOS facility	Description
Fast critical regions		
sys_arch_protect	portENTER_CRITICAL	Enter critical region
sys_arch_unprotect	portEXIT_CRITICAL	Exit critical region
Task creation		
sys_thread_t	xTaskHandle	Task data type
sys_thread_new	xTaskCreate	Create a task
Binary semaphores		
sys_sem_t	xSemaphoreHandle	Semaphore data type
sys_sem_new	vSemaphoreCreateBinary	Create a semaphore
sys_sem_free	vSemaphoreDelete	Delete a semaphore
sys_arch_sem_wait	xSemaphoreTake	Semaphore P()
sys_sem_signal	xSemaphoreGive	Semaphore V()
Mailboxes		
sys_mbox_t	xQueueHandle	Mailbox data type
sys_mbox_new	xQueueCreate	Create a mailbox
sys_mbox_free	vQueueDelete	Delete a mailbox
sys_mbox_post	xQueueSend	Post a message
sys_mbox_trypost	xQueueSend	Nonblocking post
sys_arch_mbox_fetch	xQueueReceive	Fetch a message
sys_arch_mbox_tryfetch	xQueueReceive	Nonblocking fetch

account that the main design goal was to make the porting layer as efficient as possible. Since it is used by LWIP as a foundation for data transfer and synchronization, any issue at this level would heavily degrade the overall performance of the protocol stack. The following, more detailed aspects of the mapping are worth mentioning.

- According to the principles presented in Chapter 5, the LWIP protocol stack uses *critical regions*, delimited by the functions sys_arch_protect and sys_arch_unprotect, to delineate and enforce mutual exclusion among regions of code. Those regions are very short, typically less than 100 machine instructions long, and do not contain calls to any blocking primitive.

  It should be noted that the most appropriate method to protect critical regions depends on how much time is spent within them, with respect to the time needed to execute the critical region entry and exit code. If a critical region is short, like in this case, using a semaphore would be inefficient, because more time would be spent executing the semaphore primitives to enter and exit the critical region, rather than within the critical region itself.

For this reason, the LWIP interface has been mapped directly on the most basic and efficient mutual exclusion mechanism of FREERTOS. This is adequate on single-processor systems and, as an added benefit, the FREERTOS mechanism already supports critical region nesting, a key feature required by LWIP. Readers interested in how the FREERTOS mechanism works will find more details about it in Chapter 10.

- The porting layer must provide a system-independent way to *create a new task*, called *thread* in the LWIP documentation, by means of the function `sys_thread_new`. This interface is used to instantiate the main LWIP processing thread discussed previously, but it can be useful to enhance portability at the network application level, too.

  When using FREERTOS, an LWIP thread can be mapped directly onto a task, as the two concepts are quite similar. The only additional function realized by the porting layer—besides calling the FREERTOS task creation function—is to associate a unique instance of a LWIP-defined data structure, `struct sys_timeouts`, to each thread. Each thread can retrieve its own structure by means of the function `sys_arch_timeouts`. The thread-specific structure is used to store information about any pending timeout request the thread may have.

- Internally, LWIP makes use of *binary semaphores*, with an initial value of either 1 or 0, for mutual exclusion and synchronization. Besides `sys_sem_new` and `sys_sem_free`—to create and delete a semaphore, respectively—the other two interfaces to be provided correspond to the classic timed `P()` and `V()` concurrent programming primitives.

  Even though nomenclature is quite different—the `wait` primitive of LWIP corresponds to the `Take` primitive of FREERTOS, and `signal` corresponds to `Give`—both primitives are provided either directly, or with minimal adaptation, by FREERTOS.

- A *mailbox* is used by LWIP when it is necessary to perform a data transfer among threads—namely to pass a pointer along—in addition to synchronization. The flavor of message passing foreseen by LWIP specifies an indirect, symmetric naming scheme, fixed-size messages and limited buffering, so it can be directly mapped onto the FREERTOS message queue facility. A non-blocking variant of the send and receive interfaces are needed (known as `trypost` and `tryfetch`, respectively), but their implementation is not an issue, because the underlying FREERTOS primitives already support this feature.

7.3 CONFIGURATION OPTIONS

According to the general design guidelines outlined previously, the LWIP protocol stack is highly configurable. Configuration takes place by means of an extensive set of object-like macros defined in the header `lwipopts.h`. Another header, which is called `opt.h` and should not be modified, contains the default value of all the

Table 7.2
Main LWIP **Protocol Enable/Disable Options**

Option	RFC	Description
LWIP_ARP	826	Enables the ARP protocol
LWIP_ICMP	792	Enables the ICMP protocol (required by RFC 1122)
LWIP_DHCP	2131	Enables the DHCP protocol
LWIP_AUTOIP	3927	Enables the AUTOIP protocol
LWIP_UDP	768	Enables the UDP protocol
LWIP_UDPLITE	—	Enables a lightweight variant of UDP (non-standard)
LWIP_TCP	793	Enables the TCP protocol
LWIP_IGMP	3376	Enables the IGMP protocol
LWIP_DNS	1035	Enables the DNS protocol
LWIP_SNMP	1157	Enables the SNMP protocol
LWIP_RAW	—	Enables access to other protocols above IP

options that LWIP supports. The default value automatically comes into effect if a certain option is not defined in `lwipopts.h`.

Interested readers should refer to the documentation found in the source code, mainly in `opt.h` for detailed information about individual configuration options. In this section, we will provide only a short summary and outline the main categories of option that may be of interest.

Table 7.2 summarizes a first group of options that are very helpful to optimize LWIP memory requirements, concerning both code and data size. In fact, each individual option enables or disables a specific communication protocol, among the ones supported by LWIP. Roughly speaking, when a protocol is disabled, its implementation is not built into LWIP, and hence, its memory footprint shrinks.

Each row of the table lists a configuration option, on the left, and gives a short description of the corresponding protocol, on the right. The middle column provides a reference to the *request for comments* (RFC) document that defines the protocol, if applicable. Even though a certain protocol has been extended or amended in one or more subsequent RFCs, we still provide a reference to the original one.

It is also worth mentioning that, even though the ICMP protocol [141] can in principle be disabled, it is mandatory to support it, in order to satisfy the minimum requirements for Internet hosts set forth by RFC 1122 [24].

A second group of options configures each protocol individually. The main configuration options for the IP, UDP, and TCP protocols are listed in Table 7.3 along with a terse description. Readers are referred to specialized textbooks, like [147], for more information about their meanings and side effects on protocol performance and conformance to the relevant standards.

The only aspect that will be discussed here—because it is more related to the way protocols are *implemented* with respect to their abstract definition—is the role

Table 7.3

Main LWIP Protocol--Specific Configuration Options

Option	Description
IP protocol	
IP_DEFAULT_TTL	Default time-to-live (TTL) of outgoing IP datagrams, unless specified otherwise by the transport layer
IP_FORWARD	Enable IP forwarding across network interfaces
IP_FRAG	Enable IP datagram fragmentation upon transmission
IP_REASSEMBLY	Enable IP datagram reassembly upon reception
IP_REASS_MAX_PBUFS	Maximum number of struct pbuf used to hold IP fragments waiting for reassembly
CHECKSUM_GEN_IP	Generate checksum for outgoing IP datagrams in software
CHECKSUM_CHECK_IP	Check checksum of incoming IP datagrams in software
UDP protocol	
UDP_TTL	Time-to-live (TTL) of IP datagrams carrying UDP traffic
CHECKSUM_GEN_UDP	Generate checksum for outgoing UDP datagrams in software
CHECKSUM_CHECK_UDP	Check checksum of incoming UDP datagrams in software
TCP protocol	
TCP_TTL	Time-to-live (TTL) of IP datagrams carrying TCP traffic
TCP_MSS	TCP maximum segment size
TCP_CALCULATE_EFF_SEND_MSS	Trim the maximum TCP segment size based on the MTU of the outgoing interface
TCP_WND	TCP window size
TCP_WND_UPDATE_THRESHOLD	Minimum window variation that triggers an explicit window update
TCP_MAXRTX	Maximum number of retransmissions for TCP data segments
TCP_SYNMAXRTX	Maximum number of retransmissions of TCP SYN segments
TCP_QUEUE_OOSEQ	Queue out-of-order TCP segments
TCP_SND_BUF	TCP sender buffer space in bytes
TCP_SND_QUEUELEN	Number of struct pbuf used for the TCP sender buffer space
TCP_SNDLOWAT	Number of bytes that must be available in the TCP sender buffer to declare the endpoint writable
TCP_LISTEN_BACKLOG	Enable the backlog for listening TCP endpoints
TCP_DEFAULT_LISTEN_BACKLOG	Default backlog value to use, unless otherwise specified
CHECKSUM_GEN_TCP	Generate checksum for outgoing TCP segments in software
CHECKSUM_CHECK_TCP	Check checksum of incoming TCP segments in software

of the checksum generation and check options, like CHECKSUM_GEN_UDP and CHECKSUM_CHECK_UDP.

Bearing in mind that checksum computation, especially when performed on the whole content of a message, is quite expensive from the computational point of view, there are two main and distinct reasons to *not* perform checksum generation and check in software.

1. Some network interfaces—previously used only on high-performance network servers but now becoming increasingly popular on embedded systems as well— can perform checksum generation and check for some popular communication protocols by themselves, either directly in hardware or with the help of a dedicated network processor.

   When this is the case, it is extremely convenient to *offload* this activity from software and leave the duty to the network interface.

   Namely, on the transmit path the protocol stack gives to the network interface outgoing frames where the checksums have been left blank and the interface fills them appropriately before actual transmission takes place.

   Conversely, when the network interface receives a frame, it will check that all checksums are correct and will forward the packet to the protocol stack only if they are.

2. For some protocols, for instance UDP, checksum is actually optional [140] and a special checksum value is reserved by the standard to indicate that the checksum has not been calculated, and hence, it must not be checked, for a certain datagram. Since UDP datagrams are eventually encapsulated in layer-2 frames, for instance, Ethernet frames, in some cases it is appropriate to disable UDP checksumming and rely only on layer-2 error detection capabilities.

   This approach significantly boosts the performance of the protocol stack, especially on low-end microcontrollers, and may not have any significant effect on communication reliability if it is possible to ensure that:

   • layer-2 error detection already protects its payload adequately, and
   • no layer-3 routing is performed, or software processing within routers is deemed to be error-free.

   As a side note, a very similar reasoning was applied when it was decided to suppress the IP header checksum in IPv6 [45].

The next group of options to be discussed is related to the all-important theme of memory management. The main options belonging to this group are listed in Table 7.4, where they are further divided into several subcategories.

The options in the first subcategory determine where the memory used by LWIP comes from and how it is allocated. Internally, LWIP adopts two main strategies of dynamic memory allocation, that is:

1. LWIP makes use of a *heap* to allocate variable-size data structures.
2. Moreover, it supports a number of memory *pools* for fixed-size data structures.

Table 7.4
Main LWIP Memory Management Configuration Options

Option	Description
Memory allocation method	
MEM_LIBC_MALLOC	Use the C library heap instead of the native heap
MEM_MEM_MALLOC	Use the heap instead of the dedicated memory pool allocator
MEM_ALIGNMENT	Alignment requirement, in bytes, for memory allocation
Network buffers	
MEM_SIZE	Size, in bytes, of the native heap (when enabled), mainly used for struct pbuf of type PBUF_RAM
MEMP_NUM_NETBUF	Number of struct netbuf
MEMP_NUM_PBUF	Number of struct pbuf of type PBUF_ROM and PBUF_REF
PBUF_POOL_SIZE	Number of struct pbuf of type PBUF_POOL
PBUF_POOL_BUFSIZE	Size, in bytes, of the data buffer associated to each struct pbuf of type PBUF_POOL
Communication endpoints and protocol control blocks	
MEMP_NUM_NETCONN	Number of struct netconn, that is, active communication endpoints
MEMP_NUM_UDP_PCB	Number of UDP protocol control blocks
MEMP_NUM_TCP_PCB	Number of TCP protocol control blocks
MEMP_NUM_TCP_PCB_LISTEN	Number of TCP protocol control blocks that listen for incoming connections
MEMP_NUM_RAW_PCB	Number of RAW protocol control blocks
MEMP_NUM_TCP_SEG	Number of TCP segments simultaneously queued
MEMP_NUM_REASSDATA	Number of IP datagrams queued for reassembly
MEMP_NUM_ARP_QUEUE	Number of outgoing IP datagrams queued because their destination IP address is being resolved by ARP
MEMP_NUM_SYS_TIMEOUT	Number of timeouts simultaneously active
Tcpip thread message passing	
MEMP_NUM_TCPIP_MSG_API	Number of messages to the tcpip thread for higher-level API communication
MEMP_NUM_TCPIP_MSG_INPKT	Number of messages to the tcpip thread for incoming packets

Both strategies are useful in principle because they represent different trade-offs between flexibility, robustness, and efficiency. For instance, the allocation of a fixed-size data structure from a dedicated pool holding memory blocks of exactly the right size can be done in constant time, which is much more *efficient* than leveraging a general-purpose (but complex) memory allocator, able to manage blocks of *any* size.

Moreover, dividing the available memory into distinct pools guarantees that, if a certain memory pool is exhausted for any reason, memory is still available for other kinds of data structure, drawn from other pools. In turn, this makes the protocol stack more *robust* because a memory shortage in one area does not prevent other parts of it from still obtaining dynamic memory and continue to work.

On the other hand, a generalized use of the heap for all kinds of memory allocation is more *flexible* because no fixed and predefined upper limits on how much memory can be spent on each specific kind of data structure must be set in advance.

Therefore, the three main options in this subcategory choose which of the supported strategies shall be used by LWIP depending on user preferences. Moreover, they also determine if the memory LWIP needs should come from statically defined arrays permanently reserved for LWIP, or from the C library memory allocator. Namely:

- The option MEM_LIBC_MALLOC, when set, configures LWIP to use the C library memory allocator (by means of the well-known functions malloc and free) instead of its own native heap implementation.
- The option MEM_MEM_MALLOC disables all memory pools and instructs LWIP to use the heap—either the native one or the one provided by the C library, depending on the previous option—for all kinds of memory allocation.
- The option MEM_ALIGNMENT specifies the architecture-dependent alignment requirements, in bytes, for memory blocks allocated with any of the methods above.

The second group of options determines the overall size of the heap, if the option MEM_LIBC_MALLOC has not been set, and specifies how many network buffers—that is, buffers used to temporarily store data being transmitted or received by the protocol stack—should be made available for use. Obviously, these options have effect only if MEM_MEM_MALLOC has not been set. More detailed information about the internal structure of the different kinds of network buffer mentioned in the table and their purpose will be given in Section 7.5.

The next group of options, shown in the middle part of Table 7.4, determines the size of other memory pools, related to communication endpoints and to protocol control blocks, specifically for each protocol supported by LWIP.

For instance, the MEMP_NUM_NETCONN option determines how big the pool for struct netconn data structures will be. These data structures are in one-to-one correspondence with communication endpoints created by any of the LWIP APIs. Moreover, one protocol control block is needed for every active communication endpoint that makes use of that protocol.

The fourth and last group of configuration option related to memory management, to be discussed here, determines the size of the memory pools that hold the data structures exchanged between the other parts of the protocol stack and the tcpip thread.

Since, as outlined in Section 7.1, this thread is responsible for carrying out most protocol stack activities, the availability of an adequate number of data structures to communicate with it is critical to make sure that application-level requests and incoming packets can flow smoothly through the protocol stack.

Table 7.5

Other Important LWIP **Configuration Options**

Option	Description
LWIP_NETCONN	Enable the netconn API (Section 7.4)
LWIP_SOCKET	Enable the POSIX sockets API (Section 7.6)
LWIP_COMPAT_SOCKETS	Enable POSIX names for sockets API functions
LWIP_POSIX_SOCKETS_IO_NAMES	Enable POSIX names for some sockets I/O functions
LWIP_SO_RCVTIMEO	Enable the SO_RCVTIMEO socket option
LWIP_SO_RCVBUF	Enable the SO_RCVBUF socket option
RECV_BUFSIZE_DEFAULT	Default receive buffer size when SO_RCVBUF is enabled
LWIP_STATS	Enables statistics collection

Last, but not least, we will present shortly a last group of important LWIP configuration options, listed in Table 7.5.

The two options LWIP_NETCONN and LWIP_SOCKET enable the two main user-level LWIP APIs, that is, the native netconn interface and POSIX sockets. They will be discussed in Sections 7.4 and 7.6, respectively. In normal use, at least one of the two APIs must be enabled in order to make use of the protocol stack.

They can be enabled together, though, to accommodate complex code in which components make use of distinct APIs. Moreover, the POSIX sockets API requires the netconn API in any case because it is layered on it. In addition, the options LWIP_COMPAT_SOCKETS and LWIP_POSIX_SOCKETS_IO_NAMES control the names with which POSIX sockets API functions are made available to the programmer. Their effect will be better described in Section 7.6.

The two configuration options LWIP_SO_RCVTIMEO and LWIP_SO_RCVBUF determine the availability of two socket options (also discussed in Section 7.6) used to specify a timeout for blocking receive primitives and to indicate the amount of receive buffer space to be assigned to a socket, respectively.

The last option listed in Table 7.5, when set, configures LWIP to collect a variety of statistics about traffic and protocol performance.

7.4 NETCONN INTERFACE

The native API supported by LWIP is called *netconn*, a short form of *network connection*, and is fully documented in Reference [52]. Table 7.6 summarizes the main primitives it provides, categorized into several functional groups, which will be discussed one by one in the following. In order to use this API, it must be enabled as described in Section 7.3. Moreover, it is necessary to include the LWIP header api.h.

Table 7.6

Main Functions of the netconn **Networking API**

Function	Description
Communication endpoint management	
netconn_new	Create a communication endpoint of a given type
netconn_close	Close an active connection on an endpoint
netconn_delete	Close and destroy an endpoint
netconn_type	Return endpoint type
netconn_err	Return last error code of an endpoint
Local endpoint address	
netconn_bind	Bind a well-known local address/port to an endpoint
netconn_addr	Return local address/port of an endpoint
Connection establishment	
netconn_connect	Initiate a connection (TCP) or register address/port (UDP)
netconn_peer	Return address/port of remote peer
netconn_disconnect	Disassociate endpoint from a remote address/port (UDP only)
netconn_listen	Make a TCP socket available to accept connection requests
netconn_listen_with_backlog	Same as netconn_listen, with backlog
netconn_accept	Accept a connection request
Data transfer	
netconn_send	Send data to registered address/port (UDP only)
netconn_sendto	Send data to explicit address/port (UDP only)
netconn_write	Send data to connected peer (TCP only)
netconn_recv	Receive data (both TCP and UDP)

Communication endpoint management

In order to perform any form of network communication using LWIP, it is first of all necessary to create a communication endpoint. Communication endpoints are represented by a pointer to a struct netconn and are created by means of the function netconn_new.

The only argument of this function is an enum netconn_type. It indicates the type of endpoint to be created and determines the underlying protocols it will use. Overall, LWIP supports the main endpoint types listed in Table 7.7, although (as indicated in the table) they may or may not be available depending on how the protocol stack has been configured. Due to space constraints, the following discussion will be limited to the NETCONN_TCP and NETCONN_UDP types only.

The return value of netconn_new is either a pointer to a newly created struct netconn or a NULL pointer if error occurred. In any case, this function performs exclusively *local* operations and does not generate any network traffic.

Table 7.7

Main LWIP netconn **Types**

netconn type	Configuration option	Description
NETCONN_TCP	LWIP_TCP	TCP protocol
NETCONN_UDP	LWIP_UDP	UDP protocol
NETCONN_UDPNOCHKSUM	LWIP_UDP	Standard UDP, without checksum
NETCONN_UDPLITE	LWIP_UDP, LWIP_UDPLITE	Lightweight UDP, non-standard
NETCONN_RAW	LWIP_RAW	Other IP-based protocol

The function netconn_close closes any active connection on the communication endpoint passed as argument and returns an err_t value that represents the outcome of the call. As for all other LWIP netconn functions, the return value ERR_OK (zero) indicates that the function completes successfully, whereas a negative value indicates error and conveys more information about the kind of error.

After a successful call to netconn_close, the communication endpoint remains available for further use—for instance, opening a new connection—unless it is deleted by means of the function netconn_delete. This function implicitly calls netconn_close if necessary and then releases any resource allocated to the communication endpoint passed as argument. As a result, the endpoint may no longer be used afterward.

The next two functions listed in Table 7.6 retrieve and return to the caller basic information about the communication endpoint passed as argument. In particular:

- netconn_type returns the type of communication endpoint, as an enum netconn_type, and
- netconn_err returns the last error indication pertaining to the communication endpoint, as an err_t.

Local socket address

The main function in this group is netconn_bind, which assigns a well-known local address to a communication endpoint. It takes three arguments as input:

- a pointer to a struct netconn, which represents the communication endpoint it must work on,
- a pointer to a struct ip_addr, which contains the local IP address to be assigned,
- an u16_t, which indicates the port number to be assigned.

The pointer to the struct ip_addr can be NULL. In this case, LWIP determines an appropriate local IP address automatically. The function returns an err_t value that indicates whether or not the function was successful.

It should be noted that it is often unnecessary to *explicitly* bind a communication endpoint to a specific local address by means of netconn_bind. This is because, as will be better described in the following, other netconn functions automatically and *implicitly* bind an appropriate, unique local address when invoked on an unbound communication endpoint.

Regardless of how a local address has been bound to a communication endpoint, it can be retrieved by means of the netconn_addr function. It takes a communication endpoint as first argument and stores its local IP address and port number into the locations pointed by its second and third arguments, which are a pointer to a struct ip_addr and a pointer to a u16_t, respectively.

Connection establishment

The function netconn_connect, when invoked on a communication endpoint, behaves in two radically different ways, depending on whether the underlying protocol is TCP or UDP.

1. When the underlying protocol is UDP, it does not generate any network traffic and simply associates the remote IP address and port number—passed as the second and third argument, respectively—to the endpoint. As a result, afterward the endpoint will only receive traffic coming from the given IP address and port, and will by default send traffic to that IP address and port, when no remote address is specified explicitly.
2. When the underlying protocol is TCP, it opens a connection with the remote host by starting the three-way TCP handshake sequence. The return value, of type err_t indicates whether or not the connection attempt was successful. TCP data transfer can take place only after a connection has been successfully established.

After a successful netconn_connect the remote address and port number associated to an endpoint can be retrieved by calling the function netconn_peer. It takes the endpoint itself as its only argument.

The function netconn_disconnect can be invoked only on a UDP endpoint and undoes what netconn_connect did, that is, it disassociates the endpoint from the remote IP address and port given previously without generating any network traffic. For TCP endpoints, it is necessary to use netconn_close, as described previously, to close the connection properly.

In order to accept incoming connection requests, a TCP endpoint must be put into the *listening* state first. In its simplest form, this is done by invoking the netconn_listen function on it.

If the LWIP configuration option TCP_LISTEN_BACKLOG has been set, the (more powerful) function netconn_listen_with_backlog performs the same action and it also allows the caller to specify the size of the connection *backlog* associated to the endpoint. If the option has not been set, and hence, the backlog is not available, netconn_listen_with_backlog silently reverts to the same behavior as netconn_listen.

Roughly speaking, the backlog size indicates the maximum number of outstanding connection requests that may be waiting acceptance by the endpoint. Any further connection request received when the backlog is full may be refused or ignored by the protocol stack. More information about how the backlog mechanism works can be found in [147].

The next function to be discussed is `netconn_accept`, which is the counterpart of `netconn_connect`. It takes as argument a TCP communication endpoint that has been previously marked as listening for incoming connection requests and *blocks* the caller until a connection request targeting that endpoint arrives and is accepted.

When this happens, it returns to the caller a newly allocated communication endpoint—represented, as usual, by a pointer to a `struct netconn`—that is associated to the originator of the request and can be used for data transfer. At the same time, the original endpoint is still available to listen for and accept further connection requests.

Data transfer

The main LWIP functions for data transfer are listed at the bottom of Table 7.6. Concerning data transmission, they can be divided into two groups, depending on the kind of communication endpoint at hand.

- The functions `netconn_send` and `netconn_sendto` can be used exclusively on UDP endpoints. They both take a pointer to a `struct netbuf`, which holds data to be sent as an UDP datagram. The main difference between the two is that `netconn_sendto` allows the caller to specify the target IP address and port number on a call-by-call basis, whereas `netconn_send` uses the address and port number previously registered into the endpoint by means of `netconn_connect`. Both functions return a value of type `err_t` that contains the usual status/error indication about the function outcome.
- The function `netconn_write` is reserved for TCP endpoints, and can be used only after a connection with a remote peer has successfully been established. Besides the communication endpoint it must work upon, the function takes two additional arguments:
  1. a `void *` pointer to a contiguous buffer in memory, containing the data to be sent, and
  2. a `u8_t` value, containing a set of flags that modify the function's behavior.

  More specifically:
  - The `NETCONN_MORE` flag indicates that the caller intends to send more data through the same endpoint in a short time. Hence, it is unnecessary to push the present data to the receiving socket immediately. As a result, the protocol stack does not set the TCP `PSH` flag in outgoing TCP segments until some data are sent *without* setting `NETCONN_MORE` in the data transfer call.

- The NETCONN_COPY flag is related to memory management. When it is set, it indicates that the caller intends to destroy or reuse the data buffer given to the function immediately after the call, and hence, the protocol stack must copy the data it contains elsewhere before returning. On the other hand, when the flag is not set, the protocol stack assumes that the data buffer and its contents will be available for an indeterminate amount of time and operates on it *by reference*. This is the case, for instance, when the buffer resides in read-only memory, and hence, its contents are immutable. Therefore, the two choices represent different trade-offs between efficiency (copying data is a quite expensive operation in most architectures) and flexibility in memory management.

On the other hand, the function netconn_recv is able to receive data from both UDP and TCP endpoints. It takes an endpoint as argument and blocks the caller until data arrive on that endpoint. Then, it returns a (non NULL) pointer to a struct netbuf that holds the incoming data. Upon error, or if the TCP connection has been closed by the remote peer, it returns a NULL pointer instead.

As can be seen from the description, on both the transmit and the receive paths data buffers are *shared* between the calling task and the protocol stack in different ways.

Before going into deeper detail about how network buffers are managed, as will be done in Section 7.5, it is therefore extremely important to appreciate how they should be managed, in order to avoid memory leaks (if a buffer is not deallocated when it is no longer in use) or memory corruption (that may likely happen if a buffer is released, and possibly reused, "too early"). In particular:

- When netconn_write is invoked without setting the NETCONN_COPY flag, the data buffer passed as argument effectively becomes property of LWIP for an indeterminate amount of time. If this is not the intended behavior, the flag must be set appropriately.
- For what concerns netconn_send and netconn_sendto, the caller is responsible for allocating the data buffer—represented by a pointer to a struct netbuf in this case—and fill it with the information to be sent before calling them, by means of the function to be described in Section 7.5. After these calls, the data buffer is shared with, and referenced by the protocol stack. In turn, LWIP will release the reference at a future time, when it no longer needs the data. As a consequence, the caller is still responsible for deleting the struct netbuf when it is no longer using it, but it is important to be aware that part of the memory associated with it will live on until LWIP clears all pending references to it.
- On the receive path, LWIP itself allocates the struct netbuf returned by netconn_recv. It is the responsibility of the calling task to delete this structure after dealing with the information it contains.

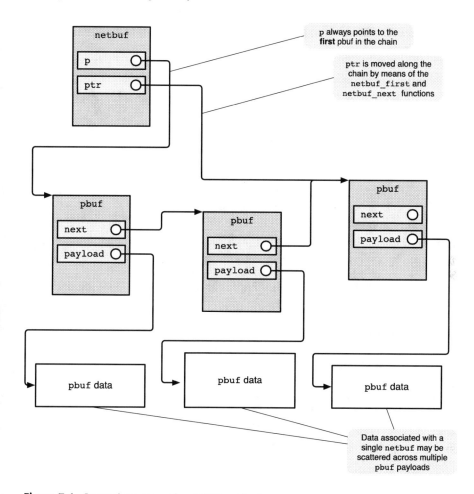

Figure 7.4   Internal structure of an LWIP `netbuf`.

7.5   NETWORK BUFFER MANAGEMENT

Even though in the previous sections the `struct netbuf` has been considered sort of a monolithic data container, it has a rather complex internal structure, outlined in Figure 7.4.

- The first aspect to be remarked is that the `struct netbuf` by itself (shown at the top left of the figure) does not hold any data. Instead, data are referenced indirectly, by means of another data structure, called `struct pbuf`.
- The `struct netbuf` references a chain of `struct pbuf` by means of pointer p. In the figure, the chain is made of three of these structures, concatenated by means of the pointer `next`. Pointer p always points to the first `struct pbuf` in the chain.

- Moreover, the additional pointer `ptr` of the `struct netbuf` points to the *current* `pbuf` in the chain and is used to scan data belonging to a `struct netbuf` sequentially. In the figure, as an example, this pointer refers to the third, and last `pbuf` in the chain.
- In turn, a `struct pbuf` does not contain any data, either. Instead, as can be seen in the figure, the data buffer associated with a certain `struct pbuf` can be accessed by means of the `payload` field of the structure. For this reason, the LWIP documentation sometimes refers to the `struct pbuf` as *pbuf header*.

As a consequence, data associated with a `struct netbuf` are not necessarily contiguous in memory and must be retrieved by following the previously mentioned pointers in an appropriate way. Fortunately, LWIP provides a rather comprehensive set of functions, listed in Table 7.8 and better described in the following, to perform those operations in an abstract and implementation-independent way as much as possible.

For the sake of completeness, it should also be noted that both the `struct netbuf` and the `struct pbuf` contain additional fields, not shown in the figure for simplicity. For instance, a `struct pbuf` contains additional fields to determine the length of the data buffer it points to.

Moreover, all `struct pbuf` are reference-counted, so that they can be referenced by multiple, higher-level data structures being certain that they will be deallocated only when no active references remain to them.

This is especially important since the `struct pbuf` is the data structure used ubiquitously by LWIP to hold network packets, as they flow through the various protocol stack layers and their ownership is transferred from one layer to another.

The first group of functions, listed at the top of Table 7.8, has to do with the *memory allocation* and *management* of a `struct netbuf` and the data buffers associated with it.

- The function `netbuf_new` has no arguments. It allocates a new `struct netbuf` and returns a pointer to it when successful. Upon failure, it returns a NULL pointer instead.
  Immediately after allocation, a `struct netbuf` has no data buffer space associated with it, and hence, it cannot immediately be used to store any data. In other words, referring back to Figure 7.4, both p and `ptr` are NULL pointers and there is no `pbuf` chain linked to it.
- In order to dynamically allocate a data buffer and assign it to a `struct netbuf`, the application must call the `netbuf_alloc` function. It takes two arguments, that is, a pointer to the `struct netbuf` it must operate upon and an integer value that indicates how many bytes must be allocated. The function allocates a single `struct pbuf` that points to a *contiguous* data buffer able to hold the requested number of bytes. Some extra space is also reserved at the beginning of it, so that LWIP can efficiently prepend

Table 7.8
LWIP **Network Buffer Management Functions**

Function	Description
Management and memory allocation	
netbuf_new	Allocate a new struct netbuf
netbuf_alloc	Allocate a data buffer for a struct netbuf
netbuf_free	Free data buffer(s) associated with a struct netbuf
netbuf_ref	Associate a data buffer to a struct netbuf by reference
netbuf_chain	Concatenate the contents of two struct netbuf into one
netbuf_delete	Deallocate a struct netbuf and any buffers associated with it
Data access	
netbuf_len	Return the total data length of a struct netbuf
netbuf_first	Reset the current data buffer pointer to the first buffer
netbuf_data	Get access to the current data buffer contents
netbuf_next	Move to the next data buffer linked to a struct netbuf
netbuf_copy	Copy data from a struct netbuf
netbuf_copy_partial	Similar to netbuf_copy, with offset
netbuf_take	Copy data into a struct netbuf
Address information	
netbuf_fromaddr	Return IP address of the source host
netbuf_fromport	Return port number of the source host

protocol headers at a later time without copying data around and without allocating any further struct pbuf for them.

Moreover, if any struct pbuf and their corresponding data buffers were already allocated to the struct netbuf in the past, they are released before allocating the new one. Upon successful completion, the function returns a pointer to the newly allocated data buffer. Else, it returns a NULL pointer.

- Conversely, the chain of struct pbuf currently linked to a struct netbuf can be released by calling the function netbuf_free. It is legal to call this function on a struct netbuf that has no memory allocated to it, the function simply has no effect in this case.

- An alternative way for an application to associate a data buffer to a struct netbuf is to allocate the data buffer by itself and then add a *reference* to it to the struct netbuf. This is done by means of the netbuf_ref function, which takes three arguments: the struct netbuf it must work on, a void * pointer to the data buffer and an integer holding the size, in bytes, of the buffer. The function returns a value of type err_t that conveys information about its outcome.

The function allocates a single `struct pbuf` that holds a reference to the application-provided data buffer. As in the previous case, if any `struct pbuf` and their corresponding data buffers were already allocated to the `struct netbuf` in the past, they are released before allocating the new one.

However, in this case, no extra space can be reserved at the very beginning of the buffer for protocol headers, and hence, subsequent LWIP processing will likely be slower.

Moreover, unlike for `netbuf_alloc`, the correct management of the data buffer becomes the responsibility of the application. In particular, it is the application's responsibility to ensure that the data buffer remains available as long as LWIP needs it. Since it is not easy to satisfy this requirement in the general case, `netbuf_ref` is most often used to reference immutable, read-only data buffers, which are always available by definition.

- The function `netconn_chain` can be used to concatenate the contents of two `struct netbuf`, passed as arguments. All data buffers are associated with the first `struct netbuf` while the second one is released and can no longer be used afterward.
- When a `struct netbuf` is no longer needed, it is important to deallocate it by means of the `netbuf_delete` function. As a side effect, this function also releases any data buffer associated with the `struct netbuf` being deallocated, exactly like `netbuf_free` does.

As shown in Figure 7.4 in the general case, the data associated to a single `struct netbuf` can be scattered among multiple, non-contiguous data buffers, often called *fragments*. Therefore, LWIP provides a set of *data access* functions to "walk through" those data buffers and get sequential access to them.

- The function `netbuf_len`, given a pointer to a `struct netbuf`, returns the *total* length of the data it contains, that is, the sum of the lengths of all data buffers associated to it by means of the `struct pbuf` chain.
- As mentioned previously, and as also shown in Figure 7.4, the `struct netbuf` contains two distinct pointers to the `struct pbuf` chain. One pointer (p) always refers to the *first* element of the chain, whereas the other (ptr) can be moved and points to the *current* element of the chain.
  The function `netbuf_first`, when invoked on a `struct netbuf`, resets pointer ptr so that it refers to the first element of the `struct pbuf` chain. As a consequence, the first element of the chain becomes the current one.
- The function `netbuf_data` is used to get access to the data buffer belonging to the current element of the `struct pbuf` chain linked to a given `struct netbuf`. In particular, it makes available to the caller a pointer to the data buffer and its length. Moreover, it returns an `err_t` indication.
- The function `netbuf_next` follows the chain of `struct pbuf` associated with a given `struct netbuf` and moves ptr so that it points to the

next element of the chain if possible. The return value indicates the outcome of the function and where the pointer is within the chain, that is:

- a *negative* return value means that it was impossible to move the pointer because it was already pointing to the last element of the chain;
- a *positive* value indicates that the current pointer was moved and now points to the last element of the chain;
- *zero* means that the pointer was moved but it is *not yet* pointing to the last element of the chain, that is, there are more elements in the chain still to be reached.

Therefore, a sequence of calls to `netbuf_first`, `netbuf_data` and `netbuf_next` (the last two repeated until `netbuf_next` returns a positive value) allows the application to get access to all data fragments associated with a `struct netbuf` and operate on it.

When the intent is simply to copy data between a `struct netbuf` and a contiguous, application-supplied buffer, a couple of specialized functions are also available, which automate the whole process. Namely:

- The function `netbuf_copy` copies data from a `struct netbuf` into a contiguous buffer in memory, up to a caller-specified maximum length. It properly handles the case in which the data associated to the `struct netbuf` are fragmented. The more powerful variant `netbuf_copy_partial` does the same, and also allows the caller to specify an initial *offset* within the `struct netbuf` data, from which the copy will start.
- The function `netbuf_take` does the opposite, that is, it copies data from a contiguous application-supplied buffer into a `struct netbuf`, scattering data among the data buffers associated to the `struct netbuf` as necessary. The function does not modify the amount of data buffer memory allocated to the `struct netbuf`, and hence, it can only copy data up to the current total length of the `struct netbuf`.

Besides data, a `struct netbuf` provided by LWIP and containing data received from the network also holds information about the IP address and port number those data come from. This information can be retrieved by means of the functions `netbuf_fromaddr` and `netbuf_fromport`. They take a pointer to a `struct netbuf` as argument and return, respectively, a pointer to a `struct ip_addr` and an `unsigned short`.

One last concept worth mentioning is that there are different kinds of `struct pbuf`, depending on their purpose and on how the memory they refer to is structured. The two main kinds are called `PBUF_RAM` and `PBUF_POOL`. They are shown in Figure 7.5.

- The most common use of a `PBUF_RAM` pbuf (depicted on the left of the figure) is to hold application-level data to be transmitted on the network. In fact, this is the kind of `pbuf` that `netbuf_alloc` allocates.

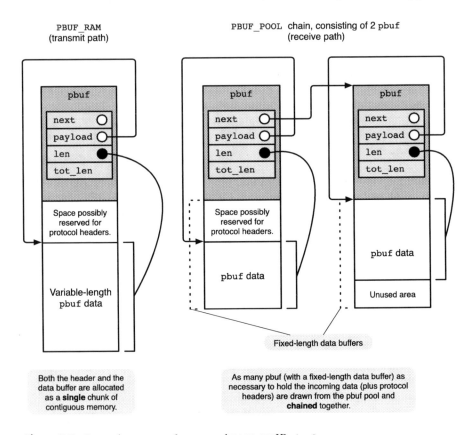

Figure 7.5   Internal structure of a RAM and POOL LWIP pbuf.

For this kind of pbuf, the memory for both the struct pbuf and the data buffer is allocated as a *single chunk* of contiguous memory from the LWIP heap memory. Therefore, the size of the data buffer is exactly tailored to the data size declared when the allocation takes place, plus some space reserved for LWIP to store protocol headers at a later time, as described previously. The overall size of the LWIP heap is controlled by a configuration option discussed in Section 7.3.

Minor padding may be necessary between the struct pbuf and the data buffer, and at the end of the data buffer, in order to satisfy the alignment constraints of the target architecture. It is not shown in Figure 7.5 for simplicity.

- On the other hand, PBUF_POOL pbuf chains are used by LWIP along the receive path and eventually passed to applications, for instance, as the return value of netconn_recv.

Unlike in the previous case, pbufs of this kind are allocated from the LWIP memory pool PBUF_POOL and have a *fixed-size* data buffer associated with them. Section 7.3 provides more details on how to configure the number of pbuf that the pool must contain and their data buffer size.

There are two direct consequences of the different memory allocation strategy used in PBUF_POOL pbufs with respect to PBUF_RAM pbufs.

1. Depending on the quantity of data to be stored, a single PBUF_POOL pbuf may or may not be sufficient to the purpose. The LWIP memory allocation module automatically allocates as many pbuf as required and links them in a chain. Hence, users of this kind of pbuf must be prepared to deal with fragmented data buffers. As an example, Figure 7.5 shows how a chain of two pbufs is structured.
2. As shown in the right part of the figure, an unused memory area may be left at the end of the data buffer associated with the last struct pbuf in the chain.

In order to keep track of how data memory is spread across a chain, each struct pbuf holds two integers, managed by LWIP itself. Namely,

- len holds the length of the data buffer linked to the struct pbuf, and
- tot_len holds the total length of all data buffers linked to the struct pbuf and to the ones that follow it in the chain.

Besides the two kinds of pbuf just mentioned, LWIP supports two others, PBUF_ROM and PBUF_REF, not shown in the figure.

Both of them differ from the ones discussed previously because they hold a *reference* to a data buffer not managed by LWIP itself. The main difference between PBUF_ROM and PBUF_REF is that, in the first case, LWIP considers the contents of the data buffer to be immutable and always available. In the second case, instead, LWIP will copy buffer contents elsewhere if it needs to use them at a later time.

It is also worth mentioning that, as a result of protocol stack processing, all kinds of pbuf can eventually be mixed and linked to a struct netbuf. Therefore, assuming that the chain of struct pbuf has got a specific structure (for instance, it contains one single pbuf) or only holds a specific kind of pbuf is a wrong design choice and makes application programs prone to errors.

7.6   POSIX NETWORKING INTERFACE

With respect to the netconn interface, described in Section 7.4, the POSIX sockets interface has two important advantages from the programmer's point of view:

1. It is backed by the renowned international standard ISO/IEC/IEEE 9945 [86]. Moreover sockets, fully described in [115], were first introduced in the "Berkeley Unix" operating system many years ago. They are now available on virtually all general-purpose operating systems and most real-time operating systems conforming to the standard provide them as well. As a consequence, most programmers are likely to be proficient with sockets and able to use them from day zero without additional effort or training.

Table 7.9

Main Functions of the POSIX sockets **API**

Function	Description
Communication endpoint management	
socket	Create a communication endpoint
closesocket	Close a communication endpoint
close	Same as closesocket [**]
shutdown	Shut down a connection, in part or completely [*]
getsockopt	Retrieve socket options [*]
setsockopt	Set socket options [*]
ioctlsocket	Issue a control operation on a socket [*]
Local socket address	
bind	Assign a well-known local address to a socket
getsockname	Retrieve local socket address
Connection establishment	
connect	Initiate a connection (TCP) or register address/port (UDP)
listen	Mark a socket as available to accept connections (only for TCP)
accept	Accept connection on a socket, create a new socket (only for TCP)
getpeername	Retrieve address of connected peer
Data transfer	
send	Send data through a socket
sendto	Like send, but specifying the target address
sendmsg	*Not implemented*
recv	Receive data from a socket
recvfrom	Like recv, also retrieves the source address
recvmsg	*Not implemented*
read	Receive data from a socket [**]
write	Send data through a socket [**]
Synchronous multiplexing	
select	Synchronous input–output multiplexing [*]
pselect	*Not implemented*
poll	*Not implemented*
FD_ZERO	Initialize a socket descriptor set to the empty set
FD_CLR	Remove a socket descriptor from a socket descriptor set
FD_SET	Add a socket descriptor to a socket descriptor set
FD_ISSET	Check whether or not a socket descriptor belongs to a set

[*] Partial implementation, see main text
[**] Only available if LWIP_POSIX_SOCKETS_IO_NAMES is set

2. With respect to the `netconn` interface, the `sockets` interface is much simpler for what concerns *memory management*. As was shown in Section 7.4, data transfer through the `netconn` interface take place by means of network buffers that are represented by a `netbuf` data structure. The internal structure of these buffers is rather complex and they are in a way "shared" between the application and the protocol stack. They must therefore be carefully managed in order to avoid corrupting them or introducing memory leaks. On the contrary, all data transferred through the `sockets` interface are stored in contiguous user-level buffers, which are conceptually very simple.

As often happens, there is a price to be paid for these advantages. In this case, it is mainly related to execution *efficiency*, which is lower for POSIX `sockets`. This is not only due to the additional level of abstraction introduced by this interface, but also to the fact that, in LWIP, it is actually implemented as an additional layer of software on top of the `netconn` interface.

In addition, as will become clearer in the following, POSIX `sockets` require more memory to memory copy operations with respect to `netconn` for commonly used functions, like data transfer.

Speaking more in general, the main advantage of `sockets` is that they support in a *uniform* way *any kind* of communication network, protocol, naming conventions, hardware, and so on, even if it is not based on the IP protocol. Semantics of communication and naming are captured by *communication domains* and socket *types*, both specified upon socket creation.

For example, communication domains are used to distinguish between IP-based network environments with respect to other kinds of network, while the socket type determines whether communication will be stream-based or datagram-based and also implicitly selects which network protocol a socket will use. Additional socket characteristics can be set up after creation through abstract *socket options*. For example, a socket option provides a uniform, implementation-independent way to set the amount of receive buffer space associated with a socket, without requiring any prior knowledge about how buffers are managed by the underlying communication layers.

On the other hand, introducing a *tailored* API for network communication is also not a new concept in the embedded system domain. For instance, the OSEK VDX operating system specification [92, 136], focused on automotive applications, specifies a communication environment (OSEK/VDX COM) less general than `sockets` and oriented to real-time message-passing networks, such as the controller area network (CAN) [91].

The API that this environment provides is more flexible and efficient because it allows applications to easily set message filters and perform out-of-order receives, thus enhancing their timing behavior. Neither of these functions is straightforward to implement with `sockets`, because they do not fit well within the general socket paradigm.

Table 7.9 summarizes the main functions made available by the POSIX `sockets` API, divided into functional groups that will be reviewed in more detail in the following. Before proceeding further it is worth noting that, as was outlined in Section 7.3, the POSIX `sockets` interface is *optional* in LWIP. Therefore, it is made available only if the configuration option `LWIP_SOCKET` is set.

However, this option does not automatically make POSIX `sockets` functions available to programmers by means of the names listed in Table 7.9, in order to keep the C language name space cleaner. Rather, the default function names are obtained by adding to the names shown in the table the prefix `lwip_`.

So, for example, when only the option `LWIP_SOCKET` is set, the function `socket` can be invoked using the name `lwip_socket`. Two notable exceptions are `closesocket` and `ioctlsocket`, for which no counterpart starting with `lwip_` is provided.

The availability of the names listed in Table 7.9, without the `lwip_` prefix, depends instead on two other LWIP configuration options. Namely:

- When the option `LWIP_COMPAT_SOCKETS` is set, all function names listed in the table become available, with the exception of `close`, `read`, and `write`. The exceptions just mentioned are made because, in the POSIX standard, these functions are in common with the file input–output subsystem, and hence, conflicts with the runtime C library may arise if it provides this feature.
- If the option `LWIP_POSIX_SOCKETS_IO_NAMES` is set, in addition to the previous one, LWIP makes `close`, `read`, and `write` available as well.

Communication endpoint management

In order to use the protocol stack for network communication, it is first of all necessary to create at least one communication endpoint, known as *socket*. This is accomplished through the invocation of the `socket` function with three arguments:

1. A *protocol family* identifier, which uniquely identifies the network communication domain the socket belongs to and operates within. A communication domain is an abstraction introduced to group together sockets with common communication properties, for example their endpoint addressing scheme, and also implicitly determines a communication boundary because data exchange can take place only among sockets belonging to the same domain.
   Since LWIP supports a single communication domain, it ignores this argument. However, in order to be portable to other, more comprehensive implementations of the POSIX standard, it is necessary to specify `PF_INET` domain, which identifies the Internet.
2. A *socket type* identifier, which specifies the communication model that the socket will use and, as a consequence, determines which communication properties will be visible and available to its user.
3. A *protocol identifier*, to select which specific protocol stack, among those suitable for the given protocol family and socket type, the socket will use—if more than one protocol is available.
   In other words, the communication domain and the socket type are first used together to determine a set of communication protocols that belong to the domain

and obey the communication model the socket type indicates. Then, the protocol identifier is used to narrow the choice down to a specific protocol within this set. The special identifier 0 (zero) specifies that a default protocol, selected by the underlying socket implementation, shall be used. It should also be noted that, in most cases, this is not a source of ambiguity, because most protocol families support exactly one protocol for each socket type.

When it completes successfully, `socket` returns to the caller a small non-negative integer, known as *socket descriptor*, which represents the socket just created and shall be passed to all other socket-related functions, in order to refer to the socket itself.

Instead, the negative value −1 indicates that the function failed, and no socket has been created. In this case, like for most other POSIX functions, the `errno` variable conveys to the caller additional information about the reason for the failure.

At the time of this writing, LWIP supports, in principle, three different socket types. Which ones are actually available for use depend on how LWIP has been configured, as explained in Section 7.3.

1. The `SOCK_STREAM` socket type provides a connection-oriented, bidirectional, sequenced, reliable transfer of a byte stream, *without* any notion of message boundaries. It is hence possible for a message sent as a single unit to be received as two or more separate pieces, or for multiple messages to be grouped together at the receiving side.
2. The `SOCK_DGRAM` socket type supports a bidirectional data flow *with* message boundaries, but does not provide any guarantee of sequenced or reliable delivery. In particular, the messages sent through a datagram socket may be duplicated, lost completely, or received in an order different from the transmission order, with no indication about these facts being conveyed to the user.
3. The `SOCK_RAW` socket type allows applications direct access to IP datagrams belonging to a specific layer-4 protocol, indicated by the *protocol identifier* argument of `socket`.

For what concerns `SOCK_STREAM` sockets, the only protocol supported by LWIP is TCP [143], and hence, the *protocol identifier* argument is ignored in this case. On the other hand, two protocols are available for `SOCK_DGRAM` sockets:

- The `IPPROTO_UDP` protocol identifier calls for the well-known UDP [140] protocol.
- The `IPPROTO_UDPLITE` identifier denotes a lighter, non-standard variant of UDP, which is LWIP-specific and does not interoperate with other protocol stacks.

The two names `close` and `closesocket` correspond to the same function, which closes and destroys a socket, given its descriptor. It must be used to reclaim system resources—mainly memory buffers—assigned to a socket when it is no longer in use.

By means of the `shutdown` function, the standard also specifies a way to shut down a socket only *partially*, by disabling further send and/or receive operations. However, the version of LWIP considered in this book only provides a partial implementation of `shutdown`, which always closes the socket completely.

Socket options can be retrieved and set by means of a pair of generic functions, `getsockopt` and `setsockopt`. The way of specifying options to these functions is modeled after the typical layered structure of the underlying communication protocols and software. In particular, each option is uniquely specified by a (`level`, `name`) pair, in which:

- `level` indicates the protocol level at which the option is defined. In addition, a separate level identifier (`SOL_SOCKET`) is reserved for the upper layer, that is, the socket level itself, which does not have a direct correspondence with any protocol.
- `name` determines the option to be set or retrieved within the level and, implicitly, the additional arguments of the functions.

It should be noted that LWIP does not implement all the options specified by the standard, and hence, the availability of a certain option must be assessed before use on a case-by-case basis. Two important socket options supported by LWIP at the `SOL_SOCKET` level are:

- `SO_RCVTIMEO`, which is used to set a timeout for blocking receive operations, expressed in milliseconds. When this timeout is set, receive operations that cannot be completed return to the caller after the specified amount of time elapses, instead of blocking the caller indefinitely.
- `SO_RCVBUF`, which determines the maximum amount of receive buffer space to be assigned to the socket.

In order to reduce memory footprint and execution overhead, both options are supported by LWIP only if it has been explicitly configured to this purpose, as described in Section 7.3.

After a socket has been created, it is possible to set and retrieve some of its characteristics, or *attributes*, by means of the `ioctl` function. Even though the POSIX standard defines a rather large set of commands that this function should accept and obey, LWIP implements only two of them. Namely:

- The `FIONREAD` command lets the caller know how many bytes of data are waiting to be received from the socket at the moment, without actually retrieving or destroying those data.
- The `FIONBIO` command allows the caller to set or reset the `O_NONBLOCK` socket flag. When this flag is set, all operations subsequently invoked on the socket are guaranteed to be *nonblocking*. In other words, they will return an error indication to the caller, instead of waiting, when the requested operation cannot be completed immediately.

Local socket address

A socket has no local address associated with it, when it is initially created by means of the `socket` function. On the other hand, a socket must have a unique local address to be actively engaged in data transfer. This is because the local address is used as the source address for outgoing data frames, and incoming data frames are conveyed to sockets by matching their destination address with local socket addresses.

The exact address format and its interpretation may vary depending on the communication domain. Within the Internet communication domain, addresses contain a 4-byte IP address and a 16-bit port number, assuming that IPv4 [142] is in use.

A local address can be bound to a socket either *explicitly* or *implicitly*, depending on how the application intends to use it.

- The `bind` function *explicitly* gives a certain local address, specified as a function argument, to a socket. This course of action gives to the caller full control on socket address.
- Other functions, `connect` for instance, automatically and *implicitly* bind an appropriate, unique address when invoked on an unbound socket. In this case, the caller is not concerned at all with local address assignment but, on the other hand, it also has no control on it.

Regardless of the way the local socket address has been assigned, it can be retrieved by means of the `getsockname` function. The function may return a failure indication when the socket has no local address, for instance, after it has been shut down.

Connection establishment

The `connect` function has two arguments:

- a socket descriptor that identifies the local communication endpoint;
- a socket address, which indicates the target remote communication endpoint.

When invoked on a *connection-oriented* socket, as those using the TCP protocol are, it sends out a connection request directed toward the target address specified in the second argument. Moreover, if the local socket is currently unbound, the system also selects and binds an appropriate local address to it beforehand.

If the function succeeds, it associates the local and the target sockets and data transfer can begin. Otherwise, it returns to the caller an error indication. In order to be a valid target for a `connect`, a socket must satisfy two conditions.

- First of all, it must have a well-known address assigned to it—because the communication endpoint willing to connect has to specify it as the second argument of the `connect` function.

- Secondly, it must be marked as willing to accept connection requests, by means of the `listen` function. The first argument of this function is, as usual, a socket descriptor to be acted upon. Informally speaking, the second argument is an integer that specifies the maximum number of outstanding connection requests that can be waiting acceptance on the given socket, known as *backlog*. It should be noted that the user-specified value is handled as a hint by the socket implementation, which is free to reduce it if necessary. Interested readers may also refer to [147] for more in-depth details about how the backlog mechanism works.

  If a new connection request is received while the queue of outstanding requests is full, the connection can either be refused immediately or, if the underlying protocol implementation supports, the request can be retried at a later time.

After a successful execution of `listen`, the `accept` function can be used to wait for the arrival of a connection request on a given socket. The function blocks the caller until a connection request arrives, then accepts it, creates a *new* socket and returns its descriptor to the caller. The new socket is connected to the originator of the connection request, while the original one is still available to wait for and accept further connection requests.

Moreover, the `accept` function also has the ability to provide to the caller the address of the socket that originated the connection request. The same information can also be obtained at a later time by means of the `getpeername`, which retrieves and returns the address of a connected peer, if any, when invoked on a connection-oriented socket.

On the other hand, *connectionless* communication is also possible. It is typical of datagram sockets, such as the ones using the UDP protocol, and does not require any form of connection negotiation or establishment before data transfer can take place.

Socket creation proceeds in the same way as it does for connection-oriented sockets, and `bind` can be used to assign a specific local address to a socket. Moreover, if a data transmission operation is invoked on an unbound socket, the socket is implicitly bound to an available local address before transmission takes place.

Due to the lack of need for connection establishment, `listen` and `accept` cannot be used on a connectionless socket. On the other hand, `connect` can still be used, albeit with different semantics.

In fact, it simply associates a destination address with the socket so that, in the future, it will be possible to use it with data transmission functions which do not explicitly indicate the destination address, for instance, `send`. Moreover, after a successful `connect`, only data received from that remote address will be delivered to the user.

The `connect` function can be used multiple times on the same connectionless socket, but only the last address specified remains in effect. Unlike for connection oriented sockets, in which `connect` implies a certain amount of network activity, connect requests on connectionless sockets return almost immediately to the caller, because they simply result in a local operation, that is, the system recording the remote address for later use.

If `connect` has not been used, the only way to send data through a connectionless socket is by means of a function that allows the caller to specify the destination address on a message-by-message case, such as `sendto`.

Data transfer

The functions `send`, `sendto`, and `sendmsg` send data through a socket, with different trade-offs between expressive power and interface complexity. Here, we will only discuss the first two, because the version of LWIP we are considering, that is, version 1.3, does not implement `sendmsg`.

- The `send` function is the simplest one and assumes that the destination address is already known to the system, as is the case when the function is invoked on a connection-oriented socket that has been successfully connected to a remote peer in the past. On the other hand, it cannot be used, for example, on connectionless sockets on which no former `connect` has been performed.

  Instead, its four arguments specify the socket to be used, the position and size of a memory buffer containing the data to be sent, and a set of *flags* that may alter the semantics of the function.

- With respect to the previous one, the `sendto` function is more powerful because, by means of two additional arguments, it allows the caller to explicitly specify a destination address, making it also useful for connectionless sockets.

The only flag currently supported by LWIP is `MSG_MORE`. This flag has effect only on TCP sockets and indicates that the caller intends to send more data through the same socket in a short time. Hence, it is unnecessary to push the present data to the receiving socket immediately. In turn, the protocol stack does not set the TCP `PSH` flag in outgoing TCP segments until some data are sent *without* the `MSG_MORE` flag set.

Symmetrically, the `recv`, `recvfrom`, and `recvmsg` functions allow a process to wait for and retrieve incoming data from a socket. Also in this case, LWIP does not implement the most complex one, that is, `recvmsg`.

Like their counterparts, these functions have different levels of expressive power and complexity:

- The `recv` function possibly waits for data to be available from a socket. When data are available, the function stores it into a data buffer in memory, and returns to the caller the length of the data just received. It also accepts as argument a set of flags that may alter the semantics of the function.
- In addition, the `recvfrom` function allows the caller to retrieve the address of the sending socket, making it useful for connectionless sockets, in which the communication endpoints may not be permanently paired.

The LWIP implementation supports two flags to modify the behavior of recv and recvfrom.

MSG_DONTWAIT: When this flag is set, recv and recvfrom immediately return an error indication to the caller, instead of waiting, if no data are available to be received. It has the same effect as setting the O_NONBLOCK flag on the socket, but on a call-by-call basis.

MSG_PEEK: When this flag is set, incoming data are returned to the application, but *without* removing them from socket buffers. In other words, they make the receive operation *nondestructive*, so that the same data can be retrieved again by a subsequent receive.

Besides the specialized functions described previously, applications can also use the read and write functions to receive and send data through a connection-oriented socket, respectively. These functions are simpler to use, but not as powerful as the others, because no flags can be specified.

Synchronous multiplexing

The socket functions described so far can behave in two different ways for what concerns *blocking*, depending on how the O_NONBLOCK flag (as well as the MSG_DONTWAIT flag for data transfer operations) has been set. Namely:

- The default behavior (when neither O_NONBLOCK nor MSG_DONTWAIT are set) is to *block* the caller until they can proceed. For example, the recv function blocks the caller until there is some data available to be retrieved from the socket, or an error occurs.
- When either flag is set, either globally (O_NONBLOCK) or on a call-by-call basis (MSG_DONTWAIT), affected socket functions return to the caller immediately and indicate whether or not they were able to perform the operation. In this way, it becomes possible to perform a periodic *polling* on each member of a set of sockets.

The default behavior in which socket functions block the caller until completion is quite useful in many cases, because it allows the software to be written in a simple and intuitive way. However, it may become a disadvantage in other, more complex situations.

Let us consider, for instance, a network server that is simultaneously connected to a number of clients and does not know in advance from which socket the next request message will arrive. In this case the server must not perform a blocking recv on a specific socket, because it would run into the risk of ignoring incoming messages from all the other sockets for an unpredictable amount of time.

On the other hand, the polling-based approach may or may not be acceptable, depending on the kind of application, because its overhead and latency grow linearly with the number of sockets to be handled. For this reason, the POSIX standard specifies a third way of managing a whole *set* of sockets at once, called *synchronous multiplexing*.

The basic principle of this approach is that a task—instead of blocking on an *individual socket* until a certain operation on it is complete—blocks until certain operations become possible on *any socket* in a set.

The standard specifies three main specialized functions for synchronous multiplexing. In order of complexity, they are `select`, `pselect`, and `poll`. Among them, we will discuss only the simplest one, that is, `select` because it is the only one implemented by LWIP at the time of this writing.

The function `select` takes as arguments three, possibly overlapping, sets of socket descriptors and a timeout value. It examines the descriptors belonging to each set in order to check whether at least one of them is ready for reading, ready for writing, or has an exceptional condition pending, respectively.

More specifically, it blocks the caller until the timeout expires or at least one of the conditions being watched becomes true. In the second case, the function updates its arguments to inform the caller about which socket descriptors became ready for the corresponding kind of operation.

It should also be noted that LWIP currently does not support the third set of socket descriptors mentioned previously. Any exceptional condition involving a certain socket is notified by informing the caller that the socket is ready for reading. The subsequent read operation will then fail, and convey more information about the nature of the error.

A set of socket descriptors is represented by the abstract data type `fd_set` and can be manipulated by means of the following function-like macros.

- `FD_ZERO` initializes a socket descriptor set to the empty set. All socket descriptor sets must be initialized in this way before use.
- `FD_CLR` and `FD_SET` remove and add a socket descriptor to a socket descriptor set.
- Finally, `FD_ISSET` checks whether or not a certain socket descriptor belongs to a socket descriptor set.

7.7 SUMMARY

This chapter provided an overview of the LWIP open-source TCP/IP protocol stack. The two main aspects described here were how to *embed* the protocol stack within a more complex software project, and how to *make use* of it from an application program.

After a short introduction to the general characteristics and internal structure of LWIP, given in Section 7.1, Section 7.2 provides information about the interface between the protocol stack and the underlying operating system, while Section 7.3 contains a short overview of the most common LWIP configuration options, which are useful to tailor LWIP to the requirements of a specific application.

The second part of the chapter, comprising Sections 7.4 through 7.6, is focused on how to use the protocol stack. Hence, it contains a thorough description of the two APIs that LWIP provides to this purpose. Moreover, Section 7.5 presents in detail the network buffer management scheme foreseen by LWIP.

Although a detailed knowledge of this topic is unnecessary when using the higher-level `sockets` API, it becomes very important for the lower-level `netconn` API. This is because, in that case, memory management responsibilities are in part shared between the protocol stack and the application program itself.

Hence, a careful application-level network buffer management is therefore of utmost importance in order to avoid corrupting them or introducing memory leaks.

8 Device Driver Development

CONTENTS

Device driver development is an important area in which general-purpose and embedded systems differ significantly. Interrupt handling and synchronization are often performed in a different way, to focus on real-time requirements. Moreover, in a device driver intended for a small, embedded operating system, the application task interface is often designed for performance instead of being fixed and standardized, as happens in a general-purpose operating system. This chapter discusses the main design principles of a real-time device driver and shows how to apply them with the help of a practical example.

8.1 GENERAL STRUCTURE OF A DEVICE DRIVER

Embedded systems can be constructed from microcontrollers. Besides the main microprocessor(s), they also include a generous suite of peripherals (namely devices), which are generally external to the microprocessor, in order to provide different functionality and serve various needs, such as network connectivity, storage, and so on.

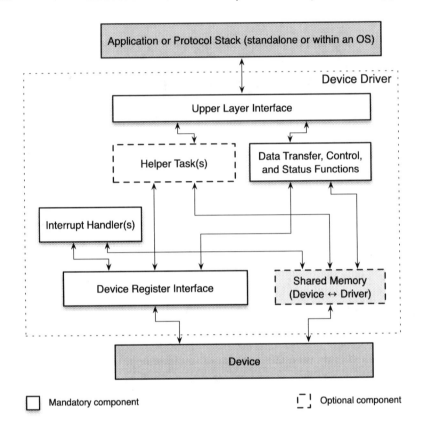

Figure 8.1 General structure of a device driver.

The device driver is essential to enable the functionality of devices and permit communication with the external world. As a consequence, it is of prominent importance in embedded system design.

Figure 8.1 demonstrates the general structure of a device driver. Whenever the application or protocol stacks (either standalone or coming as a component of an operating system) such as the TCP/IP protocol stack, USB protocol stacks, and so on would like to access a device, or vice versa when a device possesses some information worth the attention of the upper layers, they are both managed by the device driver. The two components that the device driver interfaces with are shown in dark gray. Downward, the device driver communicates with a device through the device register interface and optionally some shared memory, whereas upward another interface should be provided between the device driver and upper layers.

The device register interface allows the device driver to access registers, which can be used for two purposes:

- It can be used by device driver software to configure the device, retrieve status information, issue commands to the device, and so on;
- For some kinds of devices, registers are also used as data buffers. More specifically, when data arrives at the device, it can be stored in the data buffer before software starts processing it, and it can also be used to hold outgoing data coming from the upper layer while the device is currently busy with transmission.

The second functionality could be replaced by shared memory on platforms that support direct memory access (DMA). The DMA-based and register-based interface will be better discussed in Section 8.3. Nevertheless, proper synchronization mechanisms should be adopted to resolve concurrent access to data buffer by the hardware device and the software driver.

In order to be reactive to the external world, events of interest, for instance data arrives at the device or data has been sent out by the device, are generally handled by interrupt handlers. Depending on whether an operating system is in the scene or not, some processing related to an event can be delegated to a helper task. This represents one main difference between device driver development for real-time embedded systems and general-purpose systems.

It is common that interrupt handlers often get a priority higher than ordinary tasks in real-time operating systems which employ task-based scheduling, for instance, rate monotonic. However, if an interrupt handler is too complex, it may impair the performance of a task with real-time requirements in a non-negligible way. More information about this and interrupt handling in general will be given in Section 8.2.

Generally, outgoing data is initiated by upper layers and propagated to the device driver through the interface in between. It can be implemented as a transmit function, which performs data transfer, retrieves status information, and controls the transmit path, executing in the context of the calling task. The complexity of this function may vary from one kind of device or protocol stack to another. If so, work can be partially delegated to one or more helper tasks as well. Both the transmit function and the helper task(s) should synchronize with hardware through the interrupt handler, when a transmission can be carried out, that is when space is available in the transmit data buffer.

Since workload can be shared among different software components, including interrupt handler, helper tasks, and transmit function, unavoidably, some shared memory is also needed among them. As a consequence, synchronization should also be handled in a proper way, which will be better discussed in Section 8.4.

8.2 INTERRUPT HANDLING

Embedded systems have to handle real-world events, for instance transmission and reception of data packets for a communication device. Communication devices often offer buffers to accommodate packets. However, the number of packets a buffer can hold may be limited. This requires that the packets should be processed within a certain amount of time. Otherwise, if the buffers are completely consumed, it could lead

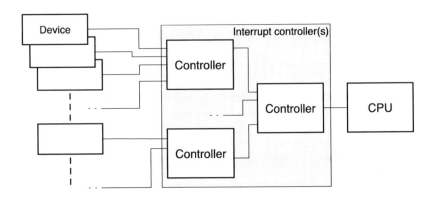

Figure 8.2 Interrupt handling path.

to packet drop for further incoming packets. Events external to the processor can be handled by *interrupts*, for instance incoming packets, whereas those detected to the processor itself are treated as *exceptions* such as data abort, instruction fetch abort, memory error, and other exceptional conditions. In this chapter, the discussion will be focused on interrupts (actually, most processors handle exceptions in a similar way), as it is essential to an embedded system to interact with the external world. Interrupt handling is also quite an important part of device driver development. Chapter 15 studies exception handling for what concerns a specific type of exception, namely, memory error, and presents different memory protection techniques to address the issue. More general information about exception can be found in Reference [161].

There are different ways to classify interrupts. One popular classification is to divide interrupts into *vectored* interrupts and *non-vectored* interrupts. The main difference between them is about how interrupts are handled. More specifically, for non-vectored interrupts, the CPU always branches to the same interrupt service routine (ISR) and from there it *polls* each device to see which may have caused the interrupt. And if there is more than one, they will be handled one by one within the same interrupt service routine. Instead, with vectored interrupts, the CPU is *notified* when there is an interrupt and where the interrupt comes from. What's more, different interrupt service routines are provided to handle different interrupts. Non-vectored interrupts are not as reactive as vectored interrupts, which makes them less suitable for some embedded system applications. As a consequence, if it is not mentioned specifically, we will focus on vectored interrupts in the following.

The basic concept about interrupt handling is that when interrupts arrive, the normal program flow of the CPU needs to be altered and context switch is required before running the interrupt handlers. Actually, the overall process of interrupt handling, especially when taking into account operations done by hardware, is much more complex. Three main hardware components involved in interrupt handling are devices, interrupt controller(s) and the CPU, as shown in Figure 8.2. In the following, they will be addressed one by one. For simplicity, we will focus on the case in which

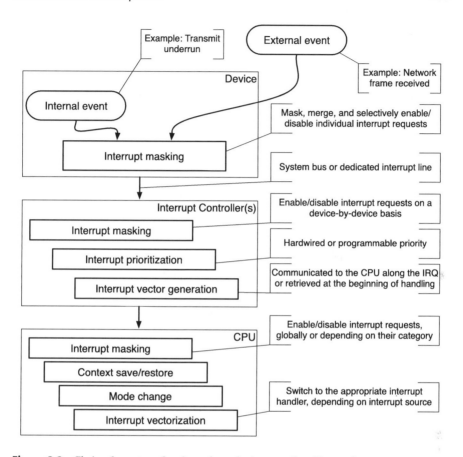

Figure 8.3 Chain of events and actions along the interrupt handling path.

just one interrupt controller is responsible for handling different sources of interrupts from different devices.

Figure 8.3 illustrates the whole process of how an interrupt is handled, starting from when it arrives at the device. Sometimes software developers are more familiar with how it should be handled in software, namely, interrupt handlers. This chapter will spend more time on how hardware addresses interrupts. By the way, interrupt handler is an alternative term for ISR and they are used in an interchangeable way in the following.

8.2.1   INTERRUPT HANDLING AT THE DEVICE LEVEL

Different devices can generate different interrupts. Interrupts for each device can be further divided into those due to external events, like frames received on the Ethernet interface, and those due to internal events, like Ethernet transmit underrun.[1] Each device can be configured to handle a certain event or not by setting appropriate device-specific registers in software. In this way, it is possible to selectively enable, disable, mask, or merge individual interrupt requests. This is called *interrupt masking*. As a result, when an event occurs, depending on the setting of the interrupt masking, it may or may not trigger an interrupt.

As we will see in the following, interrupt masking can be applied also in other places along the interrupt handling path, including the interrupt controller(s) and the CPU, in order to be flexible at selectively serving interrupt requests in different scopes. More specifically, interrupt masking set in the interrupt controller enables or disables interrupt requests on a *device-by-device* basis, while its setting at the CPU level has a more *global* effect. Generally, interrupt masking can be done by configuring proper registers in software and then it will take effect when hardware runs.

If an interrupt is generated, the corresponding interrupt signal will be propagated to the next module, namely the interrupt controller(s), through either the system bus or dedicated interrupt lines. It is quite common that each device just has a single interrupt request output to the interrupt controller through one interrupt line or the system bus. This indicates that interrupts corresponding to the same device are aggregated on a single interrupt line. This way of implementation is useful to simplify hardware design. In order to determine the origin of the interrupt, the interrupt service routine, when started, needs to query registers related to the device which activates its interrupt line. Besides, it is also possible that more than one device share the same interrupt line.

8.2.2 INTERRUPT HANDLING AT THE INTERRUPT CONTROLLER LEVEL

When an interrupt corresponding to a certain device arrives at the interrupt controller(s), depending on the interrupt masking set at this level, this interrupt may be either ignored or its processing keeps going. In other words, the interrupt controller continuously examines the interrupt request sources and checks against the value of its interrupt mask register to determine if there are active requests.

Another important concept handled in the interrupt controller(s) is *prioritization*, in which if more than one interrupt is active at the same time, the one with the highest priority will be served first. Generally, priorities are assigned to interrupts in a per-device manner, that's to say, interrupts coming from the same device share the same priority level. As we can see, this is also aligned with the hardware design. When more than one interrupt comes at the same time, the higher the priority of a specific interrupt is, the sooner it will be served. With the notion of priority, it becomes more convenient to implement real-time embedded systems by giving a higher priority to interrupts of more concern.

---

[1]This happens when the Ethernet transmitter does not produce transmit data, for example, the next fragment in a multi-fragment transmission is not available.

Priority can be either *hardwired* or *programmable*. Interrupt lines connected to the same interrupt controller are pre-assigned with a fixed and unique priority, which is generally an integer number. This type of priority is called hardwired priority. Another type of priority, namely programmable priority, can also be associated with interrupt lines. As its name tells, it can be configured in software. These two types of priority are *independent* from each other. Programmable priority offers more *flexibility* to software development as the order in which concurrent interrupts will be serviced can be changed. If programmable priority is used, it takes precedence over hardwired priority. Otherwise, the fixed hardwired priority will be adopted to define the order.

For what concerns the programmable priority, the number of priority levels available is *not* necessarily the same as the number of interrupt lines. It could be less; for instance the ARM PrimeCell vectored interrupt controller [7] has 32 interrupt lines, whereas just 16 programmable priority levels are available. What's more, an interrupt can be set to any eligible priority level. As a result, it is possible that more than one interrupt has the same value of programmable priority. When they arrive at the interrupt controller at the same time, the hardwired priorities corresponding to them are used to determine which one will be served first.

No matter which kind of priority is used, the interrupt controller is always able to select the interrupt with the highest priority among currently active interrupt requests to serve. The difference is that, if only hardwired priority is used, the interrupt sources and the associated interrupt handler names, namely the *interrupt vector table*, should be specified in the system *startup* code. Only in this way, can the interrupt controller figure out which ISR to branch to when interrupts arrive.

Instead, for what concerns programmable priority, it is not mandatory to provide this information too much in advance. Also as its name indicates, it can be done in application software, in particular the device driver, before hardware starts running. This requires that the interrupt controller should provide a minimal set of functions which allows the software to set the priority level, specify the interrupt handler, and so on. Anyway, for both hardwired and programmable priority, implementation of the interrupt handler itself is the programmers' responsibility.

After prioritization, the interrupt to be served next together with the address of the interrupt handler will be communicated to the CPU, or this information will be retrieved at the beginning of interrupt handling. This process is called *interrupt vector generation*.

The interrupt vector table is a table using the interrupt number as the index and with each element of it storing the starting address of the corresponding interrupt service routine, namely *ISR*. This permits *efficient* interrupt vector generation. Generally, the system startup code provides a default interrupt handler for each interrupt source. The default handlers are defined as `weak`, for instance,

```
weak void ETH_IRQHandler(void);
```

The above code declares an interrupt handler to serve interrupt requests coming from the Ethernet interface. As explained in Section 9.4, when a symbol is declared as a weak symbol and if a (non-weak) declaration and definition of the same symbol is provided elsewhere, like in the device driver, the latter takes precedence and will be used. The default handlers are generally quite simple and are normally implemented as a forever loop. On the other hand, when a real implementation of an interrupt handler is not available, the default handler will be executed and the system will not fall into any undefined state.

The following listing shows an example of how to set the priority level and ISR when programmable priority is adopted, on the NXP LPC24xx microcontroller [126] which is built around an ARM7 CPU core [8]:

```
#define VICVectAddr(x)      (*(ADDR_REG32 (0xFFFFF100 + (x)*4)))
#define VICVectPriority(x)  (*(ADDR_REG32 (0xFFFFF200 + (x)*4)))
#define VICIntEnable        (*(ADDR_REG32 (0xFFFFF010)))

VICVectAddr(ENET_VIC_VECTOR) = (portLONG) ENET_IRQHandler_Wrapper;
VICVectPriority(ENET_VIC_VECTOR) = ENET_VIC_PRIORITY;
VICIntEnable = (1 << ENET_VIC_VECTOR);
```

On LPC24xx, the interrupt controller supports 32 vectored IRQ slots and provides a set of registers for each individual slot. The VICVectAddr(x) register holds the address of the interrupt service routine corresponding to slot x, while VICVectPriority(x) register allows the software to configure the priority level for interrupt source identified by slot entry x. In addition, by writing to the corresponding bit in the 32-bit VICIntEnable register, it is possible to enable a certain interrupt.

As shown in the example, the macro ENET_VIC_VECTOR specifies the entry for the Ethernet interrupt in the 32 slots. Symbols ENET_IRQHandler_Wrapper and ENET_VIC_PRIORITY indicate the address of ISR and the priority level set for the Ethernet interrupt, respectively.

8.2.3   INTERRUPT HANDLING AT THE CPU LEVEL

When the CPU is notified that an interrupt needs to be handled, three actions should be carried out before the ISR code is executed, namely,

- Context switch
- Mode change
- Interrupt vectorization

Next we will explain them one by one.

An interrupt may come up at any time, during the execution of a task or even when another ISR is running if nested interrupts are supported. In order to keep the explanation as general as possible, we will first focus on the situation that an interrupt occurs while a task is running, which is the most common case. Then we will explain how nested interrupts are addressed based on the general concept.

If it occurs while a task is running, the execution of the task will be suspended and the corresponding ISR will be executed. After the ISR completes its job, the execution of the task will be resumed. However, this requires that the OS keeps an

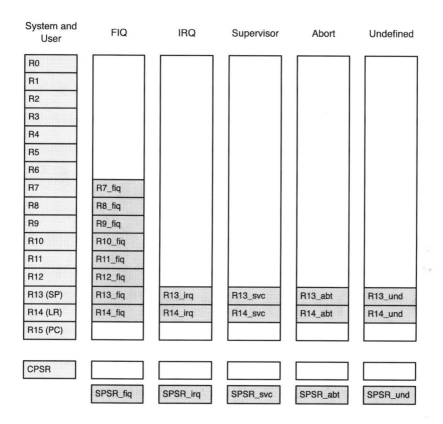

Figure 8.4 Operating modes on ARM7.

image of the state of the task when it is suspended so that the OS can resume its execution from the same place afterward. This process is referred to as *context switch* in Chapter 4, which includes first saving the context before switching to the ISR and then restoring the context, correspondingly.

Moreover, the context or, in other words, the processor state, to be saved and restored depends on several different things, including the CPU in use, the OS, and any optimization enabled for a program. For instance, it can vary from one CPU to another even if they are within the same family, because some additional registers may be available on one CPU but not the other. At a minimum, the context includes the general purpose registers, in particular the program counter register. When an OS is present, context switch also involves updating OS lists and other internal data structure. Last but not least, for systems with memory management unit (MMU), the page table will be updated as well.

With respect to context switch between tasks, which is discussed in Chapter 4, interrupt handling also requires to change the *operating mode* of the CPU when switching to an ISR or returning from it. The number of modes in which a micro-

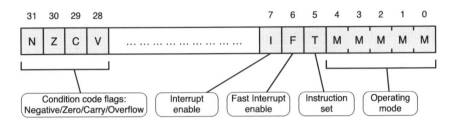

Figure 8.5   Current program status register (CPSR) on ARM7.

controller can work may vary from architecture to architecture. For example, the ARM7 CPU supports seven different operating modes as shown in Figure 8.4. The application code is running in *user* mode. And the *system* mode is a privileged mode for privileged tasks, which are often related to the operating system, or part of the operating system.

It is worth mentioning that the ARM7 CPU accepts two kinds of hardware interrupts, namely, the general purpose interrupt (IRQ) and fast interrupt (FIQ). And they are corresponding to two different operating modes, namely the *IRQ mode* and *FIQ mode*. Strictly speaking, among existing interrupt sources, only one of them can be selected as FIQ source so that the processor can enter the FIQ mode and start processing the interrupt as fast as possible. In this book, we will mainly focus on IRQ, rather than FIQ.

When there are software interrupts or when the system is reset, the CPU should work in *supervisor* mode, while the *abort* mode is for both the data access and instruction fetch memory abort. The CPU enters into the *undefined* mode when undefined instructions are executed. The availability of different modes may permit the CPU to service interrupts or exceptions more efficiently.

Each operating mode, except the system mode, has some registers *private* to its own use, in addition to the general purpose registers common among different modes. Generally, the application code runs in the user mode, where it has access to the register bank R0-R15 as well as the current program status register (CPSR), as shown in Figure 8.4.

As shown in the same figure, R0-R12 and R15 remain accessible to the IRQ mode, while R13 (link register) and R14 (stack pointer) are replaced by a pair of registers unique to the IRQ mode. This indicates that, in IRQ mode, the CPU has its own link register and stack. Besides, each mode has its own saved program status register (SPSR), except the user mode and the system mode.

In Figure 8.4, the registers available for the system and user modes are used as a reference and are shown in light gray. Registers private to a certain mode, with respect to those available for the system and user modes, are highlighted in dark gray, whereas those registers in common are omitted for clarity.

Figure 8.5 depicts the CPSR register provided on ARM7. As we can see, it contains several flag bits such as negative, zero, carry, and overflow which report the result status of a data processing operation, as well as multiple control bits which can be set to change the way the CPU behaves, including the operating mode, enabling/disabling IRQ and so on.

When saving the context and switching from one mode to another, the general purpose registers common to both modes, especially those that will be used in the second mode, are saved onto the stack of the first mode, pointed to by its R14 register. In addition, the R15 (program counter) is saved to the link register of the second mode and the value of the CPSR register is saved in the SPSR register of the later mode. Then the mode can be changed by writing to the CPSR register. The opposite will be done when the CPU needs to go back to the previous mode. The content of SPSR and the link register will be copied back to the CPSR and the program counter of the previous mode, respectively. And the register values stored in the stack will be written back to the corresponding registers.

By the way, as can be seen from the figure, the FIQ mode also has its own R7-R12 registers. This means that, when entering the FIQ mode, there is no need to save the value of those registers of the previous mode because they will not be reused in FIQ mode. This leads to more efficient handling of FIQs.

The last thing to do for interrupt handling is to perform *interrupt vectorization*, which is to jump to the appropriate interrupt handler, depending on the interrupt source.

More specifically, taking again the ARM7 architecture as an example, each operating mode has its own vector table at a predefined address, except the system and user modes. Instructions which can be used to reach and access each individual vector table are kept in an upper level vector table, namely the *exception vector table (EVT)*. More precisely, only the starting address of where the instructions for a certain mode are stored is recorded in the exception vector table.

For example, as shown in Figure 8.6, the instructions, which can be used to access the *interrupt vector table (IVT)*, are stored at a certain address, for instance, Add_B. As a consequence, the entry in the EVT corresponding to IRQ mode will be filled with Add_B. It is worth mentioning that, by convention, those instructions are often referred to as *first level IRQ handler*.

When the CPU changes from one mode to another, for example to the IRQ mode, the program counter (PC) will be filled with Add_B. By following the pointer, the CPU will execute instructions stored there, namely the first level IRQ handler. Depending on the interrupt number (or interrupt source), the first level IRQ handler is able to locate the correct entry in the interrupt vector table which stores the starting address of the associated ISR. For example, the Ethernet interrupt has an entry in the interrupt vector table and its ISR is stored at Add_C.

Then, the first level IRQ handler will alter the PC to point to Add_C. Afterward, the Ethernet interrupt handler will be executed.

For the sake of efficiency, it is not necessary that these three activities, namely context switch, mode change and interrupt vectorization, are carried out exactly one

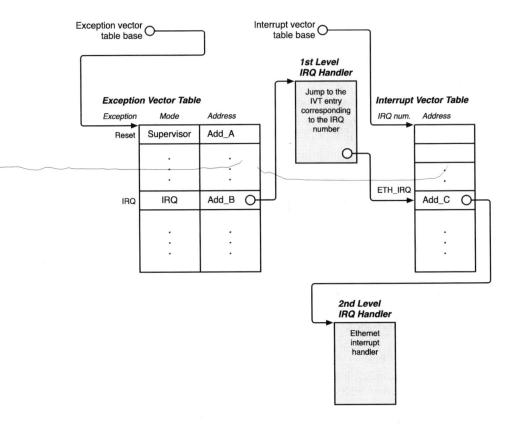

Figure 8.6 Exception and interrupt vector tables.

after the other. It may vary from one architecture to another. For example, for ARM7 architecture, the operations performed at the CPU level to service interrupt are summarized in the following:

- Save the content of the program counter into the link register of the IRQ mode;
- Save the content of CPSR into the SPSR of the IRQ mode;
- The program counter is pointed to the entry corresponding to the IRQ mode in the upper level vector table;
- Write to the CPSR register to change the operating mode to the IRQ mode;
- After entering the IRQ mode, follow the PC to jump to the right interrupt service routine;
- At the beginning of the ISR, before processing the interrupt, the content of those registers that are to be reused in ISR will be saved on the stack of the user mode (since we are considering a running task is interrupted);
- Then proceed with interrupt processing.

8.2.4 SOME REMARKS

As we can see, from when an interrupt occurs at the device level until it is handled by the CPU, a quite complex procedure is followed. The amount of time spent on performing those operations should *not* be neglected or taken for granted.

Moreover, the whole process is summarized taking ARM7 as an example. It should be noted that the general concept also applies to other architectures, except that the software/hardware boundary may vary, regarding those operations.

On some CPUs, nested interrupts are supported. More specifically, while an interrupt handler is running, if another interrupt with a priority higher than the current one occurs, it will interrupt the current one and the CPU will start serving the other interrupt request. For example, this is the case for LPC1768 [128, 127] which is built around an ARM Cortex-M3 processor core [9]. Generally, a specialized interrupt controller, like the nested vectored interrupt controller (NVIC) used on LPC1768, could be adopted to resolve the new scenario.

Briefly speaking, the interrupt controller will adopt a different mechanism to determine how interrupts should be masked. And it is also able to decide whether a newly arrived interrupt request corresponds to a higher priority or not, and if so, forward the decision together with the address of the ISR related to the new interrupt request to the CPU. The CPU simply obeys the decision and performs context switch again in order to serve the new interrupt request. More interested readers could refer to Reference [171] for more detailed information.

The interrupt service routines, namely ISRs, are suggested to be implemented as *short* as possible. This is because generally they have a higher priority than normal tasks. Interrupts may come at any time during the execution of a task. When it arrives, the execution of the task will be suspended so that the interrupt will be served. This indicates that interrupts will introduce unwanted interference to tasks. The more complex ISRs are, the larger the interference.

As a result, it is recommended that an ISR just does the bare minimum work so that the hardware device can continue its operation. For example, retrieving data from the hardware buffer, reconfiguring device registers to prepare the device for future incoming data, enabling interrupts and so on. Complex processing concerning the interrupt can be deferred to a later point. For example, the remaining processing can be delegated to a normal task. Afterward, it is possible to fine tune the task depending on its real-time requirements, for instance, by proper priority assignment. In this way, more flexibility is given to the programmers and it may permit better real-time performance of the overall system.

8.3 DEVICE DRIVER INTERFACES

As shown in Figure 8.1, a device driver interfaces with both the hardware device and other software components like the application or a protocol stack. As mentioned in Section 8.1, the interface to the device could be either register-based or DMA-based for what concerns data transfer.

The register-based interface works in a quite straightforward way, namely, device registers are not only used to control and retrieve status information of the device but also for data storage. Instead, DMA-based interface uses shared memory for data storage. Moreover, it indicates that a peripheral device is capable of directly accessing memory to retrieve data from it and store data to it, without sorting help from the processor.

One main goal of enabling DMA for some peripherals is to ensure high performance. Hence, it makes sense that a device and the memory it can reach are connected to the same bus in order to permit fast access. For example, as mentioned in Chapter 2, on LPC2468, the Ethernet controller and the SRAM memory bank reserved for its use are connected to the same advanced high-performance bus (AHB).

DMA-based interface requires that some data structures that will be referred by both hardware and software should be set up properly before use. For example, information like the starting address of the data buffer, its size, and so on should be communicated to the device through device registers so that the peripheral device knows where and how to retrieve and store data. Actually, depending on the complexity of the data structure, other more sophisticated information may be needed. For instance, in the example presented in Section 8.5, the indexes used to access the ring buffer should be updated in a proper way.

As we can see, a DMA-capable device is able to store incoming data to the shared memory and retrieve data to be transmitted from the shared memory *autonomously*. The device driver software is mainly responsible for configuring the device, delivering the data in the shared memory to upper layers for processing, and moving data that comes from the upper layers to the shared memory to prepare for transmission.

On the contrary, with register-based interface, incoming and outgoing data are both managed by the processor through the device driver. Data flows back and forth between the device driver and the corresponding device without an intermediate stop at the memory.

DMA-based interface can offer better performance because some data processing (for example, header parsing, filtering, packet classification) as well as data transfer can be done in parallel with CPU operations. As a result, there is no need to wait until CPU is free.

DMA can offload expensive memory operations, such as large copies or scatter&gather operations from the processor to a dedicated DMA engine. For example, the size of an Ethernet frame is variable and could be up to 1500 bytes. If registers are used to store them, a large number of registers would be needed and the cost is too high. What's more, the processor is fully occupied when performing I/O operations with the register-based interface. At this time, if another task with stricter real-time requirements needs to run while an I/O is ongoing, it needs to wait until I/O is completed. This impairs its real-time performance.

The upper layer interface should provide at least a minimum set of functions, which permits either the application or other protocol stacks to initialize the hardware device, retrieve received data for further processing and prepare data for transmission.

For example, the LwIP protocol stack uses the *netif* data structure to represent all network interfaces. This data structure contains several fields, for instance, `netif->input` and `netif->linkoutput`, which point to functions that should be implemented in the device driver and can be used to pass a received packet up the protocol stack and send a packet on the interface, respectively.

Moreover, device driver should also follow the requirements for data structures adopted in higher layers. For instance, internally, LwIP makes use of the `pbuf` data structure to hold incoming and outgoing packets, as discussed in Chapter 7. As a consequence, data should be moved back and forth between the shared memory (or registers) and pbuf in the device driver.

For what concerns embedded systems, the primary goal of the design and implementation of a device driver and its interface to other components is *performance* rather than being standard as for general purpose systems.

More specifically, general purpose operating systems tend to provide a standard device driver interface. In particular, all devices are categorized into just a few classes and devices belonging to the same class share the same interface to the upper layers. For example, Linux only differentiates three classes of devices, namely, block devices, character devices, and network devices. And the main interface functions for character devices include only a handful of high-level, abstract functions like `open`, `close`, `read`, `write`, `ioctl`.

Being generic, on the one hand, has the advantage of abstracting the development of application software away from hardware details and makes them as well as the OS more portable. On the other hand, sometimes it is not easy at all to map existing device features/capabilities directly onto a limited set of interface functions. For example, graphics cards are classified as character devices. However, the `read` and `write` functions are not so meaningful for this category of devices because they communicate with devices character-by-character in a sequential manner. At the end, specific `ioctl`s are defined in order to exploit the functionality of graphics cards.

What's more, in order to offer a standard device driver interface to the upper layers, some sacrifices are unavoidable. For instance, a mapping layer is needed to map the generic interface functions like those listed above to specific functions implemented within each individual device driver. The introduction of an extra layer not only adds extra software complexity, but may also affect performance.

Instead, for what concerns embedded systems, there is not any device that must be supported by all platforms; in other words, there is no standard device. As a consequence, unlike general purpose operating systems, device drivers are not deeply embedded into the RTOS, except a few commonly used ones like the UART used for early debugging. On the contrary, device drivers are generally implemented on top of the RTOS. Afterward, applications and middleware such as protocol stacks can be implemented directly on top of appropriate drivers, rather than using a standardized API like in the case of the general purpose OS.

For the same reason, the design and implementation of different protocol stacks by different people do not assume the existence of a unified device driver interface. It is possible that different protocol stacks may abstract various types of devices in quite different ways. As a consequence, device driver interfaces differentiate from each other at a more fine-grained level, with respect to general purpose OS. Most drivers (even for similar devices) are developed separately for different RTOSs and applications/middleware.

Even though this paradigm may hinder the easy replacement of hardware, it does offer more flexibility to the development of the device driver so that they can be better tailored according to the characteristics of hardware as well as the features provided by RTOSs in order to achieve better performance, which is the primary goal of embedded systems.

Some effort has been made in the past to unify the interfaces for developing device drivers for embedded systems, for instance, the real-time driver model [101] specified and implemented on top of Xenomai [67].

Last but not least, it is also worth mentioning that when the applications developed for general purpose operating systems intend to access the device, this generally should be done through the system call interface, as the device driver module resides in kernel whereas application code runs in user space. As a consequence, when data is going to be exchanged between the device and application, one extra memory copy from the kernel space to user space or vice versa is hard to avoid.

Instead, this is not necessarily the case for device drivers designed and implemented for embedded systems. Because most of the time, application tasks, real-time operating systems, device drivers are executing within the same mode. Hence, the device driver just needs to store the data required by upper layers in appropriate data structures for use and memory copy due to change of execution mode can be prevented. This is of paramount importance as embedded systems generally lack memory and memory management is generally costly.

8.4   SYNCHRONIZATION ISSUES

As discussed in Chapter 5, the set of communication and synchronization primitives available to interrupt handlers is, in many cases, more limited with respect to regular tasks. Namely, in most operating systems, the invocation of any potentially *blocking* primitive from an interrupt handler is forbidden.

As a consequence, the ways an interrupt handler can communicate and synchronize with other components of a device driver are severely constrained. As shown in Figure 8.7, even the relatively simple task of implementing critical regions, to ensure mutually exclusive access to shared data, cannot be done in the most straightforward way—that is, by means of a mutual exclusion semaphore—because the $P()$ semaphore primitive cannot be used on the interrupt handler side.

Resorting to a nonblocking variant of $P()$, which most operating systems do provide and can indeed be used from an interrupt handler, is quite intuitive, but it may not solve the problem in a satisfactory way. In this case, in fact, the interrupt handler must be prepared to receive an error indication from the semaphore primitive, signifying that it was not possible to get exclusive access to shared data. When this happens, the interrupt handler is left with the choice of aborting the access (and pos-

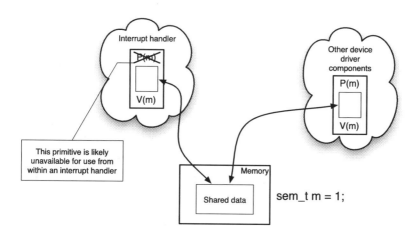

Figure 8.7 Synchronization issues in interrupt handlers.

sibly lose some data from the device because it was impossible to store them into the shared data structure) or retrying it.

The second option does not lead to loss of data because, at least by intuition, the semaphore operation will eventually succeed, but it has nonetheless two important drawbacks that hinder its applicability, especially in a real-time execution environment:

1. Retrying the critical region access multiple times represents an overhead, because each invocation of the nonblocking P() requires a certain amount of processing to be performed.

   This overhead may severely impair the schedulability scenario of the system as a whole because, as already mentioned in Section 8.2, in many operating systems interrupt handlers implicitly (and unavoidably) have a priority greater than any regular task in the system.
2. In principle, there is no upper bound on the number of retries needed to gain access to shared data. Actually, if retries are implemented in a careless way, access may *never* be granted anyway.

Regarding the second drawback let us consider, for example, a monolithic operating system running on a single-core processor. Due to the operating system architecture, as long as the interrupt handler keeps using the processor to retry the nonblocking P(), no other tasks will ever be executed. As a consequence, the device driver component that is currently engaged in its critical region—and is preventing the interrupt handler from entering its own critical region—will never be able to proceed and exit.

It must also be noted that introducing some delay between P() retries is not as easy as it looks at first sight, because:

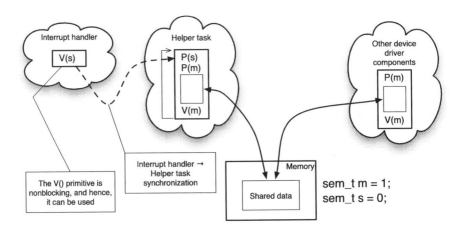

Figure 8.8 Synchronization by means of a helper task.

- an *active* delay loop does not solve the problem, and
- a *passive* delay is unfeasible because any operating system timing primitive surely counts as a blocking primitive by itself.

Finally, the significance of both drawbacks also depends on the *probability* of other device driver components being in their critical regions when the interrupt handler invokes the nonblocking P(), and the amount of *time* they spend within it.

A straightforward solution to this problem, consists of disabling interrupts (either globally or from a specific device) for the entire length of any critical regions executed by tasks in order to communicate with the interrupt handler. On the other side, the interrupt handler is allowed to enter its critical regions, and hence, access shared data directly. In the past, the method was adopted in most general-purpose operating systems, for instance 4.4BSD [115], and it is still popular nowadays.

At least on a single-core system, this approach is obviously able to ensure mutual exclusion, because the interrupt handler can never be executed while a task is within a critical region, but it also introduces several side effects:

- Depending on the amount of time spent within critical regions, interrupts may be disabled for a relatively long time. Especially on hardware architectures in which it is impractical to disable interrupts selectively, this may have a severe impact on interrupt handling latencies of the whole system.
- The method, by itself, does not work on most multi-core systems, in which disabling interrupts locally, within a single core, is still possible but disabling them globally, across all cores, is generally an extremely inefficient operation.

Concerning the second side effect, when interrupts are only disabled locally, nothing prevents an interrupt handler from executing on a certain core, while a regular task or another instance of the same interrupt handler executes within a critical region on another core. In this way, mutual exclusion is no longer guaranteed unless interrupt disabling is complemented by a different locking technique, for instance, the usage of spin locks, like it is done in modern multi-processor (MP) Linux kernels [112]. More information about this and other issues associated with multi-core processing can be found in [122, 71], and they will not be further discussed here.

Another way to address this issue, outlined in Figure 8.8 and often more appropriate for a real-time system, consists of *delegating* most of the activities that the interrupt handler must perform to a dedicated *helper task*. As shown in the figure, the helper task consists of an infinite loop containing a blocking synchronization point with the interrupt handler. Synchronization is implemented by means of a semaphore s, initialized to zero.

Within the infinite loop, the helper task blocks when it invokes $P(s)$, until the interrupt handler wakes it up by means of the corresponding $V(s)$. The last primitive is nonblocking, and hence, it can be safely invoked from the interrupt handler. At this point, the helper task interacts with the device and then makes access to the shared data structure in the usual way. At the end of the cycle, the helper task executes $P(s)$ again, to wait for the next interrupt. On the other hand, if the interrupt already occurred before $P(s)$, the helper task proceeds immediately.

With respect to the previous approach, the introduction of helper task brings the following main advantages:

- The interrupt handler becomes extremely short and often contains only a single $V()$ primitive, because most of the interrupt-related processing is delegated to the helper task. As a consequence, the amount of time spent executing in the interrupt context—with its priority assignment issues highlighted in Section 8.2—is greatly reduced.
- The priority of the helper task can be chosen at will, also depending on the requirements of the other system components. All the schedulability analysis techniques described in Chapter 6 can be applied to it, by considering it a sporadic task.
- Critical regions are implemented without disabling interrupts for their entire length. Besides improving interrupt-handling latency, especially when critical regions have a sizable execution time, this also makes the code easier to port toward a multi-core execution environment.

A third, more sophisticated approach to interrupt handler synchronization revolves around the adoption of *lock-free* and *wait-free* data structures. A thorough discussion of these kinds of data structures is beyond the scope of this book, but a simple example of their underlying concepts will be given in Section 8.5. There, a wait-free data structure is used at the interface between the device controller hardware and the device driver, but the same concepts are valid for software-only interactions, too.

Informally speaking, the main difference between traditional, lock-based object sharing versus a lock-free or wait-free data structure is that the latter guarantees the consistency of the shared object *without* ever forcing any process to wait for another.

Interested readers may refer, for instance, to References [70, 72, 2] for more information. In addition, References [4, 3] are useful starting points to learn how lock-free objects can be profitably adopted, within a suitable implementation framework, in a real-time system. The recent development of open-source libraries containing a collection of lock-free data structures such as, for example, the concurrent data structures library (libcds) [100] is also encouraging to bring lock-free data structures into mainstream programming.

8.5   EXAMPLE: ETHERNET DEVICE DRIVER

This section presents the implementation of an Ethernet device driver as an example. The Ethernet block together with the software device driver offer the functionality of the media access control (MAC) sublayer of the data link layer in the OSI reference model [85]. Briefly speaking, they are responsible for transmitting/receiving frames to/from the network and provide them to the upper layers of the protocol stack for processing.

In this example, the same device driver is assumed to work on both LPC2468, which encompasses an ARM7 microprocessor, and LPC1768, which is built around a Cortex-M3 core. Hence, differences between the two kinds of architecture as well as the hardware diversity should be taken into account when implementing a *portable* device driver. For instance, LPC1768 supports nested interrupts, whereas LPC2468 does not. In addition, even though the Ethernet controller adopted on both LPC2468 and LPC1768 is the same, its registers are mapped onto different physical memory addresses.

8.5.1   ETHERNET BLOCK ARCHITECTURE

Figure 8.9 shows the internal architecture of the Ethernet block, which can be found on both LPC2468 and LPC1768. As shown in Figure 2.3 and here, the Ethernet block as well as a SRAM bank which can be used for data storage by the Ethernet controller are connected to the same advanced high-performance bus (AHB).

The Ethernet block contains a full featured 10/100 Mbps Ethernet media access controller which has been designed to provide optimized performance through the use of direct memory access (DMA). Features include a wide set of host registers, flow control, hardware acceleration for transmit retry, receive packet filtering, and so on, even though they will not be fully explored in this example for conciseness. More specifically, the Ethernet block mainly consists of:

- Two DMA engines. They permit the transfer of frames directly to and from memory with little support from the microprocessor, while at the same time they off-load CPU processing/computation significantly.

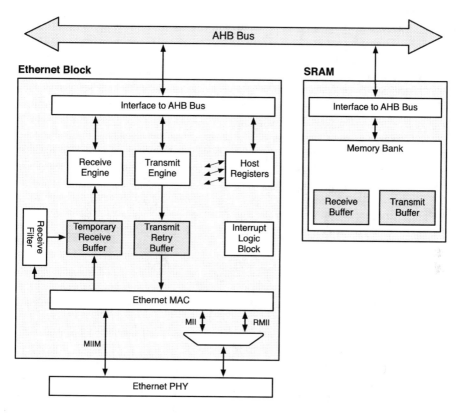

Figure 8.9   Ethernet block diagram.

Moreover, part of the on-chip SRAM is reserved for Ethernet communication. Hence, independent receive and transmit buffers can be mapped to
the Ethernet SRAM to hold incoming and outgoing packets. The Ethernet
SRAM is connected to the AHB as well.
The DMA to AHB interface allows the Ethernet block to access the Ethernet SRAM in order to store the received packets and pick up packets for
transmission.

- The Ethernet block transmits and receives Ethernet packets through an external Ethernet PHY. The Ethernet MAC is responsible for talking with
  the PHY via either media independent interface (MII) or reduced MII
  (RMII), which can be selected in software. They differ from each other
  in terms of clock frequency and communication data width. PHY registers
  can be accessed through the on-chip media independent interface management (MIIM) serial bus.
- The Ethernet MAC can also parse part of the header of incoming Ethernet
  frames and examine the frame type as well as Ethernet destination address.

This information can be exploited by the receive filter to carry out filtering. The temporary receive buffer between Ethernet MAC and the receive DMA engine implements a delay for the received packets so that the receive filter can work upon them and filter out certain frames before storing them back to memory.

- The transmit retry buffer also works as a temporary storage for an outgoing packet. It can be exploited to handle the Ethernet retry and abort situations, for example, due to collision.
- The Ethernet block also includes an interrupt logic block. Interrupts can be masked, enabled, cleared, and set by the software device driver as aforementioned. The interrupt block keeps track of the causes of interrupts and sends an interrupt request signal to the microprocessor through either the VIC (LPC2468) or NVIC (LPC1768) when events of interest occur.
- Host registers. The host registers module provides a set of registers accessible by software. The host registers are connected to the transmit and receive path as well as the MAC. As a result, they can be used to manipulate and retrieve information about network communication.

The receive path consists of the receive DMA engine, the temporary receive buffer, receive filter as well as the Ethernet MAC. Similarly, the transmit path is made up of the transmit DMA engine, transmit retry module, transmit flow control as well as the Ethernet MAC.

Referring back to Chapter 7 and Figure 7.3, the Ethernet device driver could interface with the LWIP protocol stack, which in turn is layered on top of the FREERTOS real-time operating system, to provide network communications.

Figure 8.10 demonstrates the general structure of the device driver. As shown in the figure, the main components of the Ethernet device driver include the interrupt handler, an Ethernet receive thread, as well as Ethernet transmit functions, shown in the dark gray rectangles. For simplicity, the initialization function is not shown in the figure. Moreover, two ring buffers are used to store incoming and outgoing data and they are shown in the middle of the figure.

As previously mentioned, the interrupt handler needs to synchronize with both the hardware and other components of the device driver. In the figure, a synchronization semaphore is represented by a light gray circle, and a dashed arrow stands for a synchronization primitive, either Take or Give. The same kind of arrow is also used to denote the activation of an IRQ handler by hardware. Instead, a solid arrow represents data flow among the Ethernet controller, device driver, and LWIP.

It is worth mentioning that, for the sake of demonstration, two interrupt handlers, which are responsible for the receive and transmit path separately, are shown in the figure. In practice, interrupt handling is implemented in a single interrupt handler.

In the following, we will go deeper into each individual component and provide more detailed information.

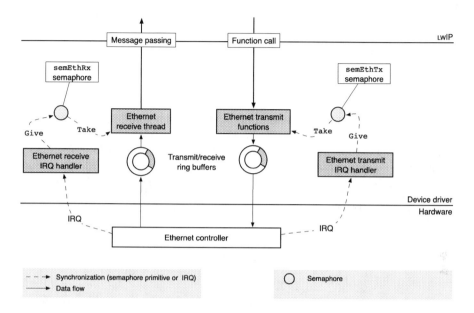

Figure 8.10   General structure of the Ethernet device driver.

8.5.2   INITIALIZATION

After reset, the Ethernet software driver needs to initialize the Ethernet block just introduced, configure proper data structures that are referred by the LWIP protocol stack, create suitable variables for synchronization, and so on. It is worth noting that interrupt handler installation is also performed during initialization. In this example, initialization is implemented in the `ethernetif_init` function shown in the following listing.

```
/* Network interface name. */
#define IFNAME0 'e'
#define IFNAME1 'n'

/* Initialize the Ethernet hardware. */
void low_level_init(struct netif *netif)

/**
 * Initialize the network interface. Internally, it calls the function
 * low_level_init() to do the actual setup of the hardware.
 */
err_t ethernetif_init(struct netif *netif)
{
  LWIP_ASSERT("netif != NULL", (netif != NULL));

  netif->name[0] = IFNAME0;
  netif->name[1] = IFNAME1;

  netif->output = etharp_output;
  netif->linkoutput = low_level_output;
```

```
/* Initialize the hardware. */
low_level_init(netif);

  return ERR_OK;
}
```

This `ethernetif_init` function is called at the beginning of the application to set up the network interface. The `netif` argument is the network interface structure used by LWIP to represent the physical network hardware. This data structure should be first allocated in the application and then passed as a parameter to this function for initialization. Its initial value should not be NULL.

More specifically, the `ethernetif_init` function is implemented as follows:

- The two character long `name` field can be set to identify the kind of network hardware and indicate the associated device driver to be used. For instance, it is set to `en` to reflect an Ethernet interface. Instead, `bt` could be used to represent a device driver for the Bluetooth network interface.
- The `output` pointer points to a function, which will be used by the IP layer when it needs to send an IP packet. This function is mainly responsible for resolving and filling in the Ethernet address header for the outgoing IP packet. At the end, it will call the function indicated by the `linkoutput` pointer to transmit the packet on the physical link.

  In theory, both functions should be implemented in the device driver. Instead, we set the `output` field directly to the `etharp_ouput` function, which is provided by the ARP module, to save an intermediate function call. You can also implement your own function, in which there is a call to `etharp_output`, if you would like to do some checks before sending the packet.
- The function `low_level_init` takes care of the Ethernet hardware initialization. We will look into this at a later time.
- Last but not least, the `netif` structure also includes an `input` pointer, which points to a function the device driver should call when a packet has been received from the Ethernet interface. This function will pass the received packet to the main LWIP task for processing.

Unlike the `output` field, which is initialized in the device driver, the `input` field can be specified through the `netif_add` function provided by the network interface abstraction layer. It is generally called in the application when we need to add a new network interface. The prototype of the `netif_add` function is shown in the following listing:

```
struct netif *
netif_add(struct netif *netif,
          struct ip_addr *ipaddr,
          struct ip_addr *netmask,
          struct ip_addr *gw,
          void *state,
          err_t (* init)(struct netif *netif),
          err_t (* input)(struct pbuf *p, struct netif *netif))
```

In the implementation of the `netif_add` function, the `input` field of the network interface structure will be set to the function indicated by the `input` argument. Moreover, the `init` argument points to the user-specified initialization function for the network interface. That's to say, the `ethernet_init` function should be given to this parameter when we call `netif_add` in the application.

The following listing demonstrates the implementation of the `low_level_init` function.

```
#define ETHARP_HWADDR_LEN          6

/* Interface Maximum Transfer Unit (MTU) */
#define EMAC_NETIF_ETH_MTU         1500

/** if set, the netif has broadcast capability */
#define NETIF_FLAG_BROADCAST       0x02U

/** if set, the netif is an device using ARP */
#define NETIF_FLAG_ETHARP          0x20U

/** if set, the interface has an active link (set by the network
 * interface driver) */
#define NETIF_FLAG_LINK_UP         0x10U

/* Number of RX and TX fragments.
   In this implementation, each fragment holds a complete frame (1536 octects)
*/
#define NUM_RX_FRAG      4             /* 4*1536= 6.0kB */
#define NUM_TX_FRAG      4             /* 4*1536= 6.0kB */

/* MAC address, from the least significant to the most significant octect */
#define MYMAC_1          0x4D
#define MYMAC_2          0x02
#define MYMAC_3          0x01
#define MYMAC_4          0xF1
#define MYMAC_5          0x1A
#define MYMAC_6          0x00

/**
 * In this function, the hardware should be initialized.
 * Called from ethernetif_init().
 */
void low_level_init(struct netif *netif)
{
  portBASE_TYPE result;
  xTaskHandle input_handle;

  /* Set MAC hardware address length */
  netif->hwaddr_len = ETHARP_HWADDR_LEN;

  /* Set MAC hardware address */
  netif->hwaddr[0] = MYMAC_6;
  netif->hwaddr[1] = MYMAC_5;
  netif->hwaddr[2] = MYMAC_4;
  netif->hwaddr[3] = MYMAC_3;
  netif->hwaddr[4] = MYMAC_2;
  netif->hwaddr[5] = MYMAC_1;

  /* Maximum transfer unit */
  netif->mtu = EMAC_NETIF_ETH_MTU;

  /* Device capabilities */
  netif->flags = NETIF_FLAG_BROADCAST | NETIF_FLAG_ETHARP | NETIF_FLAG_LINK_UP;

  semEthTx = xSemaphoreCreateCounting(NUM_TX_FRAG-1, NUM_TX_FRAG-1);
  semEthRx = xSemaphoreCreateCounting(NUM_RX_FRAG-1, 0);
```

```
if(semEthTx == SYS_SEM_NULL) {
  LWIP_DEBUGF(NETIF_DEBUG,("Creation of EMAC transmit semaphore failed\n"));
}

if(semEthRx == SYS_SEM_NULL) {
  LWIP_DEBUGF(NETIF_DEBUG,("Creation of EMAC receive semaphore failed\n"));
}

/* Initialize the Ethernet hardware. */
Init_EMAC();

/* Separate task for handling incoming Ethernet frames. */
result = xTaskCreate(ethernetif_input,
                     (const signed char *)"EMAC_RX_TASK",
                     (unsigned portSHORT)EMAC_RX_TASK_STACK_SIZE,
                     (void *)netif,
                     EMAC_RX_TASK_PRIORITY,
                     &input_handle);

if(result == pdFAIL) {
  LWIP_DEBUGF(NETIF_DEBUG,("Creation of EMAC receive task failed\n"));
}
}
```

- The `netif` structure also includes fields which represent hardware-related information, such as the link level hardware address of this interface (that is the MAC address for Ethernet interface), maximum transfer unit (MTU), and so on. They will be configured in this function. The MAC address set here will be used to fill the source MAC address of the Ethernet header when transmitting. The most significant byte of the MAC address is stored in the first entry of the array indicated by the `hwaddr` field. And it will be transmitted first with respect to the other bytes of the MAC address, as the byte order of an Ethernet frame transmission is big-endian. The total length of the MAC address is specified in the `hwaddr_len` field.

  The `flags` field describes the capabilities of the corresponding network interface. In this example, it is configured to use ARP and support broadcast. It also indicates that the Ethernet interface has an active link.

  It is worth mentioning that, in this example, only a subset of the fields defined in the `netif` structure are configured in the device driver, which permits the network interface to function properly. The reader can refer to the LWIP specification [51] and its implementation for more information.

- It is worth mentioning that, on LPC24xx, separate ring buffers are used to store Ethernet frames for both the transmitting and receiving paths. For simplicity, only the one used in the transmitting path is shown in Figure 8.11 and, as an example, it contains four elements. The number of elements available in the ring buffers used for transmission and reception are specified with the `NUM_TX_FRAG` and the `NUM_RX_FRAG` macros, respectively. The two counting semaphores `semEthTx` and `semEthRx`, created with the `xSemaphoreCreateCounting` primitive provided by FREERTOS, are first of all used to synchronize the hardware and the software. For instance, they are used to notify the software when the hardware is ready for transmission or an Ethernet frame has been received. Secondly, they are also

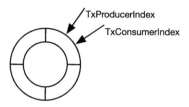

Figure 8.11 Ring buffer used for the transmission path.

used to manage resources, namely the ring buffers in this example. The two arguments indicate the maximum count value that a semaphore of this type can reach and its initial value, respectively.

The `semEthTx` semaphore is used for the transmission path. However, as shown in the listing above, the maximum count value of the `semEthTx` semaphore is set to `NUM_TX_FRAG-1`. It is configured in this way, because it is important for the hardware, namely the Ethernet controller, to be able to distinguish between an empty buffer and a full buffer. When the buffer is full, the software should stop producing new frames until hardware has transmitted some frames. Otherwise, it can keep going.

On the LPC2468, the check of buffer status is implemented in the hardware and carried out with the help of two indexes associated with the buffer, namely `TxProducerIndex` and `TxConsumerIndex`. Concerning the transmission path, the software is responsible for inserting new frames into the buffer for transmission, whereas the hardware will remove frames from the buffer and send them onto the link. As a consequence, they are considered as the *producer* and the *consumer* of the shared resource, respectively. Moreover, they are responsible for updating the corresponding index after working on the buffer.

`TxProducerIndex` always points to the next free element in the ring buffer to be filled by the device driver, whereas `TxConsumerIndex` indicates the next buffer element which contains the frame to be transmitted. Since this is a ring buffer, the two indexes are wrapped to 0 when they reach the maximum value.

Internally, the Ethernet controller considers the buffer as *empty* when `TxProducerIndex` equals `TxConsumerIndex`, and as *full* when `TxProducerIndex` equals `TxConsumerIndex-1`, as shown in Figure 8.12. As we can see, in this way, the buffer can accommodate up to `NUM_TX_FRAG-1` elements at the same time. It is also worth remarking that all elements of the buffer will still be used by the controller anyway, just not all at the same time. For clarity, the check for full buffer when `TxConsumerIndex` gets a larger value than `TxProducerIndex` is shown here. The check can be done with a slight update when it is the

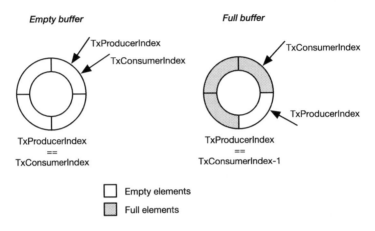

Figure 8.12 Empty and full ring buffer check.

opposite, for instance, when `TxConsumerIndex` wraps around to 0. For simplicity, it is not shown here.

As a consequence, the initial count value of the `semEthTx` semaphore, which indicates the number of free elements in the transmission ring buffer, is also set to `NUM_TX_FRAG-1` at the very beginning.

The same reasoning applies to the reception path as well, except that the initial value for the `semEthRx` semaphore should be 0, as the `semEthRx` tells how many frames are available in the reception ring buffer for software processing. By the way, if the creation of the two semaphores fails, some debugging information is printed out.

- The initialization of the hardware, including the external Ethernet PHY chip, are performed by the `Init_EMAC` function. This generally requires to work at the register level. The main things that need to be done during initialization include pin routing, initializing the PHY, configuring the transmit and receive DMA engines, enabling the transmit and receive paths, as well as enabling interrupts and installing the interrupt handler. We will further explain them one by one in the following.

- After the Ethernet hardware is successfully initialized, a separate task is created with the FREERTOS `xTaskCreate` primitive. This task is responsible for handling incoming Ethernet frames. According to the definition of `xTaskCreate` given in Chapter 5, the `ethernetif_input` function represents the entry point of the new task, namely the function that the task will start executing from. More information about this function can be found in Section 8.5.4.

It is now time to have a deeper look at the specific operations carried out during hardware initialization, namely, within the `Init_EMAC`.

- One thing typically done during the hardware initialization is pin routing. It brings the signals back and forth between the Ethernet controller and the right I/O pins of the CPU. In this way, data and control information can be properly exchanged between the controller and the PHY. It is worth mentioning that each I/O pin of the CPU can be configured for different functions, which are predefined and can be found in the related manual or data sheet [126, 128].

- As shown in Figure 8.9, the Ethernet block transmits and receives Ethernet frames through an external off-chip PHY, namely the Ethernet physical layer. It should also be initialized before being used. One main thing is to set up the link and decide the link speed as well as the duplex mode, which also depend on the node at the other end of the connection. A convenient way to perform this is through autonegotiation, which can be enabled by configuring, for example for the DP83843 PHY device supported by LPC2468, the basic mode control register of the PHY as follows:

```
write_PHY (PHY_REG_BMCR, PHY_AUTO_NEG);
```

The `write_PHY` function is responsible for writing a value to one of the PHY registers and waiting until the operation completes. After that, depending on the negotiation result, we can set the corresponding registers accordingly. In addition, the RMII or MII interface can be enabled by configuring the Ethernet command register. The following code enables the RMII interface:

```
/* Command register. */
#define MAC_Command (*(ADDR_REG32 (0xFFE00100)))

MAC_Command = CR_RMII;
```

- As mentioned at the very beginning of this section, the Ethernet block contains DMA engines for both transmission and reception. Besides, the way software, namely the device driver, and hardware interact with the ring buffers, which is also part of the DMA engines, has already been explained above. Instead, during initialization, it is necessary to configure the DMA engines, including setting up the ring buffers for use. For instance, the size of the ring buffers as well as the initial value of the indexes, which are used to access the ring buffers, should be specified here.

- Another thing that is done during the initialization phase is to enable interrupts at the device level, whose general concept has been discussed in Section 8.2.1. As mentioned there, each device can support multiple interrupt sources and it can choose to enable them in a selective way by interrupt masking, which in turn can be done by configuring proper registers. The following listing shows how it is done in the Ethernet driver:

```
/* Interrupt enable register. */
#define MAC_IntEnable (*(ADDR_REG32 (0xFFE00FE4)))

/* Enable interrupts. */
MAC_IntEnable = INT_RX_DONE | INT_TX_DONE;

/* Reset all interrupts. */
MAC_IntClear  = 0xFFFF;
```

As shown above, interrupt masking can be set through the interrupt enable register. Among the interrupts that could be generated by the Ethernet module, we are interested in the interrupts corresponding to successful reception and transmission of a frame.

After that, we also reset interrupts by writing to the interrupt clear register. It will clear all the bits in the interrupt status register. Individual bits in the interrupt status register are set by the hardware when an interrupt corresponding to that bit occurs. During initialization, they should be reset to avoid spurious interrupt requests.

- The receive and transmit mode of the Ethernet module can be enabled through the command register, as shown in the following listing. What's more, by setting the receive enable bit of the MAC configuration register 1, we allow incoming frames to be received. This is necessary, because internally the MAC synchronizes the incoming stream to this control bit.

```
/* Command register. */
#define MAC_Command (*(ADDR_REG32 (0xFFE00100)))

/* MAC configuration register 1. */
#define MAC_MAC1    (*(ADDR_REG32 (0xFFE00000)))

MAC_Command  |= (CR_RX_EN | CR_TX_EN);
MAC_MAC1     |= MAC1_REC_EN;
```

- During initialization, it is also necessary to install the Ethernet interrupt handler and enable the Ethernet interrupt at the *interrupt controller* level. In this way, when the Ethernet module starts, it is clear to the processor which interrupt handler to call when interrupts arrive. As mentioned at the beginning, one difference between LPC2468 and LPC1768 is that the former one includes a vectored interrupt controller (VIC), whereas the latter contains a nested vectored interrupt controller (NVIC). However, the installation of the interrupt handler is done in a quite similar way, with minor variety, as shown in the following listing.

```
/* Ethernet interrupt handler. */
void ENET_IRQHandler ( void );

#ifdef GCC_ARM7_EA_LPC2468
/* The interrupt entry point is naked so we can control the context
saving. */
void ENET_IRQHandler_Wrapper( void ) __attribute__ ((naked));
#endif
```

```
#ifdef GCC_ARM7_EA_LPC2468
/**
 * Interrupt handler wrapper function.
 */
void ENET_IRQHandler_Wrapper( void )
{
    /* ARM7: Save the context of the interrupted task. */
    portSAVE_CONTEXT();

    /* Call the handler to do the work.  This must be a separate
    function to ensure that the stack frame is set up correctly. */
    ENET_IRQHandler();

    /* ARM7: Restore the context of whichever task will execute next. */
    portRESTORE_CONTEXT();
}
#endif

#ifdef GCC_ARM7_EA_LPC2468
VICVectAddr(ENET_VIC_VECTOR) = ( portLONG ) ENET_IRQHandler_Wrapper;
VICVectPriority(ENET_VIC_VECTOR) = ENET_VIC_PRIORITY;
VICIntEnable = (1 << ENET_VIC_VECTOR);
#endif

#ifdef GCC_CM3_UN_LPC1768
NVIC_SetPriority(ENET_IRQn, ENET_INTERRUPT_LEVEL);
NVIC_EnableIRQ(ENET_IRQn);
#endif
```

First of all, as we can see, the interrupt handler is not specified here for LPC1768. This is because a default vector table is provided by the system startup file used on LPC1768. It includes an entry where the interrupt handler to be used for Ethernet interrupts is indicated. The ISR is named ETH_IRQHandler and it is defined as a weak symbol. What's more, a default implementation of the interrupt handler is provided as well. Instead, this is not the case for LPC2468. As we will see in the following, for LPC1768, an interrupt handler with the same symbol name is defined and implemented. It will take precedence over the default one and we do not need to install the interrupt handler again. For what concerns LPC2468, the interrupt handler is specified explicitly.

Secondly, another difference is that, context switch needs to be done explicitly in software on LPC2468, whereas LPC1678 provides hardware support for it so that the general purpose and status registers are saved automatically in the stack when an interrupt arrives.

In the above listing, portSAVE_CONTEXT and portRESTORE_CONTEXT are two function-like macros, which implement context switch by saving and restoring the entire execution context. More detailed implementation information of them can be found in Chapter 10. Since context is saved and restored explicitly with these two macros in the function body, the ENET_IRQHandler_Wrapper function should be declared or defined with the *naked* attribute. Otherwise, some processor registers would be saved in the prologue and restored in the epilogue sequence of this function when the compiler generates code for it. This will result in redundant

operations. With the *naked* attribute, the compiler will no longer generate
any prologue or epilogue for it. For more information about the *naked* attribute, readers can refer to Chapter 9.

The processing concerning the interrupt itself is implemented in the
ETH_IRQHandler function. It is recommended to implement this function as a separate function with respect to the wrapper so that the stack
frame can be set up correctly.

Thirdly, on LPC1768, library functions are used to set the priority and enable the interrupt. Different macros are used to specify the Ethernet interrupt number and the priority level assigned to it. This is simply because
they are different from the hardware point of view, namely they correspond
to different interrupt lines on the two platforms, and also due to the use of
the library function.

We can observe that the Ethernet device driver for LPC2468 and the one for
LPC1768 have a major part in common, except, for instance, the registers
map, external PHY in use, interrupt controller. As a consequence, when
porting the device driver from one platform to the other, there is no need
to rewrite the device driver. It is more convenient to just make the code
corresponding to the parts different from each other executed conditionally,
as shown in the above listing.

8.5.3 INTERRUPT HANDLER

```
/**
 * Ethernet interrupt handler.
 * On LPC2468, the interrupt handler function must be separate from
 * the entry function to ensure the correct stack frame is set up.
 */
void ENET_IRQHandler ( void )
{
  u32_t status = MAC_IntStatus;

#ifdef GCC_ARM7_EA_LPC2468
  signed portBASE_TYPE TaskWokenByRx = pdFALSE;
  signed portBASE_TYPE TaskWokenByTx = pdFALSE;
#endif
#ifdef GCC_CM3_UN_LPC1768
  /* Added static to avoid
     internal compiler error: in expand_expr_addr_expr_1, at expr.c:6925
     with CodeSourcery toolchain arm-2010.09-51-arm-none-eabi.bin
  */
  static signed portBASE_TYPE TaskWokenByRx;
  static signed portBASE_TYPE TaskWokenByTx;
  TaskWokenByRx = pdFALSE;
  TaskWokenByTx = pdFALSE;
#endif

  /* RxDoneInt */
  if(status & INT_RX_DONE) {
    xSemaphoreGiveFromISR(semEthRx, &TaskWokenByRx);
  }

  /* TxDoneInt */
  if(status & INT_TX_DONE) {
    xSemaphoreGiveFromISR(semEthTx, &TaskWokenByTx);
  }
```

```
/* Clear the corresponding interrupt bits of the IntStatus
 * register. */
MAC_IntClear = status;

/* If a task with a priority higher than the interrupted task is
 * unblocked, rescheduling is needed. */
if(TaskWokenByRx || TaskWokenByTx)
{
#ifdef GCC_ARM7_EA_LPC2468
    portYIELD_FROM_ISR();
#endif
#ifdef GCC_CM3_UN_LPC1768
    vPortYieldFromISR();
#endif
}

#ifdef GCC_ARM7_EA_LPC2468
/* On LPC2468, this register must be written with any value at the
   very end of an ISR, to update the VIC priority hardware.
*/
VICAddress = 0;
#endif
}
```

The above listing shows how the interrupt handler is implemented. As mentioned in Section 8.2, it is recommended that the interrupt handler just does the minimum to permit the hardware to keep working correctly. In this way, leaving the main part of the interrupt processing to a task could help to improve the real-time performance.

The interrupt status register, that is MAC_IntStatus, keeps a record of which interrupts corresponding to the Ethernet device are triggered. During initialization, it is specified that two kinds of interrupt are of interest. The first one corresponds to the event that an Ethernet frame has been received correctly in the Ethernet interface, while the second one indicates that a frame has been transmitted to the Ethernet link and more space is available in the transmit ring buffer.

If either interrupt occurs, the same operation will be taken. The interrupt handler releases the corresponding semaphore with a call to the FREERTOS xSemaphoreGiveFromISR primitive. In this way, if previously there are some tasks blocked on one of the semaphores, they will be woken up.

As discussed in Chapter 5, with respect to the xSemaphoreGive primitive, xSemaphoreGiveFromISR can be invoked from an interrupt handler since it never blocks the caller. Moreover, it returns to the caller an indication—stored by the driver in the variable TaskWokenByRx or TaskWokenByTx—of whether or not a task with a higher priority than the interrupted task has been unblocked. If it is, the unblocked task should be executed after the ISR completes, instead of returning to the interrupted task. This requires to run the FREERTOS task scheduling algorithm before exiting the interrupt handler. As shown in the listing, this is achieved by the portYIELD_FROM_ISR function on LPC2468 and vPortYieldFromISR function on LPC1768.

Besides dealing with architectural differences, the two functions have a lot in common. A possible implementation of portYIELD_FROM_ISR is provided in Chapter 9.

Writing a "1" to bits in the interrupt clear register, that is MAC_IntClear in the listing, clears the corresponding status bit in the interrupt status register. This is to mark the related interrupt as processed. Otherwise, it will be served again when we enter the ISR next time. The way to clear an interrupt status bit may vary from one device to another, for instance, some devices support *clear on read*.

According to References [126, 125], on LPC2468, the vector address register (indicated by `VICAddress` in the above listing) contains the address of the ISR for the interrupt that is to be serviced. What's more, this register must be written, with any value, at the very *end* of an ISR in order to update the vectored interrupt controller (VIC) priority hardware. Writing to the register at any other time can cause incorrect operation. Instead, this is not needed for LPC1768.

By the way, it is also worth mentioning that registers are referred to by name in this example. They are mapped to different physical memory addresses for LPC2468 and LPC1768 in separate header files.

8.5.4 RECEIVE PROCESS

```c
/**
 * This function should be called when a packet is ready to be read
 * from the interface. It uses the function low_level_input() to
 * handle the actual reception of bytes from the network interface.
 */
static void
ethernetif_input(void *parameter)
{
    struct eth_hdr *ethhdr;
    struct pbuf *p;
    struct netif *netif;
    netif = (struct netif *)parameter;

    for(;;)
    {
        /* Wait until a packet is arrived at the interface. */
        if(xSemaphoreTake(semEthRx, portMAX_DELAY) == pdTRUE)
        {
            /* When another interrupt request is generated within the
               EMAC while another request is still pending, the
               interrupt status register is updated, but no additional
               requests are sent to the CPU.

               This may happen, for instance, when two frames are
               received in rapid succession.  In this case, semEthRx
               is signaled only once, even if there is more than one
               frame to process, and we need to loop until the Rx ring
               buffer is empty.
             */
            while(CheckFrameReceived())
            {
                /* Move received packet into a new pbuf. */
                p = low_level_input(netif);

                /* No packet could be read, silently ignore this */
                if (p == NULL) continue;

                /* Point to the pbuf payload. It contains the packet
                   and it starts with the Ethernet header. */
                ethhdr = p->payload;
```

```
                    switch (htons(ethhdr->type))
                    {
                        /* IP or ARP packet? */
                    case ETHTYPE_IP:
                    case ETHTYPE_ARP:
#if PPPOE_SUPPORT
                        /* PPPoE packet? */
                    case ETHTYPE_PPPOEDISC:
                    case ETHTYPE_PPPOE:
#endif /* PPPOE_SUPPORT */

                        /* No matter what type it is, the full packet is
                           sent to the tcpip_thread to
                           process. netif->input is normally set in the
                           application */
                        if (netif->input(p, netif)!=ERR_OK) {
                           LWIP_DEBUGF(NETIF_DEBUG,
                                       ("ethernetif_input: IP input error\n"));
                           pbuf_free(p);
                           p = NULL;
                        }
                        break;
                    default:
                        pbuf_free(p);
                        p = NULL;
                        break;
                    }
                }
            }
        }
}
```

As mentioned in Section 8.5.2, this function is the entry point of the task created during initialization. When a frame has been received in the Ethernet interface, this function is responsible for retrieving it from the hardware ring buffer and delivering it to the higher layer protocol for processing.

As we can see, all operations are performed within an infinite loop. This is because, first of all, this task should prepare to handle any number of frames whenever they arrive at the interface. Secondly, as required by FREERTOS, tasks should be implemented to never return.

It blocks on the semEthRx semaphore indefinitely, unless this semaphore is signaled elsewhere, for instance, the interrupt handler. In other words, it waits until a frame arrives at the interface. Then it moves the received frame, which resides in the hardware buffer, to a software pbuf using the low_level_input function. In this way, more space in the hardware ring buffer is available for future incoming frame. Detailed implementation information of low_level_input will be shown later. By the way, as discussed in Section 7.5, the pbuf data structure is the internal representation of packet in LWIP.

The low_level_input function returns to the caller a pointer to the pbuf structure, whose payload field points to the received Ethernet frame (including the Ethernet header) stored in the pbuf pool.

In theory, it is possible to check the type of the received frame, whether it corresponds to an IP packet, an ARP packet, or other types of frame, and then react accordingly. Instead here, the whole packet is simply passed to the main tcpip thread of the LWIP protocol stack for input processing by means of the function indicated by the `input` field of the `netif` data structure. As mentioned in Section 8.5.2, this field is initialized in the application when a new network interface is added, with a call to the `netif_add` function. In LWIP, this is handled by the `tcpip_input` function. Consequently, the `input` field should point to this function.

The `CheckFrameReceived` function is used to check whether there is any frame remaining in the receive ring buffer. It is performed by simply comparing the two indexes used to access the ring buffer and checking whether the buffer is empty or not. This is not just a redundant test, because in some extreme cases it might happen that more than one frame has been received while just one interrupt request is handled by the processor.

For example, this is possible when two frames are received in rapid succession. In this case, when the processor is handling the interrupt request corresponding to the frame received first, that is when the interrupt service routine is running, another interrupt request is generated by the Ethernet device. Concerning the second interrupt request, the interrupt status register is updated. However, no additional request is sent to the CPU.

This is because, for what concerns the ARM7 architecture, interrupts are disabled at the CPU level while it is serving an interrupt request. They will be enabled again when the interrupt handler returns. What is more, as shown in Section 8.5.3, the content of the interrupt status register will be cleared in the interrupt handler. As a consequence, there is not a second chance that the later interrupt request will be propagated to the processor. Another side effect is that the `semEthRx` semaphore is signaled only once, even if there is more than one frame to process. To work around this issue, the receive task needs to check whether this aforementioned situation happens or not and keep receiving, if so.

This also somehow explains why the interrupt handler should be kept as short as possible. Otherwise, due to the way hardware works, it is possible that some interrupts are missed when interrupt rate is too high.

Instead, in the normal case, when two incoming frames are apart from each other, the receive interrupt is enabled again before the next frame is coming.

The situation is slightly different on the Cortex-M3 architecture, as it supports nested interrupts. However, an interrupt at a lower or equal priority is not allowed to interrupt the one that is currently being served. As we know, interrupts corresponding to the same device share the same priority level.

As can be seen from the above listing, the `low_level_input` function may return to the caller a `NULL` pointer, which indicates that no packet could be read. The caller just silently ignores this error indication. This may happen due to memory errors. It is recommended to attempt receiving at a later time because it is possible that in the meanwhile other tasks may free some memory space.

The following code shows the implementation of the `low_level_input` function.

```
/**
 * Allocate a pbuf and transfer the bytes of the incoming packet from
 * the interface to the pbuf.
 */
struct pbuf *
low_level_input(struct netif *netif)
{
  struct pbuf *p, *q;
  s16_t len;

  /* Obtain the size of the incoming packet. */
  len = StartReadFrame();

  if(len < 0)
  {
      /* StartReadFrame detected an error, discard the frame */
      EndReadFrame();

      /* Update statistics */
      switch(len)
      {
      case SRF_CHKERR:  /* CRC Error */
          LINK_STATS_INC(link.chkerr);
          break;

      case SRF_LENERR:  /* Length Error */
          LINK_STATS_INC(link.lenerr);
          break;

      case SRF_ERR:  /* Other types of error */
          LINK_STATS_INC(link.err);
          break;
      }

      /* In any case, the packet is dropped */
      LINK_STATS_INC(link.drop);

      /* Return a null pointer to the caller */
      p = NULL;
  }
  else
  {

#if ETH_PAD_SIZE
      /* Allow room for Ethernet padding */
      len += ETH_PAD_SIZE;
#endif

      /* Allocate a chain of pbufs from the pool for storing the
       * packet. */
      p = pbuf_alloc(PBUF_RAW, len, PBUF_POOL);

      if (p != NULL) {

#if ETH_PAD_SIZE
          /* Drop the padding word. Find the starting point to store
           * the packet. */
          pbuf_header(p, -ETH_PAD_SIZE);
#endif

          /* Iterate over the pbuf chain until the entire packet is
             read into the pbuf. */
          for(q = p; q != NULL; q = q->next) {
              /* Read enough bytes to fill this pbuf in the chain. The
                 available data in this pbuf is given by the q->len
                 variable. */
              CopyFromFrame_EMAC(q->payload, q->len);
          }
```

```
              /* Notify the hardware that a packet has been processed by
               * the driver. */
              EndReadFrame();

#if ETH_PAD_SIZE
              /* Reclaim the padding word. The upper layer of the protocol
               * stack needs to read the data in the pbuf from a correct
               * location. */
              pbuf_header(p, ETH_PAD_SIZE);
#endif

              LINK_STATS_INC(link.recv);
        }
        else
        {
              /* Notify the hardware but actually nothing has been done
               * for the incoming packet. This packet will just be ignored
               * and dropped. */
              EndReadFrame();
              LINK_STATS_INC(link.memerr);
              LINK_STATS_INC(link.drop);
        }
    }

    return p;
}
```

More specifically, the `StartReadFrame` function is used to locate the next Ethernet frame to be read from the hardware ring buffer as well as its size. Actually, the next frame to be read can be accessed through the `RxConsumerIndex` index. Normally, the return value of this function simply indicates the length of the frame to be read, represented with positive values. Regarding erroneous frames, a negative length is used to distinguish different types of error.

If the frame corresponds to an incorrectly received frame, the `EndReadFrame` function should be called. It just updates the `RxConsumerIndex` to point to the next element in the receive ring buffer. In this way, the erroneous frame is simply discarded.

Instead, if the next frame to be retrieved from the ring buffer represents a correctly received frame, the following actions will be performed:

First of all, room will be made for Ethernet padding, depending on the configuration. This is because for some architectures, word access has to be aligned to 4 byte boundary. If this is the case, the Ethernet source address of the Ethernet header structure will start at a non-aligned (4 byte) location. `ETH_PAD_SIZE` is used to pad this data structure to 4 byte alignment.

After that, a chain of pbufs is allocated from the pbuf pool to accommodate the Ethernet frame, by means of the `pbuf_alloc` function. Since each pbuf is of fixed size, as discussed in Chapter 7, one or more pbufs may be needed to store the whole frame. If there is not enough memory and the required amount of pbufs cannot be allocated, the frame is simply dropped and some statistic information is updated, namely, `link.memerr` and `link.drop`.

On the other hand, upon successful memory allocation, the pbuf chain allocated is pointed to by the p pointer. It is worth noting that the padding field is prepended before the Ethernet header. In other words, the Ethernet header starts after the padding field. When storing the received packet, the pointer should be moved forward along the pbuf chain to find the right starting point. This is achieved by means of the pbuf_header function implemented in the LWIP protocol stack. The second argument of this function specifies which direction to move and how much. If it is a negative value, the pointer will move forward, otherwise backward.

Then the Ethernet frame can be copied from the hardware ring buffer to the pbuf chain by iterating through the pbuf chain and copying enough bytes to fill a pbuf at a time until the entire frame is read into the pbuf by means of the CopyFromFrame_EMAC function. After that, the EndReadFrame function is called to update the RxConsumerIndex. In this case, the hardware is notified that a packet has been processed by the driver and more space is available for further incoming packets.

At the end, before returning the pointer to the pbuf chain to the caller, it should be adjusted to include the padding field so that the upper layer of the protocol stack can operate on it correctly.

8.5.5   TRANSMIT PROCESS

The transmission of a packet is performed by the low_level_output function, which is shown in the following listing. Actually, transmission is initiated by the upper layer of the protocol stack. The packet to be transmitted is contained in the pbuf (chain) that is passed to this function. This function returns ERR_OK if the packet could be sent, otherwise an error indication.

```
/**
 * This function performs the actual transmission of a packet.
 */
static err_t
low_level_output(struct netif *netif, struct pbuf *p)
{
  struct pbuf *q;

/* Drop the padding word. */
#if ETH_PAD_SIZE
  pbuf_header(p, -ETH_PAD_SIZE);
#endif

  /* Wait until space is available in the transmit ring buffer. */
  if(xSemaphoreTake(semEthTx, portMAX_DELAY) == pdTRUE) {

    /* Set up a transmit buffer to the outgoing packet which is stored
       at p->payload and with a length of p->tot_len.
    */
    RequestSend(p->tot_len, p->payload);

    /* Move the data from the pbuf to the ring buffer, one pbuf at a
       time. The size of the data in each pbuf is kept in the ->len
       field.
    */
    for(q = p; q != NULL; q = q->next) {
      CopyToFrame_EMAC_Start(q->payload, q->len);
    }

    /* Notify the Ethernet hardware that a packet is ready to be sent
     * out. */
    CopyToFrame_EMAC_End();
  }
```

```
/* Reclaim the padding word. The upper layer of the protocol stack needs
   the pointer to free the memory allocated for pbuf *p from a correct
   position.
*/
#if ETH_PAD_SIZE
  pbuf_header(p, ETH_PAD_SIZE);
#endif

  LINK_STATS_INC(link.xmit);

  return ERR_OK;
}
```

First of all, the padding word should be dropped before starting a transmission because it is just used within the LWIP protocol stack. There is no need to send it onto the link. The real Ethernet frame starts after it.

Moreover, transmission cannot be started if there is no space available in the hardware transmit ring buffer. If this is the case, the caller will be blocked on the semEthTx semaphore created and initialized in low_level_init indefinitely. Otherwise, the next free element in the buffer is indicated by TxProducerIndex. The RequestSend function will set up that buffer for the outgoing frame. Since the size of each transmit buffer is the same as the maximum Ethernet frame size, any outgoing Ethernet frame will just require one single transmit buffer.

Then the data can be copied from the pbuf (chain) to the ring buffer, one pbuf at a time, by means of the CopyToFrame_EMAC_Start function, until the whole Ethernet frame is stored in it. When it is done, the TxProducerIndex is updated in the CopyToFrame_EMAC_End function to point to the next free element in the ring buffer. In this way, the Ethernet hardware will also find out that a packet is ready to be sent out.

Last but not least, the pointer to the pbuf (chain) should be adjusted to reclaim the space allocated for the padding field so that when the upper layer of the protocol stack is about to free the memory of the pbuf, the right amount of memory will be freed.

8.6 SUMMARY

This chapter discussed the design and implementation of device driver for embedded systems, which are quite different from those on general purpose operating systems. The most significant differences mainly related to the concept of interrupt handling, application task interface, as well as synchronization.

The chapter starts with a general description of the internal structure of typical device drivers, presented in Section 8.1. Device driver does nothing different than moving data back and forth between peripheral devices and other software components. The main processing is generally done within an interrupt handler and one or more helper tasks in order to be reactive to external events while at the same time permit better real-time performance.

Section 8.2 illustrates the whole process of how an interrupt is handled. In particular, it focuses on interrupt handling at the hardware level, from when an interrupt arrives at the device until it is accepted by the processor, since this part may be less familiar to embedded software developers.

Moreover, in order to interact with both the hardware and other software components, interfaces to them should be provided by the device driver, which is discussed in Section 8.3. For what concerns hardware interface, it could be either register-based or DMA-based. With respect to general purpose operating systems, the application task interface is more performance oriented rather than being standardized.

Last but not least, synchronization is another major topic in concurrent programming. It becomes even trickier when interrupt handler(s) are added into the picture. As a simple example, any potentially blocking primitive is forbidden in an interrupt handler. To this purpose, synchronization issues between interrupt handlers and upper layers have been analyzed and addressed in Section 8.4

At the end, a practical example about the Ethernet device driver is presented in Section 8.5, which demonstrates how the main issues related to the three topics just mentioned are addressed in a real implementation.

9 Portable Software

CONTENTS

The previous chapters set the stage for embedded software development, starting from the general requirements of embedded applications and proceeding by presenting the main software development tools and runtime components. The practical aspects of the discussion were complemented by theoretical information relating to real-time execution models and concurrent programming.

In this chapter, the focus is now on how to make the best profit from the software development activity, by producing software that can be easily migrated, or *ported*, from one processor architecture to another and among different projects, without sacrificing its efficiency.

This aspect is becoming more and more important nowadays because new, improved processor architectures are proposed at an increased rate with respect to the past and it becomes exceedingly important to adopt them in new projects as quickly and effectively as possible. At the same time, the time to market of new projects is constantly decreasing, putting even more focus on quick prototyping and development.

Further portability issues may arise if some important software components—most notably, the operating system and the protocol stacks—or the software development toolchain itself may also change from one project to another and from one target system to another. In this case, an appropriate use of abstract application program interfaces and porting layers becomes of primary importance.

9.1 PORTABILITY IN EMBEDDED SOFTWARE DEVELOPMENT

Modern software is more and more often built from a multitude of modular components, each realizing a specific function of interest. This plays a key role, first of all, to keep the ever-increasing software complexity under control. Furthermore, it also encourages designers and programmers to *reuse* existing software components multiple times.

For instance, major system-level components like the software development toolchain and the language runtime libraries (both presented in Chapter 3), the operating system (Chapter 5), and the TCP/IP protocol stack (Chapter 7) are almost never written from scratch for every new project.

The same is true also at the application layer, where some components—for instance graphics user interface engines and libraries—are indeed largely independent from the specific application at hand.

This approach brings several important advantages for what concerns the software design and development process. Among them, we recall:

- *Faster time to market,* because software design and development must mainly deal only with the highest level of application software—which is often significantly different from one project to another—rather than with the software system as a whole.

  This aspect becomes even more significant when projects are closely related to each other, for instance, when they pertain to the same application domain. This is because, in this case, there is more shared functionality and more opportunities for software reuse arise.

- *Rapid software prototyping* becomes more affordable even when the software development team is small because a working prototype of a new software, albeit admittedly not yet ready for prime time, can be assembled by "putting together" existing components in limited time.

  Rapid prototyping has been heavily studied in the recent past and is an important tool to give project designers and developers valuable user feedback about a new project as early as possible [42].

  In turn, this helps to validate user requirements and refine software specifications early in the software development cycle, when changes can still be made with limited impact on project budget and time span. This fact is supported not only by intuition, but also by well-established works on software engineering methodologies [44, 68].

- *Higher reliability* because existing software components have already been tested and debugged thoroughly, especially when they have already been adopted successfully in previous projects. If they have been designed correctly, migrating those components from one project to another is unlikely to introduce new defects.

  As a consequence, the time needed to test the component and repair any newly found defects becomes much shorter, and the probability of having latent bugs in the system after software release is lower.

- *Reduced software development cost,* due to the synergy of the above-mentioned advantages, because reusable software components often account for a significant part of the overall software size, also taking into account that they often implement the most complex functions in the system.

- Last, but not least, the fast pace of contemporary *hardware evolution* becomes easier to follow. This aspect is especially important in some areas—for instance, consumer appliances—in which being able to quickly incorporate the latest advances in microcontroller technology into a new product often plays a key role in its commercial success.

However, it is possible to take full advantage of an existing software component in a new project *only if* it can be easily migrated, or *ported* from one project to another. In other words, the amount of effort needed to migrate the component must be significantly lower than the effort needed to develop the component anew.

In turn, this is possible only if the component is designed and built to be as much as possible independent from the environment in which it executes, from at least two different points of view.

1. The component itself must have been designed and implemented in a portable way. In turn, this encompasses two different aspects. The first one concerns how the code belonging to the component itself has been *written* at the programming language level.

   Namely, an improper use of the C programming language and its compiler may easily lead to portability pitfalls, which will be summarized in Section 9.2. On the other hand, at least in some cases, the use of some compiler-specific language extensions, such as the ones presented in Section 9.4, becomes unavoidable.

   Sometimes, the standard C language may simply be inadequate to express the operation to be performed—for instance, accessing the contents of some processor registers directly—or implementing the operation using only standard language constructs would have unacceptable, adverse effects on performance.

   In this case, to limit the impact on portability, it becomes essential to organize the code in a proper way and enclose non-standard language constructs in well-defined code modules, often called *porting layers*, to be kept as small as possible. The second aspect concerns how the component *interfaces* with other components and, symmetrically, how the application programming interface it makes available to others is architected.

   As will be better described in Section 9.3, choosing the right interface when more than one is available (when using another component) and providing a proper interface (when designing a component) are key factors to reach a proper trade-off between execution efficiency and portability.

2. Some software components interface heavily with hardware. For instance, as described in Chapter 4, the operating system needs to leverage an extremely deep knowledge of the processor architecture in order to perform some of its critical functions, for instance, context switch.

   Therefore, when the operating system is ported from one processor architecture to another, it becomes necessary to extend this knowledge. As for the usage of non-standard language constructs, the goal is conveniently reached by means of a *porting layer*. A detailed example of how a porting layer is architected and implemented will be given in Chapter 10.

Sometimes, hardware dependencies are limited to the interface toward a few specific devices. For instance, the TCP/IP protocol stack obviously needs to have access to at least one network device, like an Ethernet controller, in order to perform most of its functions.

In this case, the problem is often solved by confining device-specific code to special software modules called *device drivers*, which have been presented in Chapter 8 and must be at least partially rewritten in order to support a new device.

For this reason, when starting a new project possibly involving new hardware (for instance, a new microcontroller) or new software (for instance, a different TCP/IP protocol stack) it's important to pay attention to code portability issues.

The same is true also in perspective, when designing a system that is likely to be ported to other hardware architectures or has to support different software components in the foreseeable future.

In order to make a more specific and practical example, Figure 9.1 depicts the typical structure of a simple, networked embedded system. The main software components that build up the system, represented as white rectangles in the picture, are:

- The language runtime libraries, which are shared among all the other components. In an open-source embedded system using the C language those libraries are typically provided by NEWLIB.
- The FREERTOS real-time operating systems, providing multitasking, timing, and inter-process communication services to all the other components.
- A number of system components, each providing additional services and interfacing with the specific devices they control. Most notably, network services are provided by the LWIP TCP/IP protocol stack interfaced with an Ethernet controller.
- Several high-level components that build up the application. They make use of all the other components mentioned so far and, in addition, interface with each other.

According to the general discussion presented above, portability issues may occur in various areas of the system and for two different reasons, mainly due to the presence of *interfaces* to other *software* or *hardware*. In the figure, these interfaces are depicted as lighter and darker gray blocks, respectively. These areas are highlighted with black dots in the figure. In particular:

1. As outlined in Chapter 3, in order to configure NEWLIB for a certain processor architecture it is necessary to provide some information about it, ranging from very basic facts (for instance, how big an `int` variable is) to more detailed architectural nuances (like which and how many registers must be saved and restored to implement `setjmp` and `longjmp` properly).
2. The NEWLIB I/O system provides the full-fledged set of I/O functions foreseen by the C language standard. However, in order to be able to perform "real" input–output operations, it must be linked to at least one hardware device. As a bare

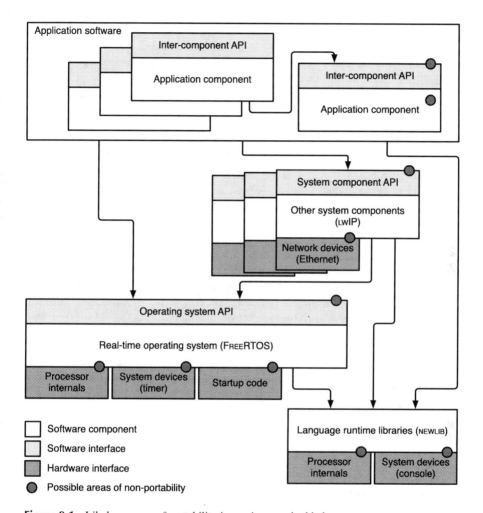

Figure 9.1 Likely sources of portability issues in an embedded system.

minimum, when printouts are used as a debugging aid, it shall be linked to the console device (usually a serial port), by means of a simple device driver.

3. The operating system contains some code to initialize the system when it is turned on or after a reset, usually called *startup code*. Some typical operations performed by the startup code are, for instance:

- Set up the processor clock appropriately, switching to a higher-quality oscillator and programming the clock generation phased-locked loops (PLLs) to achieve a faster speed with respect to the default configuration used by the microcontroller upon system startup, which is often extremely conservative.
- Make the memory external to the chip accessible by configuring the external memory controller with parameters appropriate to the hardware and memory characteristics, such as external bus width, speed, and timings.

- Prepare the system for executing code written in C, by setting up the initial processor stack as well as the interrupt-handling stack when required, and initializing the data segment.

As can be seen, in order to implement any of those operations, it is strictly necessary to have an intimate knowledge of the hardware characteristics of the microcontroller in use, as well as how it has been configured and integrated into the embedded board. For obvious reasons, all these aspects are hardly portable from one project to another and some of them cannot even be expressed in a high-level language without using non-standard extensions, which will be outlined in Section 9.4.

It must also be noted that, strictly speaking, the responsibility of carrying out these operations is somewhat "shared" between the language runtime libraries and the operating system. Therefore, even though the startup code is shown to be part of the operating system in this example, it is not uncommon to find part or all of the startup code within those libraries as well.

4. Even after the system has been initialized, the operating system still needs to perform operations that are inherently architecture-dependent, most notably *context switch* and *interrupt masking*. They are implemented in a dedicated code module, that constitutes the operating system porting layer. Due to its importance, the operating system porting layer will be discussed in more detail in Chapter 10.

5. One important function of the operating system is to provide timing references to all tasks in the system. This feature is implemented with the help of a hardware *timer* that raises an interrupt request at predetermined intervals. Even though timers are a relatively simple component, their way of working highly depends on the microcontroller in use. For this reason, the interface toward the timer is implemented as part of the operating system porting layer, too.

6. Additional system components are often the most critical for what concerns portability because, as shown in Figure 9.1, they must deal with multiple interfaces toward other software and hardware components.

   For instance, the LWIP protocol stack, besides making use of the C language support library, is concerned with three other important interfaces.

   - At the bottom, LWIP interfaces with an Ethernet controller, by means of a device driver presented in Section 8.5.
   - For timing and synchronization—both internal and with respect to the application tasks—LWIP leverages the operating system services through the operating system programming interface.
   - At the top, LWIP offers two different programming interfaces to application tasks. Both these interfaces and the previous one have been discussed in Chapter 7.

7. When a program component interacts with another it does so by means of a well-defined application programming interface (API). In some cases a component implements and offers multiple APIs, which represent different trade-offs between execution efficiency and portability of the code that makes use of them.

   As was mentioned earlier, both aspects are of great importance in an industrial applications context. Therefore, choosing the most appropriate API to be used in a certain project it not just a matter of habit or style, as it is often considered by programmers, but it may also have important consequences to the project's present (execution efficiency) and future (portability).

8. The code itself may be easier or harder to port to a new project, depending on how it has been implemented. From this point of view, the way the programming language is used plays an important role, as explained in Section 9.2.

This is usually not a concern when adopting existing open-source software components, because they have most often been explicitly designed and implemented to be as portable as possible. However, it may become a concern for application-level software built in-house, especially when portability was not considered to be a design goal right from the beginning.

As a final remark before discussing portability issues and the ways of addressing them in greater detail, it is worth remarking that, although the areas that may be problematic seem to cover a large part of Figure 9.1, they actually represent a minor fraction of the code. For instance, a typical porting layer for FREERTOS consists of only about 1000 (heavily commented) lines of code with respect to about 17000 lines for the architecture-independent part of the operating system.

Furthermore, may popular open-source components—like FREERTOS itself—come with a number of industrial-grade porting layers that are already available for use or, at least, represent a good starting point to develop a new porting layer more quickly and with less effort.

9.2 PORTABILITY ISSUES IN C-LANGUAGE DEVELOPMENT

As was outlined in the previous section, designing and developing portable code involves several different language-related facets that will be further elaborated upon here.

The C programming language is backed by a well-established international standard [89], which has been in effect for a long time and is quite stable, even though it is still being improved further. Moreover, the GCC compiler supports many different architectures and, in many cases, switching from one architecture to another seems to be just a matter of acquiring the right compiler version and changing several command-like flags related to code generation. However, writing portable code also requires programmers to conform to several rules during software development, which go beyond just adhering to the standard. In fact, C language portability has been a long-standing topic in computer programming [95].

The first aspect of the C programming language that may lead to portability issues lies in the area of data types. Even restricting the discussion to integer data types—the simplest and most common ones—it must be noted that, historically, the language standard only specified their *minimum range*, rather than their *actual width* in bits, as shown in Table 9.1.

The object-like macros listed in the table, containing the minimum and maximum value allowed for a certain data type, are made available in the system header file

Table 9.1

Minimum Range of Basic Integer Data Types Specified by the C Standard

Type	Macro	Minimum Value	Macro	Maximum Value
signed char	SCHAR_MIN	-127	SCHAR_MAX	127
unsigned char	—	0	UCHAR_MAX	255
char	CHAR_MIN	(a)	CHAR_MAX	(b)
short	SHRT_MIN	-32767	SHRT_MAX	32767
unsigned short	—	0	USHRT_MAX	65535
int	INT_MIN	-32767	INT_MAX	32767
unsigned int	—	0	UINT_MAX	65535
long	LONG_MIN	-2147483647	LONG_MAX	2147483647
unsigned long	—	0	ULONG_MAX	4294967295

(a): either SCHAR_MIN or 0

(b): either SCHAR_MAX or UCHAR_MAX

<limits.h>. It is therefore possible, at *compile time*, to check them and define a user data type that corresponds to the most suitable data type listed in the table.

However, the exact characteristics of a data type are still mostly unknown at software *design time*. For instance, the only information known at this stage about the int data type is that it is *signed* and it can hold *at least* values within the range -32767 to 32767. Moreover, it is known from the standard that its width, in bits, is the "natural size suggested by the architecture of the execution environment" [89]. Other important information about the data type is not specified, for instance:

- The *actual* width of an int, in bits.
- The integer representation.

About the integer representation it must be remarked that, even though most recent architectures adopt the two's complement representation, this is not mandated in any way by the standard, which also supports the sign and magnitude, as well as the one's complement representation.

As a side note for curious readers, this is the reason why the minimum int range does not include the value -32768.

In some cases, knowing those details is indeed unimportant. For instance, when using an int variable to control a for loop, the main goal of the programmer usually is to obtain code that executes as fast as possible.

This goal is fulfilled by the standard because choosing a "natural size" for int, and making the variable fit in a machine register, is beneficial for machine code efficiency. The only check that careful programmers should perform is to compare the maximum value that the control variable can assume by design with INT_MAX.

Table 9.2

Extended Integer Data Types Specified by the C99 Standard

Type	Width (b)	Properties
int8_t	8	Signed, two's complement, exact width
int16_t	16	Signed, two's complement, exact width
int32_t	32	Signed, two's complement, exact width
int64_t	64	Signed, two's complement, exact width
uint$n$_t	n	As above, but unsigned
int_least8_t	≥ 8	Signed, minimum guaranteed width
int_least16_t	≥ 16	Signed, minimum guaranteed width
int_least32_t	≥ 32	Signed, minimum guaranteed width
int_least64_t	≥ 64	Signed, minimum guaranteed width
uint_least$n$_t	$\geq n$	As above, but unsigned
int_fast8_t	≥ 8	Signed, minimum guaranteed width, fastest
int_fast16_t	≥ 16	Signed, minimum guaranteed width, fastest
int_fast32_t	≥ 32	Signed, minimum guaranteed width, fastest
int_fast64_t	≥ 64	Signed, minimum guaranteed width, fastest
uint_fast$n$_t	$\geq n$	As above, but unsigned
intmax_t	—	Signed, maximum width
uintmax_t	—	Unsigned, maximum width
intptr_t	—	Signed, can hold a pointer to void
uintptr_t	—	Unsigned, can hold a pointer to void

Unfortunately, in many other scenarios—and especially in embedded programming—details matter. This is especially important, for instance, when an integer value must be written into some device registers, and hence, interpreted by hardware. In this case, knowing exactly how wide the integer is and how negative values are represented clearly becomes crucial for correctness.

A related issue is to consider pointers and int variables to be the same size and freely convert from the one to the other. Even though this may work on some architectures, the standard offers no guarantee at all about this property.

Before the introduction of the C99 international standard [89], the only way of addressing this shortcoming was to refrain from using basic integer data types in those cases, and replace them with user data types. Then, user data types were defined in terms of appropriate basic data types in an architecture-dependent way, usually within a header file to be customized depending on the target architecture.

When using a C99 compiler, the selection of appropriate data types becomes easier because the standard introduced a new system header <stdint.h> that defines a set of integer data types having well-defined width and representation.

As summarized in Table 9.2, <stdint.h> provides data type definitions in several different categories:

Table 9.3

Minimum and Maximum Values of C99 Integer Data Types

Macro	Description
INTn_MIN	Minimum value of int$n$_t, exactly -2^{n-1}
INTn_MAX	Maximum value of int$n$_t, exactly $2^{n-1}-1$
UINTn_MAX	Maximum value of uint$n$_t, exactly $2^{n}-1$
INT_LEASTn_MIN	Minimum value of int_least$n$_t
INT_LEASTn_MAX	Maximum value of int_least$n$_t
UINT_LEASTn_MAX	Maximum value of uint_least$n$_t
INT_FASTn_MIN	Minimum value of int_fast$n$_t
INT_FASTn_MAX	Maximum value of int_fast$n$_t
UINT_FASTn_MAX	Maximum value of uint_fast$n$_t
INTMAX_MIN	Minimum value of intmax_t
INTMAX_MAX	Maximum value of intmax_t
UINTMAX_MAX	Maximum value of uintmax_t
INTPTR_MIN	Minimum value of intptr_t
INTPTR_MAX	Maximum value of intptr_t
UINTPTR_MAX	Maximum value of uintptr_t

1. Integer data types having an *exact* and known in advance width, in bits.
2. Integer data types having *at least* a certain width.
3. *Fast* integer data types having at least a certain width.
4. Integer data types having the *maximum* width supported by the architecture.
5. Integer data types able to hold *pointers*.

In addition, the same header also defines the object-like macros listed in Table 9.3. They specify the minimum and maximum values that the data types listed in Table 9.2 can assume. The minimum values of unsigned data types are not specified as macros because they are invariably zero.

Another common source of portability issues is related to which version of the C language standard a certain module of code has been written for. In fact, as summarized in the right part of Table 9.4, there are three *editions* of the standard, ratified in the past 25 years, plus one intermediate version introduced by means of an amendment to the standard.

For the sake of completeness, we should mention that full support for the most recent edition of the standard [90], commonly called C11 because it was ratified in 2011, is still being added to the most recent versions of GCC and is incomplete at the time of this writing.

In each edition, the language evolved and new features were introduced. Although every effort has been put into preserving backward compatibility, this has not always been possible. The problem is further compounded by the presence of several language *dialects*, typical of a specific compiler.

Table 9.4

Main Compiler Flags Related to Language Standards and Dialects

Flag	Description
-std=c89	Support the first edition of the C standard, that is, ISO/IEC 9899:1990 [87], commonly called C89 or, less frequently, C90. Support includes the technical corrigenda published after ratification. The alternative flag -ansi has the same meaning.
-std=iso9899:199409	Support the ISO/IEC 9899:1990 standard plus "Amendment 1" [88], an amendment to it published in 1995; the amended standard is sometimes called C94 or C95.
-std=c99	Support the second edition of the C standard, that is, ISO/IEC 9899:1999 [89], commonly called C99. Support includes the technical corrigenda published after ratification.
-std=c11	Support the third edition of the C standard, that is, ISO/IEC 9899:2011 [90], commonly called C11.
-std=gnu89	Support the C89 language plus GCC extensions, even if they conflict with the standards.
-std=gnu99	Support the C99 language plus GCC extensions, even if they conflict with the standard.

In many cases, these dialects were historically introduced to work around shortcomings of earlier editions of the standard, which were addressed by later editions. However, they often became popular and programmers kept using them even after, strictly speaking, they were made obsolete by further standardization activities.

For this reason, GCC offers a set of command-line options, listed in Table 9.4, which specify the language standard the compiler shall use. By default, these options do not completely disable the GCC language dialect.

Namely, when instructed to use a certain standard, the compiler accepts all language constructs conforming to that standard *plus* all GCC dialect elements that *do not conflict* with the standard itself. When strict checks against the standard are required, and all GCC-specific language constructs shall be disabled, it is necessary to specify the -pedantic option, too, as discussed in the following.

Another group of flags is related to code portability in two different ways. We list the main ones in Table 9.5. Namely:

1. Some flags direct the compiler to perform additional checks on the source code. Those checks are especially useful in *new code*, in order to spot potential issues that may arise when it is ported to another architecture in the future.
2. Other flags "tune" the language by affecting the default behavior of the compiler or relaxing some checks, with the goal of making it easier to compile *existing code* on a new architecture successfully.

As shown in the table, the main flag to disable GCC-specific language extensions and enable additional source code checks is -pedantic. The two additional flags -Wall and -Wextra enable additional warnings.

Table 9.5
Main Compiler Flags Related to Code Portability

Flag	Description
Extra source code checks	
-pedantic	Issue all the warnings required by the ISO C standard specified by means of the -std= option (see Table 9.4) and reject GCC-specific language extension, even though they do not conflict with the standard.
-Wall	Often used in conjunction with -pedantic, this option enables extra warnings about questionable language constructs even though, strictly speaking, they do not violate the standard.
-Wextra	Enables even more warnings, beyond the ones already enabled by -Wall.
Language tuning	
-fno-builtin	Instruct the compiler to *neither* recognize *nor* handle specially some *built-in* C library functions. Instead, they will be handled like ordinary function calls.
-fcond-mismatch	Allow conditional expressions, in the form a ? b : c with mismatched types in the second (b) and third (c) arguments. The expression can be evaluated for side effects, but its value becomes void.
-flax-vector-conversions	Relax vector conversion rules to allow implicit conversions between vectors with differing number of elements and/or incompatible element types.
-funsigned-char	Sets the char data type to be unsigned, like unsigned char.
-fsigned-char	Sets the char data type to be signed, like signed char.
-funsigned-bitfields	Sets bit-fields to be unsigned, when their declaration does not explicitly use signed or unsigned. The purpose of this option and of the following one is similar to -funsigned-char and -fsigned-char above.
-fsigned-bitfields	Sets bit-fields to be signed, when their declaration does not explicitly use signed or unsigned.
-fshort-enums	Allocate the minimum amount of space to enumerated (enum) data types, more specifically, only as many bytes as required for the range of values they can assume.
-fshort-double	Use the same size for both float and double floating-point data types.
-fpack-struct	When this flag is specified *without* arguments, it directs the compiler to pack structure members tightly, without leaving any padding in between and ignoring the default member alignment. When the flag is accompanied by an argument in the form =n, n represents the maximum default alignment requirement that will be honored by the compiler.

However, it should be remarked that this flag does *not* enable a thorough set of conformance checks against the C language standard. Rather, only the non-conforming practices for which the standard *requires* a diagnostic message, plus a few selected others, are reported.

Other tools, like the static code analysis tool mentioned in Chapter 16, are often able to perform even more accurate checks concerning code portability, although this is not their primary goal.

For what concerns the language tuning flags:

- The `-fno-builtin` flag instructs the compiler to *neither* recognize *nor* handle specially some *built-in* standard C library functions. When built-in functions are enabled, the compiler generates special code for them, which is more efficient than a function call. For instance, it may replace a call to `memcpy` with a copying loop.

  Moreover, the compiler may also issue more detailed warning and error messages when it recognizes a built-in function. For instance, when `printf` is handled as a built-in function, the compiler is able to check that its optional arguments are consistent with the output format specification.

  When built-in functions are handled specially, the resulting code is smaller and executes faster in most cases. On the other hand, the program becomes harder to debug. This is because there are no actual calls to built-in functions in the code, and hence, for instance, it becomes impossible to set a breakpoint on them.

  Moreover, it is also impossible to replace a built-in function with an alternate implementation, for instance, by linking the program against a different library that takes priority over the standard C library. This possibility is sometimes used by legacy code and makes it difficult to port.

- Some flags, like `-fcond-mismatch` and `-flax-vector-conversions`, relax language rules, as denoted in the table, with the main goal of accommodating legacy code without errors. They should be used with care (or, if at all possible, not be used at all) when developing new code, because relying on those relaxed rules may, in turn, introduce latent issues in future ports.

- The two flags `-funsigned-char` and `-fsigned-char` determine whether `char` variables that are not explicitly qualified with `unsigned` or `signed` are unsigned or signed, respectively.

  According to the standard, programs should not depend on `char` being signed or unsigned. Instead, they should explicitly use `signed char` or `unsigned char` when they depend on signedness.

  However, many programs and libraries (especially older ones) use plain `char` variables and expect them to be signed or unsigned, depending on the architecture and compiler they were originally designed for.

  These options obviously help to port code designed in this way to a new architecture. However, it is also important to remark that these options generally apply to the *whole program*. In fact, it is usually unwise to use different settings for individual code modules because this would result in the same data type, char, having two different meanings depending on the code module it's encountered in.

- The flags -funsigned-bitfields and -fsigned-bitfields have a similar meaning, but they apply to bit-fields that are not explicitly qualified as either unsigned or signed.

- The two flags -fshort-enums and -fshort-double determine the amount of storage the compiler reserves for enumerated and floating-point data types. As a consequence, they make it possible to port legacy code that contains hidden assumptions about these aspects without extensive modifications. However, it makes the resulting code incompatible with code generated without these options. Hence, it must be used with care, especially when mixing legacy and newly developed code.

- Similarly, the flag -fpack-struct controls how structure members are packed. When used alone, that is, without arguments, the flag directs the compiler to pack structure members as tightly as possible, that is, without leaving any padding between them, regardless of the default or "natural" alignment members may have.

  Thus, for instance, an int member that follows a char member will be allocated immediately after the char, even though its default, natural alignment (on a 32-bit architecture) would be to an address which is a multiple of 4 bytes.

  The flag also accepts an optional integer argument, n, which is specified as -fpack-struct=n. In this case, the compiler will introduce some padding, in order to honor default alignments up to a multiple of n bytes, but no more than that.

  For example, when the flag -fpack-struct=2 is in effect, the compiler will use padding to align members up to a multiple of 2 bytes if their default alignment says so. However, continuing the previous example, int members will still be aligned only to a multiple of 2 bytes, instead of 4.

  This flag may sometimes be useful to define a data structure whose layout is exactly as desired, for instance, when the data structure is shared with a hardware device. However, it makes the generated code incompatible with other code modules that are not compiled with exactly the same flag setting. Moreover, it also makes the generated code less efficient than it could be. This is because, on most recent architectures, accessing a memory-resident variable that is not naturally aligned requires extra instructions and, possibly, additional working registers.

As a final remark, it is important to mention that writing portable code, being able to reuse existing code easily, and code efficiency are often conflicting goals. This concerns not only the use of non-standard language extensions, described in Section 9.4, but also (at least to some extent) aggressive code optimization, to be discussed in Chapter 11.

9.3   APPLICATION PROGRAMMING INTERFACES

As was mentioned previously, the choice of the best application programming interface (API) to be used between software components depends on a rather complex set of criteria. For instance:

- Execution *efficiency* is usually better when using a streamlined, specific interface instead of a more general one.
- Code *portability* is easier to achieve if the API conforms to a standard (either *de facto* or backed by an international standardization body) that is implemented by more than one product of the same kind.
- When interfacing with existing software components, a quite obvious constraint comes from which APIs they make available and their suitability for use, depending on the *application* requirements.

Since the criteria just mentioned are not easy to express formally, and it is even harder to evaluate quantitatively how much a software component adheres to a certain criterion—for instance, how can we measure how portable a piece of code is, especially when it is complex—in the following we will discuss a small set of real-world examples to give readers an informal feeling of which trade-offs they may have to confront in practice, and which main aspects they should consider.

- At the C library level, the most intuitive way of producing some output, for instance, to roughly trace program execution during software development, is to call the standard library function `printf`.

  Assuming that the C library is properly configured, as described in Chapter 3, the resulting string of character will eventually be transferred to a suitable console I/O device on the target board and, from there, reach the programmer.

  On the other hand, many runtime libraries offer alternate ways of producing output. For instance, the LPCOpen software development platform offered by NXP for use on their range of LPC microcontrollers [129] provides several macros, whose name starts with _DB, with the same purpose.

  When considering which functions shall be used in a certain project, it is important to fully realize in which aspects they differ from each other. In our specific case—but this scenario is rather typical—it is important to consider that the two approaches are rather far away from each other for what concerns the *abstraction level*.

  In fact, the `printf` function is very complex and implements a wide range of data conversion and formatting facilities, analogous to what is available on a general-purpose computer. On the contrary, the _DB functions offer just a few formatting options.

  Namely, they can output only single characters, character strings, as well as 16- and 32-bit integer variables. For integer variables, the only conversion options are to the decimal and hexadecimal representations. In turn, this brings some important differences from the practical point of view.

1. On the one hand, `printf` and other closely related functions, such
    as `fprintf`, `fputs`, and so on, provide a *general* and very flexible
    input–output framework. For instance, by just calling `fprintf` with
    one stream or another as an argument, it is easy to direct the output to
    another device rather than to the console.

    If the C runtime library is appropriately configured, it is also possible
    to direct the output to an entity belonging to a quite different level
    of abstraction with respect to a physical device, for instance, a file
    residing on a storage medium.

    On the other hand, the `_DB` functions are able to operate only on a
    single device, that is, the console, which is chosen at configuration
    time and cannot be changed afterward.

2. The generality and flexibility of `printf` comes at the expense of
    memory footprint, because `printf` necessarily depends on many
    other lower-level library functions that implement various parts of its
    functionality.

    Due to the way most embedded applications are linked, those depen-
    dencies imply that calling `printf` in an application brings many ad-
    ditional library modules into the executable image besides `printf`
    itself, even though most of them are not actually used at runtime.

    For instance, `printf` is able to convert and format floating-point
    variables. In most cases, the (rather large) portion of code that im-
    plements this capability will be present in the executable image even
    though the application does not print any floating-point variables
    at all.

    On the other hand if, for instance, the program never calls the `_DBD32`
    macro (that is responsible for printing a 32-bit integer in decimal), the
    corresponding conversion and formatting code will not be present in
    the executable image.

3. For the same reasons, `printf` also has a bigger execution time over-
    head with respect to other functions that are less general and flexible.
    Continuing the previous example, scanning and obeying even a rela-
    tively simple format specification like `"%-+5d"` (which specifies that
    `printf` should take an `int` argument and print it in decimal, left-
    aligned, within a 5-character wide output field, and always print its
    sign even though it's positive) is much more complex than knowing
    in advance that the argument is a 32-bit integer and print it in full
    precision with a fixed, predefined format, like `_DBD32` does.

4. Due to its internal complexity, `printf` also requires a relatively large
    amount of stack space (for example, at least 512 bytes on a Cortex-
    M3 using the NEWLIB C library). In turn, this may make the function
    hard to use when stack space is scarce, for instance, within an interrupt
    handler.

5. Last, but not least, `printf` works only if the C library input–output
    subsystem has been properly configured. This may or may not be true

in the initial stages of software development, further limiting the usability of `printf` in these cases.

- At the protocol stack level, LWIP offers two distinct APIs, namely, a standard POSIX *sockets* API [86] and a custom API called *netconn*, both discussed in Chapter 7.

  With respect to POSIX `sockets`, the `netconn` API has the advantage of being smaller in terms of footprint, simpler for what concerns application design and development, and more efficient regarding execution speed. Moreover, the `netconn` API is often expressive and powerful enough for the range of network applications typically found in small, embedded systems.

  On the other hand, a clear shortcoming of the `netconn` API with respect to the POSIX `sockets` API is that it makes the port of existing software more difficult, because it is not based on any international standard. Development of new software may be impaired or slowed down, too, because the `netconn` API is not as well known to programmers as the POSIX `sockets` API.

  Both aspects are of great importance when porting an existing piece of software, for instance a higher-level protocol stack, especially in an industrial applications context.

  On the contrary, making use of the `netconn` API makes sense in the following scenarios:

  - When the easiness of porting *existing* software is not of concern, because a brand *new* software component is being developed and it is reasonable to assume that LWIP will still be used as a protocol stack in future projects.
  - If the project design aims at the best possible performance regardless of any other factors, the only constraint being to use LWIP and its proven implementation of the TCP/IP protocol for communication, in order to streamline and simplify implementation.

- Even though this is not the case for FREERTOS, due to its simplicity, even at the operating system level there may be more than one choice of API.

  For instance, RTEMS [131], another popular open-source real-time system in common use for embedded applications, supports three different APIs:

  1. A *native* API, which is defined and thoroughly described in the RTEMS documentation [130].
  2. The *POSIX* API [133], conforming to the international IEEE standard 1003.13 [79].
  3. The *ITRON* API [132], conforming to ITRON version 3.0, an interface for automotive and embedded applications, mainly popular in Japan [148].

As can be expected, the native API provides the richest set of features, at the expense of portability because it is available only on RTEMS, and not on any other operating system. It is the most efficient one, too, because the other two are *layered* on top of it, at the cost of some performance.

At the other extreme, the POSIX API, being backed by an international standard, makes software development easier and quicker, also because many programmers are already familiar with it and are able to use it effectively from day zero.

It also offers the best guarantee against software obsolescence because, even though the standard will continue to evolve, backward compatibility will always be taken into consideration in any future development.

The ITRON API, instead, is an example of interface provided for yet another reason. In this case, the emphasis is mainly on encouraging the adoption of a certain operating system in a very specific application area, where existing software plays a very important role. This can be obtained in two different, symmetric ways:

- Port existing applications to the new operating system by modifying their porting layer.
- Informally speaking, do the opposite, that is, port the operating system to those applications by providing the API they expect.

Of course, the second alternative becomes appealing when the relative weight of existing software and operating system within applications is strongly unbalanced toward the former.

In this case, it was considered more profitable to write an API adaptation layer once and for all, and then make applications pass through two layers of code to reach the operating system (their own porting layer plus the API adaptation layer just described), rather than updating existing code modules one by one.

9.4   GCC EXTENSIONS TO THE C LANGUAGE

Although in the previous sections of this chapter we highlighted the importance of having a *standard* language specification and to adhere to that specification as much as possible when developing application code, for good reasons, it is now time to briefly introduce several useful *extensions* to the C programming language that compilers usually make available. As always, the open-source GCC toolchain discussed in Chapter 3 will be used as an example.

Even though it may seem counter-intuitive to discuss language extensions after stressing the importance of standards from the point of view of portability, it should also be noted that software development, especially when considering embedded systems, it is always a trade-off between portability and other factors, most notably execution efficiency.

From this point of view, the opportunity (or even the need) for using non-standard language extensions stems from two distinct main reasons:

1. When portability is achieved by means of a porting layer, as described in Section 9.1, within that layer it is convenient to take advantage of all the features the

compiler can offer to the maximum extent possible, mainly to improve execution efficiency.

This opportunity becomes even more appealing because, in this particular case, there are no adverse side effects. In fact, informally speaking, there is no reason to have a "portable porting layer" and we can accept that the code that belongs to the porting layer is non-portable "by definition."

2. In some cases, within porting layers, the use of some extensions to the C language is sometimes necessary. In other cases, even though in principle they could be avoided, their use is indeed extremely convenient to avoid developing standalone assembly language modules in the project.

Two examples, which we will further elaborate upon in the following, are assembly-language *inserts* within C code—to express concepts that are not supported by high-level programming languages—as well as object *attributes* to direct the toolchain to place some special code and data structures into specific areas of memory.

9.4.1 OBJECT ATTRIBUTES

Object attributes are attached to variable and function declarations (and, in some cases, to data type definitions) in order to provide additional information to the compiler, beyond what is made possible by the standard C language syntax.

Among the diverse purposes and uses of object attributes, the main ones are listed in the following.

- They drive the compiler's code optimizer in a more accurate way and improve optimization quality.
- When applied to a function, they modify the code generation process in order to meet special requirements when the function is invoked from particular contexts, like interrupt handling.
- Some attributes are propagated through the toolchain and may affect other toolchain components as well, for instance, how the linker resolves cross-module symbol references.

Before proceeding further, it is important to remark that object attributes, being a non-standard language feature, may change their meaning from one compiler version to another. Moreover, new attributes are introduced from time to time, whereas old ones may become obsolete. In addition, object attributes are sometimes related to very specific details of a certain processor architecture, and hence, are available for that architecture only.

In the following description, we will focus on GCC version 4.3 for Cortex-M3 processors that, due to its stability, was a popular choice for embedded industrial application development at the time of this writing. For the sake of conciseness, we will only present a subset of the available attributes, namely, the ones that we consider most stable and that are available, possibly with minor variations, for other architectures, too.

Table 9.6

Main Object Attributes, Cortex-M3 GCC Toolchain

Name	Kind of object	Purpose
`aligned(n)`	Data & Code	Specifies that the object address must be aligned at least to a multiple of n bytes, in addition to any existing alignment constraints already in effect for the object. As the `packed` attribute discussed next, `aligned` can be applied either to a data structure as a whole or to individual fields.
`packed`	Data	Specifies that the object it refers to must have the smallest possible alignment, that is, one byte for a variable and one bit for a bit field.
`hot`	Code	Specifies that a function is a "hot spot" of the program, that is, it is frequently used and/or critical from the performance point of view. It tunes compiler optimization to privilege execution speed, and code placement to improve locality.
`cold`	Code	Specifies that the function is "cold," that is, is used infrequently. As a consequence, it will be optimized for size instead of execution speed. Code generation within functions that call cold function is also optimized by assuming that such calls are unlikely to occur.
`short_call`	Code	Indicates that calls to the function it is attached to must use the default calling sequence, in which the offset to the function entry point is encoded directly in a "branch with link" `BL` instruction. This calling sequence is very efficient but the offset must be in the range from -16777216 to 16777214.
`long_call`	Code	Indicates that the function it is attached to must be invoked by means of a more expensive calling sequence (in terms of code size and execution time) with respect to `short_call`. In return, this calling sequence is able to reach the function regardless of where it is allocated in memory.
`interrupt`	Code	Indicates that the function it is attached to is an interrupt handler, thus affecting code generation in the areas of prologue and epilogue sequences generation, as well as stack pointer alignment assumptions.
`naked`	Code	Indicates that the compiler must not generate any prologue or epilogue sequences for the function it is attached to, because these sequences will be provided explicitly by the programmer within the function body.
`section("s")`	Data & Code	Specifies that an object must be allocated in section `"s"` instead of the default section for that kind of object. This information is propagated through the toolchain, up to the linker.
`weak`	Data & Code	Primarily used in library function declarations, this attribute specifies that the symbol corresponding to the function or variable name must be "weak," that is, it can be silently overridden at link time by a normal, "strong" declaration of the same symbol appearing in another object module.

As such, the main goal of the discussion will be to provide a general idea of which concepts object attributes are able to convey, and how they can profitably be used to develop a porting layer, like the one presented in Chapter 10, in the most effective way. Interested readers should refer to the compiler documentation [159, 160] for up-to-date, architecture-specific, and thorough information.

The general syntax of an object attribute is

```
__attribute__ ((<attribute-list>))
```

where `<attribute-list>` is a list of attributes, separated by commas. Individual attributes can take several different syntactic forms. The two main ones are:

1. An attribute can be made up of a single word. For instance, referring to Table 9.6, which lists all the attributes that will be discussed in the following, `cold` is a valid attribute.
2. Some attributes are more complex and take one or more parameters. In these cases, parameters are listed within parentheses after the attribute name and are separated by commas. For instance, `section("xyz")` is a valid `section` attribute with one parameter, the string `"xyz"`.

In some cases, not discussed in detail in this book, attribute parameters have a more complex syntax and can be expressions instead of single words. In addition, it must be noted that it is also possible to add `__` (two underscores) before and after an attribute name, without embedded spaces, without changing its meaning. Therefore, the meaning of `cold` and `__cold__` is exactly the same.

Attributes may appear at several different syntactic positions in the code, depending on the object they refer to. In the following, we will discuss only attributes attached to a variable, function, or data type definition, even though other cases are possible, too.

- The typical position of attributes within a variable definition is after the name of the variable and before the (optional) initializer. For instance,

  ```
  int v __attribute__((<attributes>)) = 3;
  ```

  attaches the specified `<attributes>` to variable v.
- Attributes can be attached to a function definition or, in some cases that will be better described in the following, to its prototype declaration. The attributes specification shall come after the function name and the list of formal parameters. For example,

  ```
  void f(int a) __attribute__((<attributes>)) {
    <function_body>
  }
  ```

  defines function f and attaches the specified `<attributes>` to it.
- When attributes are attached to a structured data type definition, their position within the definition affects the object they are attached to. For

instance, the position of an attribute specification within a `struct` definition determines whether it affects the structure as a whole or an individual
field.

For instance, in the example that follows,

```
struct s {
    int a __attribute__((<attributes_field>));
    ...
} __attribute__((<attributes_struct>));
```

the list of attributes `<attributes_struct>` applies to the structure as a
whole (and to all its fields), whereas `<attributes_field>` applies only
to field a.

A first group of attributes affects the layout of data and code in memory, namely,
for what concerns alignment. The most important attributes in this group are:

- The `aligned(n)` attribute, which specifies the *minimum* alignment of the
  corresponding object in memory. In particular, the object address will be *at
  least* an integer multiple of n bytes. The alignment specification is considered to be *in addition* to any existing alignment constraints already in effect
  for the object.

  It is also possible to specify just `aligned`, without giving any arguments.
  In this case, the compiler will automatically choose the maximum alignment factor that is ever used for any data type on the target architecture,
  which is often useful to improve execution efficiency.

- Informally speaking, the `packed` attribute has the opposite effect, because
  it specifies that the corresponding object should have the smallest possible
  alignment—that is, one byte for a variable, and one bit for a bit field—
  disregarding the "natural" alignment of the object.

  Besides the obvious benefits in terms of storage space, it is necessary to
  consider that the use of `packed` often has negative side effects on performance and code size. For instance, on many architectures, multiple load
  operations and several mask and shift instructions may be needed to retrieve
  a variable that is not aligned naturally in memory.

The second group of attributes can be attached only to function definitions. These
attributes tune code optimization toward improving execution speed or reducing code
size on a function-by-function basis. Namely:

- The `hot` attribute specifies that a certain function is frequently used and/or
  it is critical from the performance point of view. The compiler will optimize
  the function to improve execution speed, possibly at the expense of code
  size, which may increase.

  Depending on the target, this attribute may also place all "hot" functions
  in a reserved subsection of the text section, so that all of them are close

together in memory. As explained in Chapter 2, this enhances locality and, in turn, may improve average cache performance.

Further performance improvements may also be expected on systems equipped with a memory management unit (MMU) and the associated translation lookaside buffer (TLB), which will be described in Chapter 15.

- On the contrary, the `cold` attribute specifies that a certain function is unlikely to be executed. As a consequence, the compiler will optimize the function to reduce code size at the expense of execution performance.

  On some targets, all "cold" functions are placed together in a reserved subsection of the text section, as is done for "hot" functions. In this way, cold functions are not interspersed with other functions and do not undermine their locality.

  In addition, when generating code for functions that contain calls to "cold" functions, the compiler will mark all execution paths leading to those calls as unlikely to be taken. Thus, code generation is tuned to privilege the execution speed of other execution paths, which are more likely to be taken.

Another group of attributes does not affect code generation within the function they are attached to, but changes the way this function is *called* from other parts of the application. As before, the two most important attributes in this category for the Cortex-M3 processor select different trade-off points between execution efficiency and flexibility.

- The `short_call` attribute attached to a function definition or to a function prototype indicates that the compiler, when calling that function, should generate a "branch with link" (`BL`) instruction and then instruct the linker to encode the offset between the `BL` instruction and the target function entry point directly in the instruction itself. The compiler makes use of this calling sequence by default, unless a special command-line option is given to it.
- The `long_call` attribute attached to a function specifies that the compiler, when calling that function, should generate a more complex sequence of instructions. The sequence of instructions takes more memory space and is slower to execute with respect to a single `BL`, but it has the important advantage of imposing no constraints on where the target function shall be in memory.

The most obvious advantage of the `short_call` calling sequence is that it is as efficient as it can possibly be, since it is composed of one single machine instruction. However, due to constraints in the `BL` instruction encoding, the legal values that the offset can assume is limited to the range from -16777216 to 16777214, and hence, it cannot span the whole address space.

Considering these aspects is especially important in embedded software development because, as discussed in Chapter 2, microcontrollers often contain different banks or areas of memory, with diverse characteristics, based at very different addresses within the processor address space.

It is therefore possible (and even likely) that a function placed in a certain bank of memory calls a function residing in a different bank, resulting in an extremely large offset that cannot be encoded in the BL instruction when using the short_call calling sequence.

Due to the way the toolchain works, as described in Chapter 3, the compiler is unable to detect this kind of error because it lacks information about where functions will be placed in memory. Instead, the error is caught much later by the linker and triggers a rather obscure message similar to the following one:

```
object.o: In function 'f':
(.text+0x20) relocation truncated to fit:
            R_ARM_PC24 against symbol 'g' ...
```

Actually, bearing in mind how the short_call calling sequence works and its restriction, it becomes easy to discern that the error message simply means that function f, defined in object module object.o, contains a call to function g.

The compiler generated a BL instruction to implement the call but the linker was unable to properly fit the required offset in the 24-bit signed field (extended with a zero in the least significant bit position) destined to it. As a consequence, the linker *truncated* the offset and the BL will therefore not work as intended, because it will jump to the wrong address. The very last part of the message, replaced by . . . in the previous listing for concision, may contain some additional information about where g has been defined.

The next two attributes listed in Table 9.6 affect the way the compiler generates two important (albeit hidden) pieces of code called function *prologue* and *epilogue*. The prologue is a short fragment of code that the compiler automatically places at the beginning of each function, unless instructed otherwise. Its goal is to set up the processor stack and registers to execute the body of the function.

For instance, if the function body makes use of a register that may also be used by the caller for another purpose, the prologue must save its contents onto the stack to avoid clobbering it. Similarly, if the function makes use of stack-resident local variables, the prologue is responsible for reserving stack space for them.

Symmetrically, the epilogue is another fragment of code that the compiler automatically places at the very end of each function, so that it is executed regardless of the execution path taken within the function body. Its aim is to undo the actions performed by the prologue immediately before the function returns to the caller.

The attributes interrupt and naked control prologue and epilogue generation for the function they are attached to, as outlined in the following.

- The interrupt attribute specifies that a certain function actually is an interrupt handler. If interrupts are enabled, interrupt requests can be accepted by the processor while it is executing anywhere in the code. As a consequence, it becomes harder to predict which registers must be saved in the interrupt handler prologue and, in order to be safe, the prologue must be generated in a more conservative way.

On some architectures, including the Cortex-M3, the invocation of an interrupt handling function differs from a regular function call in other aspects, too, most notably about which assumptions the compiler may make about the machine state when the function is invoked.

For instance, on the Cortex-M3 processor it is safe to assume that, upon a normal function call, the stack pointer will always be aligned to a multiple of 8 bytes, and hence, the compiler may generate optimized code leveraging this assumption.

On the contrary—depending on architectural details that will not be discussed in detail here—it is possible that, when an interrupt handling function is invoked, the stack pointer is only aligned to a multiple of 4 bytes.

In the last case, any code generated according to the stricter alignment assumption may or may not work properly. The `interrupt` attribute, when attached to a function, informs the compiler that it cannot rely on that assumption and must adjust its code generation rules accordingly.

- The `naked` attribute specifies that the compiler must *not* generate any prologue or epilogue for a certain function. As those pieces of code are needed in all cases, it is up to the programmer to provide them.

  Since C language statements are by themselves unable to manipulate registers directly, this can usually only be obtained by means of assembly language inserts, to be outlined in Section 9.4.

  The `naked` attribute is sometimes needed when developing low-level code, for instance, interrupt handlers or functions belonging to the operating system porting layer. In fact, in these cases, the compiler may not have enough knowledge to generate a correct prologue and epilogue, because it is not aware of the somewhat special scenarios in which those functions may be invoked.

  A rather intuitive way to illustrate this point is to consider the operating system context save function. Since, informally speaking, the responsibility of this function is to save all processor registers somewhere in memory in order to preserve them, it becomes useless (and unnecessarily time consuming) to have an automatically generated function prologue that saves a subset of the same registers on the calling task stack.

The last two attributes to be discussed here are peculiar because they affect only marginally the compiler's behavior, but they are propagated to, and take effect during, the linking phase.

The `section` attribute allows programmers to control exactly where a certain object, consisting of either code or data, will be placed in memory. On many architectures this is important, for instance, to achieve maximum performance in the execution of critical sections of code.

In other cases, this may even be necessary for correctness. For example, as described in Chapter 8, when developing a device driver, the shared buffers between software and hardware must be allocated in a region of memory that is accessible to both.

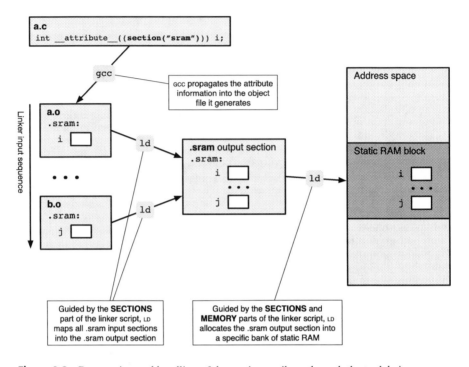

Figure 9.2 Propagation and handling of the section attribute through the toolchain.

Due to design constraints, mainly related to the on-chip interconnection architecture (described in Chapter 2) hardware components may have rather severe constraints on which regions of memory they have access to, and software must necessarily meet those constraints.

As shown in Figure 9.2, the propagation and handling of this attribute through the toolchain is rather complex and its correct implementation requires cooperation between the two main toolchain components, that is, the compiler GCC and the linker LD.

Namely, the `section` attribute is handled in several stages, or phases. In order to better appreciate the description that follows, readers should also refer to Chapter 3, which provides thorough information about how the toolchain components involved in the process being discussed work.

1. The `section` attribute is specified in a C source file. It takes a mandatory argument, `"sram"` in the example, which is a character string containing a section name.
2. The main role of the compiler is to *propagate* the attribute to the linker by storing the same information in the object file it generates. In the example, we suppose that variables `i` and `j`, defined in source modules `a.c` and `b.c`, have been marked

with `section("sram")`. Accordingly, in the corresponding object files a.o and b.o, these variables will have the same information attached to them.

3. When the linker LD maps input sections (coming from the input object files and library modules) into the output sections specified in the linker script, it follows the instructions it finds in the linker script itself—in particular, within the SECTIONS part of the script—and gathers all `sram` input sections into the same output section. In our example, we imagine that the relevant SECTIONS part of the linker script looks like this.

```
SECTIONS {
  .sram = { *(.sram) } > staticram
}
```

In this case, the output section has the same name as input sections, a rather common choice.

4. Afterward, the linker itself allocates the output section into a specific area, or bank, of memory—again, following the directives contained in the linker script. In particular:

   • The SECTIONS part of the script establishes into which area of memory the `sram` output section will go. Referring back to the previous listing, this is done by means of the `> staticram` specification. In our example, as the bank name suggests, we imagine that the `sram` output section must be allocated within `staticram`, which is a bank of static RAM.

   • The MEMORY part of the script specifies where the bank of static RAM is, within the processor address space, and how big it is. The linker makes use of the last item of information to detect and report any attempt to overflow the memory bank capacity. For instance, the following specification:

```
MEMORY {
  staticram: ORIGIN=0x1000, LENGTH=0x200
}
```

   states that the `staticram` memory bank starts at address `0x1000` and can hold up to `0x200` bytes.

The `weak` attribute requires the same kind of information propagation through the toolchain, but it has a simpler meaning. As shown in Figure 9.3, when this attribute is attached to an object within a source file, the compiler will mark it in the same way in the corresponding object file.

In the link phase, if the linker has to choose between referring to a "normal" object or a "weak" one, it will invariably prefer the first one and disregard the second, without emitting any error or warning message about duplicate symbols.

The main use of the mechanism just described is to provide a default version of a function or a data structure—possibly within a library—but still provide the ability to override it and provide a different implementation elsewhere—for instance, in the application code. An example of use of this technique, in the context of interrupt handling, has been given in Chapter 8.

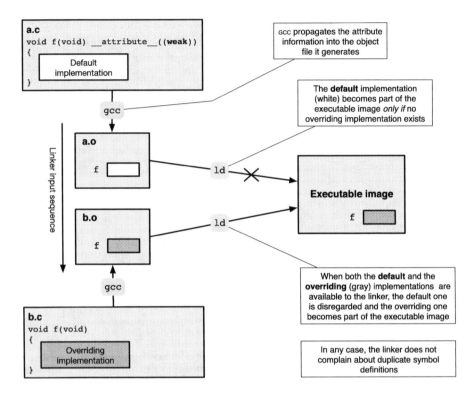

Figure 9.3　Meaning of the weak attribute.

9.4.2　ASSEMBLY LANGUAGE INSERTS

Assembly language *inserts* are an extension to the C language that, as their name says, allows programmers to insert a fragment of assembly language code into the body of a function. They are inherently non-portable, because the assembly language used to write the inserts themselves is highly dependent on the target processor architecture.

On the other hand, they are also extremely useful to express concepts that are not supported by high-level programming languages without introducing a full-fledged assembly language module in the software project.

For instance, when performing a context switch it is necessary to save and restore all processor registers to and from a memory area. How can we do this in the C language that—like virtually all other high-level programming languages—does not even make processor registers visible to the programmer?

Moreover, including an assembly language insert within a function rather than putting it in an assembly language module on its own and calling it, is often beneficial for what concerns execution performance and code size.

From the syntax point of view, assembly language inserts are enclosed in an "envelope," introduced by the keyword __asm__, delimited by a pair of parentheses, and closed by a semicolon, as shown below.

```
__asm__ (
  <assembly_language_insert>
);
```

Depending on the compiler version and command-line options, a pair of alternate keywords (namely, asm and __asm) may also be accepted.

Within the envelope, the insert is composed of four parts:

1. A sequence of assembly instructions within a character string. They must be written according to the syntax of the assembly language of the target processor, described in the assembler documentation [55], except for what concerns *operands*. In fact, the compiler provides a mechanism to automatically "map" C language expressions and variables into assembly instruction operands.

   In this way it becomes possible, for instance, to use the result of an expression as an operand of an assembly instruction, or store the result of an assembly instruction into a C variable, without guessing which registers or memory locations the compiler has reserved to this purpose.

2. A list of zero or more *output* operand descriptions, separated from the previous part by means of a colon (:). If there is more than one operand description, they must be separated by commas.

   Each operand description refers to an output operand and is composed of:
   - An optional symbolic operand *name*, delimited by brackets.
   - An operand *constraint* specification, a character string that provides key information about the operand itself, for instance, its data type and its storage location.
   - The name of a C variable or expression that corresponds to the operand, between parentheses. In order to be valid, the variable or expression must be a legal *lvalue*, that is, it must be a legal target of a C assignment operator.

3. A list of zero or more *input* operand descriptions, with the same syntax as the output operand descriptions. As before, the first input operand description must be separated from the last output operand description by means of a colon. Unlike for output operands, the C expression associated with an input operand description may not be a valid *lvalue*.

4. A list of comma-separated character strings, called *clobber* list. It indicates which processor registers are manipulated by the assembly instructions, even though they are not mentioned in the above-mentioned operand descriptions. The compiler will ensure that these registers do not contain any useful information when the execution of the assembly language insert begins, because it is assumed that the insert destroys their contents.

   As before, the first element of this list must be separated from the last input operand description by means of a colon.

In order to establish the association between an assembly instruction operand and the corresponding operand description, the assembly instruction operand is expressed by means of a string like %[<name>], in which <name> is the symbolic name given to the operand in the output or input descriptions.

An alternate, older syntax is also possible, in which operands are referred to by *number* instead of by *name*. In this case, assembly instructions operands are expressed as %<number>, where <number> is the operand number, and symbolic operand names become unnecessary. Operand numbers are assigned according to the order in which they appear in the operand descriptions, starting from 0.

As an example let us consider the following assembly language insert, taken from [160]:

```
__asm__ (   "fsinx %[angle],%[output]"
         : [output] "=f" (result)
         : [angle] "f" (angle)
);
```

- The insert contains one assembly instruction, fsinx, which calculates the sine of an angle and has two operands. Namely, the first operand is the angle and the second one is the result.
  In the example, these operands are mapped to two C-language expressions, with symbolic names [angle] and [output], respectively.
- After the first colon, there is a single output operand description that refers to operand [output]. The description specifies, according to the constraint specification "=f", that it is a write-only (=) floating-point (f) operand. The operand corresponds to the C variable result.
- After the second colon, there is the description of the input operand [angle]. It is a real-only floating-point operand and corresponds to the C variable angle.

Table 9.7 lists a subset of the characters that can appear in a constraint specification. It should be noted that many of them are architecture-dependent, because their goal is to specify the key properties of interest of an operand from the point of view of the assembly instruction that makes use of it.

For this reason, in the table we only list the most commonly used ones, which are available on most architectures supported by GCC. Interested readers should refer to the compiler documentation [159, 160] to obtain a comprehensive list of constraint letters and modifiers pertaining to the specific architecture they are working on.

The compiler makes use of the constraints specifiers to generate the code that surrounds the assembly language insert and ensure that all constraints are met or, when this is impossible, emit an error message. However, it is worth noting that the checks the compiler can perform are strictly limited to what is specified in the operand descriptions, because it parses the first part of the assembly language insert (that is, the list of assembly instructions) in a very limited way.

For instance, the compiler is unaware of the meaning of assembly instructions themselves and cannot ensure they are given the right number of arguments.

Table 9.7
Main Assembly Operand Constraints Specifiers

Name	Purpose
Simple constraints	
m	The operand can be any memory-resident operand, reachable by means of any addressing mode the processor supports.
r	No constraints are posed upon the type of operand, but the operand must be in a general-purpose register.
i	The operand must be an immediate integer, that is, an operand with a constant integer value. It should be noted that an operand is still considered immediate, even though its value is unknown at compile time, provided the value becomes known at assembly or link time.
g	The operand can be either in memory or in a general-purpose register; immediate integer operands are allowed, too, but the operand cannot be in a register other than a general-purpose register.
p	The operand must be a valid memory address, which corresponds to a C-language pointer.
f	The operand must be a floating-point value, which is stored in a floating-point register.
X	No constraints are posed on the operand, which can be any operand supported by the processor at the assembly language level.
Constraint modifiers	
=	The operand is a write-only operand for the instruction, that is, its previous value is not used and is overwritten by the instruction with output data.
+	The operand is both an input and an output operand for the instruction, that is, the instruction uses the previous operand value and then overwrites it with output data; operands that are not marked with either = or + are assumed to be input-only operands.
&	Indicates that the instruction modifies the operand before it is finished using the input operands; as a consequence, the compiler must ensure that the operand does not overlap with any input operands to avoid corrupting them.

Referring again to Table 9.7, a constraint specification is composed of one or more simple constraint characters, and zero or more constraint modifiers. Some modifier characters must appear in specific parts of the specification. For instance = or +, when needed, must be the very first character of the specification itself.

In addition, an input operand constraint specification may also refer to an output operand by using either an operand name between brackets or an operand number. This gives rise to a so-called *matching* constraint, which specifies that a certain input operand must reside at exactly the same location as an output operand when the execution of the assembly language insert begins. This cannot be ensured by merely using the same C language expression for both operands, due to code optimization.

The very last part of an assembly language insert, that is, the clobber list, plays a very important role that should not be forgotten for what concerns code correctness. In fact, assembly language inserts often appear "in between" C-language statements. In order to generate the surrounding code correctly, the compiler must be made aware of all their possible *side effects*. This is especially important when the compiler, as usually happens, is also asked to optimize the code. More specifically:

- The most immediate way to indicate side effects is through output operand specifications. Each specification explicitly indicates that the assembly language insert may modify the operand itself, thus destroying its previous contents.
- Some assembly instructions make implicit use of some registers and destroy their previous contents even though, strictly speaking, they are not used as operands. These registers must be specified in the clobber list, to prevent the compiler from storing useful information into them *before* the assembly language insert and use their content *afterward*, which would obviously lead to incorrect results.
- In many architectures, some assembly language instructions may modify *memory* contents in a way that cannot be described by means of output operands. This fact must be indicated by adding the keyword memory to the clobber list. When this keyword is present, the compiler will not cache memory values in registers and will not optimize memory load and store operations across the assembly language insert.
- Virtually all processor architectures have a *status* register that holds various pieces of information about the current processor state, which can affect future instruction execution.

  For instance, this register is called program status register (xPSR) on the Cortex-M3 architecture [8, 9] and is divided into 3 subregisters. Among those subregisters, the application program status register (APSR) holds a set of flags that arithmetic and test instructions may set to indicate their outcome.

  The processor uses these flags at a later time to decide whether or not conditionally executed instructions must be executed and whether or not conditional branch instructions will be taken.

  In GCC terminology, these pieces of information are called *condition codes*. If the assembly language insert can modify these condition codes, it is important to make the compiler aware of this fact by adding the keyword cc to the clobber list.

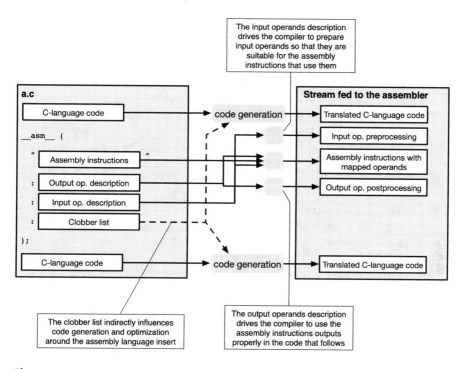

Figure 9.4 Handling of assembly language inserts by the compiler.

- Last, but not least, it is important to remark that the compiler applies its ordinary optimization rules to the assembly language insert, considering it as an indivisible unit of code. In particular, it may opt for *removing* the insert completely if (according to its knowledge) it can prove that none of its outputs are used in the code that follows it.

  As discussed previously, a common use of output operands is to describe side effects of the assembly language inserts. In this case, it is natural that these output operands are otherwise unused in the code. If, as it happens in some cases, the assembly language insert must be retained in any case, because it has other important side effects besides producing output, it must be marked with the `volatile` keyword, as shown below:

```
__asm__ volatile(
  <assembly_language_insert>
);
```

An insert marked with `volatile` may be removed by the compiler only if it can prove that it is *unreachable*, that is, there is no execution path that may lead to it.

To conclude this section, Figure 9.4 summarizes in a graphical form how the compiler handles assembly language inserts.

9.5   SUMMARY

The focus of this chapter was on *portable software*. As shown in Section 9.1, this is a complex topic that involves virtually all software components of an embedded system, from the interface between software and hardware, up to the application programming interfaces adopted to connect software modules together.

In order to reach the goal, it is first of all important to realize where the main sources of portability issues in C-language software development may be and how the compiler and other tools can help programmers to detect them as early as possible. This was the topic of Section 9.2, whereas the all-important aspect of choosing the right application programming interfaces was briefly considered in Section 9.3.

However, especially when working on low-level software component, such as device drivers, resorting to non-portable constructs of the C language is sometimes highly convenient, or even unavoidable. Section 9.4 provided an overview of the features provided, in this respect, by the GCC-based open-source toolchain.

To conclude the discussion, a real-world example of porting layer to achieve architecture-independence, and hence, portability at the operating system level will be presented in the next chapter.

10 The FREERTOS Porting Layer

In this chapter we will have a glance at how a typical porting layer is designed and built. The example taken as a reference is the FREERTOS porting layer, which contains all the architecture-dependent code of the operating system.

For what concerns device driver development, another important method to achieve portability and hardware independence, a comprehensive example has been presented and discussed as part of Chapter 8.

10.1 GENERAL INFORMATION

As was outlined in the previous chapter, to enhance their portability to different processor architectures, software development systems, and hardware platforms, most modern operating systems, including FREERTOS, specify a well-defined interface between the operating system modules that do not depend on the architecture, and the architecture-dependent modules, often called porting layer or, even more specifically, hardware abstraction layer (HAL).

As their name suggests, those modules must be rewritten when the operating system is ported to a new architecture, and must take care of all its peculiarities. Moreover, they often include driver support for a limited set of devices needed by the operating system itself and the language support library. For instance, FREERTOS needs a periodic timer interrupt to work properly.

As an example, we will now discuss the main contents of the FREERTOS porting layer, referring specifically to the ARM Cortex-M3 port of FREERTOS when

Table 10.1
Object-Like Macros to Be Provided by the FREERTOS Porting Layer

Group/Name	Purpose
Basic data types	
portCHAR	8-bit character
portFLOAT	Single-precision, 32-bit floating point
portDOUBLE	Double-precision, 64-bit floating point
portLONG	Signed, 32-bit integer
portSHORT	Signed, 16-bit integer
portSTACK_TYPE	Task stack elements, used to define StackType_t
portBASE_TYPE	Most efficient signed integer, used to define BaseType_t
Time representation	
portMAX_DELAY	Highest value that can be represented by a TickType_t, special value used to indicate an infinite delay.
portTICK_PERIOD_MS	Tick period, approximated as an integral number of milliseconds.
Architectural details	
portSTACK_GROWTH	Direction of stack growth, −1 is downward (toward lower memory addresses), +1 is upward (toward higher addresses).
portBYTE_ALIGNMENT	Alignment required by critical data structures, namely, task stacks and dynamically allocated memory.

concrete examples and code excerpts are needed. More information about this architecture can be found in References [8, 9].

The FREERTOS version taken as a reference is V8.0.1. Since this operating system is still evolving at a fast pace, there may be minor differences in the porting layer when referring to other versions of the source code.

When another architecture is needed as a comparison and to show how the porting layer shall be adapted to a different processor architecture, we will also make reference to the port of the same operating system for the Coldfire-V2 family of microcontrollers [66].

Both families are typical representatives of contemporary, low-cost components for embedded applications and, at the same time, they are simple enough so that the reader can gain a general understanding of how the porting layer works without studying them in detail beforehand. The example has a twofold goal:

1. It complements the higher-level information about FREERTOS provided in Chapter 5, showing how abstract concepts—for instance, context switch—are mapped into concrete sequences of machine instructions for a certain processor architecture.

Table 10.2
Function-Like Macros to Be Provided by the FREERTOS Porting Layer

Group/Name	Purpose
Context switch requests	
portYIELD	Requests a context switch to be performed at the earliest opportunity, that is, as soon as the processor is not handling an interrupt.
portYIELD_WITHIN_API	Same semantics as portYIELD, but called from within FREERTOS itself; if not explicitly defined by the porting layer it defaults to portYIELD.
portYIELD_FROM_ISR	Conditionally requests a context switch from an interrupt service routine, depending on the value of its Boolean argument.
portEND_SWITCHING_ISR	Alternate definition of portYIELD_FROM_ISR provided by some ports.
Critical sections and interrupt handling	
portSET_INTERRUPT_MASK_FROM_ISR	Sets interrupt mask from an *interrupt service routine* context to disable interrupt sources whose handler may interact with FREERTOS, returns previous value.
portCLEAR_INTERRUPT_MASK_FROM_ISR	Resets interrupt mask to a previous value from an interrupt service routine context, the value is usually obtained from portSET_INTERRUPT_MASK_FROM_ISR.
portDISABLE_INTERRUPTS	Disables interrupt sources whose handler may interact with FREERTOS, called from a *task* context.
portENABLE_INTERRUPTS	Enables all interrupt sources, called from a task context.
portENTER_CRITICAL	Opens a critical region delimited by disabling interrupts.
portEXIT_CRITICAL	Closes a critical region opened by portENTER_CRITICAL, possibly re-enabling interrupts depending on the current critical region nesting level.

2. It illustrates the typical structure and contents of a porting layer and gives a summary idea of the amount of effort it takes to write one.

The bulk of the port to a new architecture is done by defining a set of C preprocessor macros (listed in Tables 10.1 and 10.2), C data types (listed in Table 10.3), and functions (listed in Table 10.4), pertaining to different categories. Figure 10.1 summarizes them graphically. Besides public functions, which are mandatory and must be provided by all porting layers, Table 10.4 also lists several *private* porting layer

Table 10.3
Data Types to Be Provided by the FREERTOS Porting Layer

Group/Name	Purpose
Basic data types	
StackType_t	Task stack elements
BaseType_t	Most efficient signed integer data type
UBaseType_t	Unsigned variant of BaseType_t
Time representation	
TickType_t	Relative or absolute time value

Table 10.4
Functions to Be Provided by the FREERTOS Porting Layer

Group/Name	Purpose
Public functions	
pxPortInitialiseStack	Initialize a task stack and prepare it for a context switch.
xPortStartScheduler	Perform all architecture-dependent activities needed to start the scheduler.
Private functions (Cortex-M3)	
vPortYield	Request a context switch at the earliest opportunity, by means of a PendSV software interrupt.
xPortPendSVHandler	PendSV interrupt handler that implements context switch, that is, the sequence: context save, next task selection (scheduling), context restoration.
vPortEnterCritical	Implementation of portENTER_CRITICAL.
vPortExitCritical	Implementation of portEXIT_CRITICAL.
ulPortSetInterruptMask	Implementation of portSET_INTERRUPT_MASK_FROM_ISR.
vPortClearInterruptMask	Implementation of portCLEAR_INTERRUPT_MASK_FROM_ISR.
prvPortStartFirstTask	Start the first task selected by the operating system for execution, by context switching to it, requests a SVC software interrupt.
vPortSVCHandler	SVC interrupt handler, performs a context restoration to start the first task in the system.
vPortSetupTimerInterrupt	Initialize and start the tick timer and enable it to interrupt the processor.
xPortSysTickHandler	Handler for the tick timer interrupt.

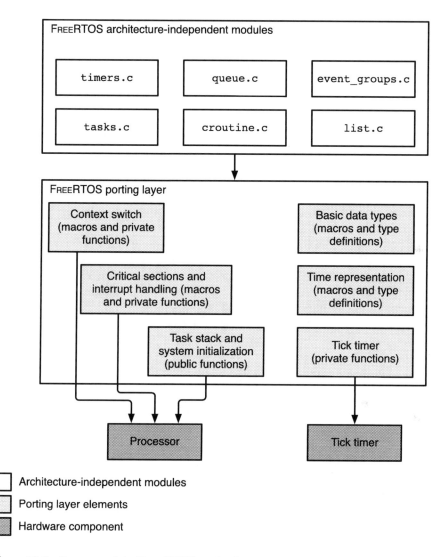

Figure 10.1 Summary of the FREERTOS porting layer.

functions described in the following. They are used as "helpers" in the Cortex-M3 porting layer and may or may not be present in other ports.

Macros and data types are defined in the header file `portmacro.h`. When the operating system is compiled, the contents of this file are incorporated by means of an `#include` directive contained in the FREERTOS header file `portable.h` and, in turn, made available for use to the operating system source code.

10.2 BASIC DATA TYPES

The first thing to be found in `portmacro.h` is a set of macros that maps some basic data types required by FREERTOS into appropriate data types supported by the compiler and the underlying processor architecture.

This is done because the C language standard does not define precisely the width in bits of predefined data types, such as `short` and `int`. Hence, for instance, the size of an `int` may change from one architecture to another and, within the same architecture, from one compiler to another.

By contrast, the FREERTOS code sometimes needs to define variables having a well-defined, fixed width across all supported architectures and compilers. For example, the FREERTOS data type `portLONG` must be a signed, 32-bit integer regardless of the architecture. Hence, it will probably be mapped to `int` on a 32-bit processor and to `long` on a 16-bit processor.

Strictly speaking, it should also be noted that these definitions are no longer used directly by FREERTOS since version 8. In fact, with the adoption of the C99 international standard [89] for the C language, the toolchain-supplied `stdint.h` header provides equivalent functionality and FREERTOS now makes use of it. For instance, the `uint32_t` data type, defined in `stdint.h`, always corresponds to an unsigned, 32-bit integer. However, the macro definitions listed in Table 10.1 are still retained in most porting layers for backward compatibility with existing code.

Two additional macro definitions convey additional information about the underlying architecture.

- The `portBASE_TYPE` macro must be defined as the most "natural" and efficient *signed* integer data type supported by the architecture, which usually corresponds to a machine word. Moreover, it must be equivalent to the type definition of `BaseType_t` and its unsigned variant `UBaseType_t`, which must also be provided by the porting layer.

  The `BaseType_t` and `UBaseType_t` data types are widely used across the FREERTOS source code. As described in Chapter 5, `BaseType_t` is also used as return value by most operating system functions.

  As for the previous macros, recent versions of FREERTOS make use only of the `BaseType_t` and `UBaseType_t` types, which are essentially defined as

  ```
  typedef BaseType_t portBASE_TYPE;
  typedef UBaseType_t unsigned portBASE_TYPE;
  ```

  The macro definition is kept for backward compatibility.
- The `portSTACK_TYPE` macro and the corresponding data type `StackType_t` are used as the base type to allocate task stacks. Namely, task stacks are allocated as an integral number of elements of type `StackType_t` and its correct definition is crucial for correct stack alignment.

  It must also be noted that also the stack size is expressed in terms of `StackType_t` elements, rather than bytes. This has an application-visible side effect when determining how big task stacks should be, in order to call `xTaskCreate` with a sensible value for the `usStackDepth` argument, as discussed in Chapter 5.

10.3 TIME REPRESENTATION AND ARCHITECTURAL DETAILS

Then, the porting layer must define that data type used by FREERTOS to represent both absolute and relative time. This data type is called `TickType_t` and holds an integral number of *ticks*, the minimum time unit that the operating system is able to represent. Its definition must depend on the FREERTOS configuration options `configUSE_16_BIT_TICKS`, which defines whether time shall be represented on 16 or 32 bits.

Although, strictly speaking, the definition of `TickType_t` is part of the porting layer, in most cases it is possible to make use of standard data types defined in `stdint.h` to obtain an architecture-independent definition for them or, better, shift the burden of defining the underlying data types in an appropriate way to the toolchain.

For instance, in both porting layers considered in this example, `TickType_t` is defined as follows.

```
#if( configUSE_16_BIT_TICKS == 1 )
        typedef uint16_t TickType_t;
        #define portMAX_DELAY ( TickType_t ) 0xffff
#else
        typedef uint32_t TickType_t;
        #define portMAX_DELAY ( TickType_t ) 0xffffffffUL
#endif
```

As can be seen, in practice the definition proceeds exactly as explained above. Namely, It depends on the value of `configUSE_16_BIT_TICKS` and relies on the `uint16_t` and `uint32_t` data types from `stdint.h`.

Since the definition of `TickType_t` affects the range of relative delay that can be represented by the operating system and used by applications, the fragment of code above also defines the `portMAX_DELAY` macro accordingly. This macro must correspond to the highest numeric value that the `TickType_t` can hold and is used to represent an infinite delay.

The last macro related to time representation to be defined by the porting layer is `portTICK_RATE_MS`. It must be set to the tick length, or period, expressed as an integral number of milliseconds. Hence, it depends on the FREERTOS configuration variable `configTICK_RATE_HZ` that establishes the tick frequency, in Hertz.

As before, even though this definition belongs to the porting layer, it can be given in an architecture-independent way most of the time, for instance, as follows.

```
#define portTICK_PERIOD_MS ( ( TickType_t ) 1000 / configTICK_RATE_HZ )
```

As a side note, it it useful to remark that this value is an integer, and hence, it may not accurately represent the actual tick length when `configTICK_RATE_HZ` is not an integral sub-multiple of 1000.

The next two macros listed in Table 10.1 provide additional details about the underlying architecture. Namely:

- The macro `portSTACK_GROWTH` can be defined as either -1 or $+1$ to indicate that, on this architecture, stacks grow downward (toward lower memory addresses) or upward (toward higher addresses), respectively.
- The macro `portBYTE_ALIGNMENT` determines the alignment of some critical objects within the operating system, most notably task stacks and dynamically allocated memory. Namely, these objects are aligned to a multiple of `portBYTE_ALIGNMENT` bytes.

For what concerns the second macro, it is worth remarking the first important difference between the two porting layers been considered in this example because, according to the corresponding data sheets [9, 66], the two architectures have different alignment requirements. More specifically, the Cortex-M3 port defines this macro as 8, whereas the Coldfire-V2 port defines it as 4.

10.4   CONTEXT SWITCH

The next group of macro definitions contains the implementation of the architecture-dependent part of the *context switch*, a key operating system's mechanism described in Chapter 4 and Chapter 8, from different points of view.

In particular, the macro `portYIELD`, when invoked, must direct the system to perform a context switch from the current task to a new one chosen by the scheduling algorithm as soon as possible, that is, as soon as the processor is not handling any interrupt.

In the Cortex-M3 port, this activity is delegated to the architecture-dependent function `vPortYield`, as shown in the code fragment below. The macros `portNVIC_INT_CTRL_REG` and `portNVIC_PENDSVSET_BIT` are used only within the porting layer, in the implementation of `vPortYield`, which will be discussed next.

```
extern void vPortYield( void );
#define portNVIC_INT_CTRL_REG  ( * ( ( volatile uint32_t * ) 0xe000ed04 ) )
#define portNVIC_PENDSVSET_BIT ( 1UL << 28UL )
#define portYIELD() vPortYield()
```

The `portmacro.h` header only contains data type and macro definitions. We have just seen that, in some cases, they map macro names used by FREERTOS, like `portYIELD`, into architecture-dependent function names, like `vPortYield`.

The implementation of those functions—along with other functions required by FREERTOS and to be discussed later—is done in at least one source module belonging to the porting layer and usually called `port.c`.

Namely, the implementation of `vPortYield` found in that file is:

```
void vPortYield( void )
{
        /* Set a PendSV to request a context switch. */
        portNVIC_INT_CTRL_REG = portNVIC_PENDSVSET_BIT;

        /* Barriers are normally not required but do ensure the code
           is completely within the specified behaviour for the
           architecture. */
        __asm volatile( "dsb" );
```

```
        __asm volatile( "isb" );
}
```

As can be seen, on the Cortex-M3 architecture a context switch is requested by setting a software interrupt request controlled by the `portNVIC_PENDSVSET_BIT` bit of the interrupt controller's control register `portNVIC_INT_CTRL_REG`. Both the register address and the value to be written into it are determined by means of port-dependent macro definitions given, as shown previously, in `portmacro.h`.

The next two instructions are *barriers*, which force the processor to commit the previous store operation (and hence, actually write into the interrupt controller's register) before proceeding with the instructions that follow. More specifically, `dsb` is a *data synchronization barrier* and stalls the processor until all memory or I/O space operations that precede it complete.

On the other hand, `isb` is an *instruction synchronization barrier*, which flushes the processor pipeline, so that all the following instructions are fetched again after `isb` itself completes.

The Coldfire-V2 porting layer shows a remarkable similarity—even though the implementation details differ—because it contains the following fragments of code, in `portmacro.h`.

```
1   #define portNOP()        asm volatile ( "nop" )
2
3   #define portYIELD() \
4       MCF_INTC0_INTFRCL = ( 1UL << configYIELD_INTERRUPT_VECTOR ); \
5       portNOP(); portNOP()
```

In particular:

- The context switch is still requested by setting a software interrupt request.
- In this case, the interrupt vector used by the interrupt request is programmable by means of a port-dependent configuration macro, `configYIELD_INTERRUPT_VECTOR`.
- Instead of barriers, two no-operation (`nop`) instructions are inserted after manipulating the interrupt controller's register to ensure that the software interrupt request is committed before any further, real instructions are executed.

On some architectures, the fine details of code to be executed to request a context switch depend on the execution context it is invoked from. Namely, three cases are possible and they are supported by different macros.

1. The context switch is requested from an application task and, in this case, the macro `portYIELD` just described is used.
2. The context switch is requested from within a FREERTOS function, using the macro `portYIELD_WITHIN_API`.
3. The context switch is requested from an interrupt service routine, using the macro `portYIELD_FROM_ISR` or, in some ports, the macro `portEND_SWITCHING_ISR`, which has the same semantics.

In the Cortex-M3 port, the macro `portYIELD_WITHIN_API` is left undefined and the operating system automatically reverts to use `portYIELD`. Instead, `portYIELD_FROM_ISR` and `portEND_SWITCHING_ISR` are implemented as follows.

```
1   #define portEND_SWITCHING_ISR( xSwitchRequired ) \
2          if( xSwitchRequired ) portNVIC_INT_CTRL_REG = portNVIC_PENDSVSET_BIT
3   #define portYIELD_FROM_ISR( x ) portEND_SWITCHING_ISR( x )
```

As can be seen, these macros contain a "simplified" version of `vPortYield` to achieve a greater execution efficiency because, in this case, no barrier instructions are needed. The Coldfire-V2 port is very similar and is not shown here for conciseness.

On both architectures, rescheduling is performed by an interrupt handler triggered by a software interrupt request. The Cortex-M3 implementation is listed in the following.

```
1   void xPortPendSVHandler( void ) __attribute__ (( naked ));
2
3   void xPortPendSVHandler( void )
4   {
5       /* This is a naked function. */
6
7       __asm volatile
8       (
9       "       mrs r0, psp                             \n"
10      "       isb                                     \n"
11      "                                               \n"
12      "       ldr     r3, pxCurrentTCBConst           \n"
13      "       ldr     r2, [r3]                        \n"
14      "                                               \n"
15      "       stmdb r0!, {r4-r11}                     \n"
16      "       str r0, [r2]                            \n"
17      "                                               \n"
18      "       stmdb sp!, {r3, r14}                    \n"
19      "       mov r0, %0                              \n"
20      "       msr basepri, r0                         \n"
21      "       bl vTaskSwitchContext                   \n"
22      "       mov r0, #0                              \n"
23      "       msr basepri, r0                         \n"
24      "       ldmia sp!, {r3, r14}                    \n"
25      "                                               \n"
26      "       ldr r1, [r3]                            \n"
27      "       ldr r0, [r1]                            \n"
28      "       ldmia r0!, {r4-r11}                     \n"
29      "       msr psp, r0                             \n"
30      "       isb                                     \n"
31      "       bx r14                                  \n"
32      "                                               \n"
33      "       .align 2                                \n"
34      "pxCurrentTCBConst: .word pxCurrentTCB          \n"
35      ::"i"(configMAX_SYSCALL_INTERRUPT_PRIORITY)
36      );
37  }
```

When the processor eventually honors the software interrupt request, it automatically saves part of the execution context onto the task stack, namely, the program status register (`xPSR`), the program counter and the link register (`PC` and `LR`), as well as several other registers (`R0` to `R3` and `R12`). Then it switches to a dedicated operating system stack and starts executing the exception handling code, `xPortPendSVHandler`.

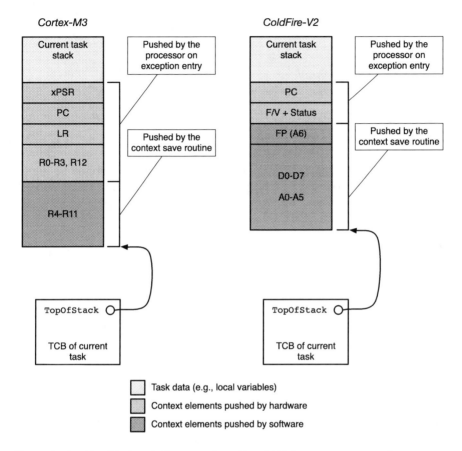

Figure 10.2 Simplified stack diagrams after a FREERTOS context save operation.

The handler first retrieves the task stack pointer PSP and stores it in the R0 register (line 9). This does not clobber the task context because R0 has already been saved onto the stack by hardware. Then, it puts into R2 a pointer to the current TCB taken from the global variable pxCurrentTCB (lines 12–13).

The handler is now ready to finish the context save initiated by hardware by pushing onto the task stack registers R4 through R11 (line 15). At last, the task stack pointer in R0 is stored into the TopOfStack field of the task control block (TCB), which is dedicated to this purpose (line 16). At this point, the stack layout is as shown on the left of Figure 10.2.

In particular,

- the stack pointer currently used by the processor, SP, points to the operating system stack;

- the PSP register points to where the top of the task stack was after exception entry, that is, below the part of task context saved automatically by hardware;
- the TopOfStack field of the current task TCB points to the top of the task stack after the context save has been concluded.

Going back to the listing of xPortPendSVHandler, the function now invokes the operating system scheduling algorithm, that is, the function vTask-SwitchContext (lines 18–24). To avoid race conditions, interrupt sources whose handler may interact with FREERTOS are disabled during the execution of this function by setting the processor base priority mask basepri [9] appropriately. The role of this register will be better described in Section 10.5 while discussing interrupt handling.

The main effect of vTaskSwitchContext is to update pxCurrentTCB so that it points to the TCB of the task to be executed next. Hence, xPortPendSVHandler dereferences pxCurrentTCB again (line 26) to get a pointer to the new TCB. From there, it extracts the TopOfStack field and stores it into R0 (line 27). Using R0 as a stack pointer, the function pops registers R4 through R11, that is, the part of context previously saved by software, from the stack of the new task (line 28). After that, the updated stack pointer is stored into the task stack pointer register PSP (line 29).

The last step of context restoration is performed by asking the hardware to restore the remaining part of the task context, which was automatically saved on exception entry. This is done by the bx instruction (line 31). The last action also restores the task PC, and thus execution continues from where it left off when the context was saved.

The ColdFire-V2 implementation of context switch functions took a different approach. Namely, instead of using assembly language inserts within C functions—as is done in the Cortex-M3 port—a sharp separation has been kept between assembly code (that has been put in the portasm.S source module) and C code (that resides in port.c).

The assembly part of the context switch function is shown below.

```
1    .macro portSAVE_CONTEXT
2
3            lea.l           (-60, %sp), %sp
4            movem.l         %d0-%fp, (%sp)
5            move.l          pxCurrentTCB, %a0
6            move.l          %sp, (%a0)
7
8            .endm
9
10   .macro portRESTORE_CONTEXT
11
12           move.l          pxCurrentTCB, %a0
13           move.l          (%a0), %sp
14           movem.l         (%sp), %d0-%fp
15           lea.l           %sp@(60), %sp
16           rte
17
18           .endm
19
20   __cs3_isr_interrupt_80:
```

```
21              portSAVE_CONTEXT
22              jsr vPortYieldHandler
23              portRESTORE_CONTEXT
```

It consists of two macros, which are instantiated within __cs3_isr_-interrupt_80. In turn, this code fragment is the handler of the software interrupt used for context switching. The execution of the software interrupt handler is triggered by the invocation of the portYIELD() macro, discussed previously.

In particular, the code corresponding to portSAVE_CONTEXT:

- Reserves space on the task stack by subtracting 60 (actually, adding −60) to the task stack (line 3).
- Moves the processor registers to be saved into the stack area just reserved by means of a move multiple instruction (line 4). This instruction saves all data registers, from D0 to D7, as well as address registers from A0 to fp, that is, to A6. Address register A7 must not be saved at this stage because, on this processor architecture, it contains the stack pointer sp.
- Loads the address of the current TCB into register A0 and stores the value of the stack pointer into its first word (lines 5–6).

On the other hand, in order to restore the task context, portRESTORE_CONTEXT proceeds in the opposite way:

- It loads the address of the TCB of the task to be restored and retrieves the stack pointer saved into its first word (lines 12–13).
- It restores processor registers from the stack by means of a move multiple instruction (line 14).
- It adjusts the stack pointer to release the memory area reserved when the context was saved (line 15)

In between, the interrupt handler calls the C function vPortYieldHandler to perform the higher-level portion of a context switch, that is, the selection of the next task to be executed. The contents of vPortYieldHandler are listed in the following.

```
1   void vPortYieldHandler( void )
2   {
3           uint32_t ulSavedInterruptMask;
4
5           ulSavedInterruptMask = portSET_INTERRUPT_MASK_FROM_ISR();
6               MCF_INTC0_INTFRCL = 0;
7               vTaskSwitchContext();
8           portCLEAR_INTERRUPT_MASK_FROM_ISR( ulSavedInterruptMask );
9   }
```

As can be seen, the similarities between the two context switch procedures on the Cortex-M3 and ColdFire-V2 are still remarkably strong, despite the architectural differences.

10.5   INTERRUPT HANDLING AND CRITICAL REGIONS

The last set of port-dependent definitions found in `portmacro.h` are concerned with in-kernel critical section management and interrupt handling.

```
1   extern void vPortEnterCritical( void );
2   extern void vPortExitCritical( void );
3   extern uint32_t ulPortSetInterruptMask( void );
4   extern void vPortClearInterruptMask( uint32_t ulNewMaskValue );
5   #define portSET_INTERRUPT_MASK_FROM_ISR() ulPortSetInterruptMask()
6   #define portCLEAR_INTERRUPT_MASK_FROM_ISR(x) vPortClearInterruptMask(x)
7   #define portDISABLE_INTERRUPTS() ulPortSetInterruptMask()
8   #define portENABLE_INTERRUPTS() vPortClearInterruptMask(0)
9   #define portENTER_CRITICAL() vPortEnterCritical()
10  #define portEXIT_CRITICAL() vPortExitCritical()
```

As usual, the implementation of the functions referenced by the macros above is found in `port.c`, as shown in the following.

```
1   static UBaseType_t uxCriticalNesting = 0xaaaaaaaa;
2
3   void vPortEnterCritical( void )
4   {
5           portDISABLE_INTERRUPTS();
6           uxCriticalNesting++;
7           __asm volatile( "dsb" );
8           __asm volatile( "isb" );
9   }
10
11  void vPortExitCritical( void )
12  {
13          configASSERT( uxCriticalNesting );
14          uxCriticalNesting--;
15          if( uxCriticalNesting == 0 )
16          {
17                  portENABLE_INTERRUPTS();
18          }
19  }
20
21  __attribute__(( naked )) uint32_t ulPortSetInterruptMask( void )
22  {
23          __asm volatile                                              \
24          (                                                           \
25          "       mrs r0, basepri                             \n"  \
26          "       mov r1, %0                                  \n"  \
27          "       msr basepri, r1                             \n"  \
28          "       bx lr                                       \n"  \
29          :: "i" ( configMAX_SYSCALL_INTERRUPT_PRIORITY ) : "r0", "r1" \
30          );
31
32          /* This return will not be reached but is necessary to
33             prevent compiler warnings. */
34          return 0;
35  }
36
37  __attribute__(( naked ))
38  void vPortClearInterruptMask( uint32_t ulNewMaskValue )
39  {
40          __asm volatile                                              \
41          (                                                           \
42          "       msr basepri, r0                             \n"  \
43          "       bx lr                                       \n"  \
44          :::"r0"                                                     \
45          );
46
```

```
47        /* Just to avoid compiler warnings. */
48        ( void ) ulNewMaskValue;
49    }
```

The macros portSET_INTERRUPT_MASK_FROM_ISR and portCLEAR_-
INTERRUPT_MASK_FROM_ISR are used by FREERTOS in order to manipulate the
processor interrupt mask from an *interrupt service routine* context. Namely:

- portSET_INTERRUPT_MASK_FROM_ISR sets the interrupt mask so that
  all interrupt sources whose handler may invoke FREERTOS primitives are
  disabled, and returns the previous value of the interrupt mask itself.
- portCLEAR_INTERRUPT_MASK_FROM_ISR resets the interrupt mask to
  a value it had previously, passed as an argument and usually obtained as
  return value of portSET_INTERRUPT_MASK_FROM_ISR, thus possibly
  re-enabling interrupts.

In the corresponding implementation, that is, in the bodies of ulPort-
SetInterruptMask and vPortClearInterruptMask, this is accomplished
by manipulating the basepri processor register. The manipulation must be done
with the help of a non-standard assembly language insert (introduced by the GCC-
specific keyword __asm described in Section 9.4) because the basepri register can
be accessed only by means of the specialized msr instruction instead of a standard
mov.

Due to the presence of the assembly language insert, it becomes necessary to
change the way the compiler generates the function entry and exit code, by means of
naked attribute, also described in the same section.

The effect of assigning a non-zero value to basepri is that all interrupt requests
with a priority lower than or equal to either the specified value or the current execution priority of the processor are not honored immediately but stay pending. On the
other hand, setting basepri to zero unconditionally re-enables all interrupt sources.
As a side effect, still taking its current execution priority into account, the processor
will also handle immediately any interrupt request that was left pending.

More specifically, as can be inferred from the code, and as specified
in the FREERTOS documentation, portSET_INTERRUPT_MASK_FROM_ISR
sets the interrupt mask to the priority defined in the configuration macro
configMAX_SYSCALL_INTERRUPT_PRIORITY and returns to the caller the original value of the interrupt mask.

Therefore, depending on the value of this macro, interrupts may or may
not be disabled completely. Interrupt requests with a priority higher than
configMAX_SYSCALL_INTERRUPT_PRIORITY are still handled normally, with
the constraint that their handler must not invoke any FREERTOS function.

The portCLEAR_INTERRUPT_MASK_FROM_ISR macro sets the basepri processor register to a specific value passed as argument. Normally, this value can be:

- Zero, as discussed previously, to re-enable all interrupt sources.
- A value obtained from portSET_INTERRUPT_MASK_FROM_ISR, to restore the interrupt mask to a past value.

In the Cortex-M3 porting layer, the two macros just described are used directly to implement portDISABLE_INTERRUPT and portENABLE_INTERRUPT, which are invoked by FREERTOS to disable and enable interrupts, respectively, from a *task context*. In the general specification of porting layers, the two sets of macros are independent from each other, as this distinction is needed on some architectures.

The last two functions related to interrupt handling, to be defined by the porting layer, are portENTER_CRITICAL and portEXIT_CRITICAL. They are used within FREERTOS to delimit very short critical regions of code that are executed in a task context, and must be protected by disabling interrupts.

Since these critical regions can be nested into each other, it is not enough to map them directly into portDISABLE_INTERRUPTS and portENABLE_INTERRUPTS. If this were the case, interrupts would be incorrectly reenabled at the end of the *innermost* nested critical region instead of the *outermost* one. Hence, a slightly more complex approach is in order.

For the Cortex-M3, as shown in previous listings, the actual implementation is delegated to the functions vPortEnterCritical and vPortExitCritical.

The global variable uxCriticalNesting contains the critical region nesting level of the current task. Its initial value 0xaaaaaaaa is invalid, to catch errors during startup. It is set to zero, its proper value, when the operating system is about to begin the execution of the first task.

The two functions are rather simple: vPortEnterCritical disables interrupts by means of the portDISABLE_INTERRUPTS macro discussed before. Then, it increments the critical region nesting counter because one more critical region has just been entered. The function vPortExitCritical, called at the end of a critical region, first decrements the nesting counter and then reenables interrupts by calling portENABLE_INTERRUPTS only if the count is zero, that is, the calling task is about to exit from the outermost critical region. Incrementing and decrementing uxCriticalNesting does not pose any concurrency issue on a single-processor system because these operations are always performed with interrupts disabled.

It should also be noted that, although, in principle, uxCriticalNesting should be part of each task context—because it holds per-task information—it is not necessary to save it during a context switch. In fact, due to the way the Cortex-M3 port has been designed, a context switch never occurs unless the critical region nesting level of the current task is zero. This property implies that the nesting level of the task targeted by the context switch must be zero, too, because its context has been saved exactly in the same way. Then it is assured that any context switch always saves and restores a critical nesting level of zero, making this action redundant.

10.6   TASK STACK INITIALIZATION

The next porting layer function to be discussed is pxPortInitialiseStack, invoked by FREERTOS when it is creating a new task. It should initialize the new task stack so that its layout is identical to the layout shown in Figure 10.2, that is, the stack layout after a context save operation. In this way, task execution can be started in the most natural way, that is, by simply restoring its execution context.

It takes as arguments the task stack pointer `pxTopOfStack`, the address from which task execution should begin `pxCode`, and a pointer to the task parameter block `pvParameters`. The return value of the function is the new value of the task pointer after the context has been saved. The listing that follows shows the Cortex-M3 implementation.

```
#define portINITIAL_XPSR ( 0x01000000UL )

#ifdef configTASK_RETURN_ADDRESS
        #define portTASK_RETURN_ADDRESS configTASK_RETURN_ADDRESS
#else
        #define portTASK_RETURN_ADDRESS prvTaskExitError
#endif

StackType_t *pxPortInitialiseStack(
    StackType_t *pxTopOfStack, TaskFunction_t pxCode, void *pvParameters )
{
        pxTopOfStack--;
        *pxTopOfStack = portINITIAL_XPSR;           /* xPSR */
        pxTopOfStack--;
        *pxTopOfStack = ( StackType_t ) pxCode; /* PC */
        pxTopOfStack--;
        *pxTopOfStack = ( StackType_t ) portTASK_RETURN_ADDRESS; /* LR */
        pxTopOfStack -= 5;          /* R12, R3, R2 and R1. */
        *pxTopOfStack = ( StackType_t ) pvParameters;   /* R0 */
        pxTopOfStack -= 8;          /* R11, R10, R9, R8, R7, R6, R5 and R4. */

        return pxTopOfStack;
}
```

By comparing the listing with Figure 10.2, it can be seen that the initial context is set up as follows:

- The initial processor status register `xPSR` is the value of the macro `portINITIAL_XPSR`.
- The program counter `PC` comes from the `pxCode` argument.
- The link register `LR` must initially contain the return address of the main task function, that is, the address from which execution must resume when the main task function returns.

  It is set to the value of the `portTASK_RETURN_ADDRESS` macro. In turn, the macro may take two different values, depending on how FREERTOS has been configured.

  1. If the port layer-dependent configuration macro `configTASK_-RETURN_ADDRESS` is set, the value of that macro is taken as the main task return address, thus allowing programmers to override the porting layer default, to be discussed next.
  2. Otherwise, the return address is set to the address of function `prvTaskExitError`, which is defined by the porting layer itself, although it is not listed here for conciseness. That function basically contains an endless loop that locks the task in an active wait and can be caught, for instance, by debuggers.

- Register `R0`, which holds the first (and only) argument of the main task function, points to the task parameter block `pvParameters`.
- The other registers are not initialized.

It is useful to remark that the ColdFire-V2 implementation (listed below) is almost identical, the only differences being dictated by the differences in the stack layout between the two architectures, which was already evident by comparing the two sides of Figure 10.2, as well as from the description of the context switch procedure given earlier.

```
1    #define portINITIAL_FORMAT_VECTOR          ( ( StackType_t ) 0x4000 )
2    #define portINITIAL_STATUS_REGISTER        ( ( StackType_t ) 0x2000 )
3
4    StackType_t *pxPortInitialiseStack(
5        StackType_t * pxTopOfStack, TaskFunction_t pxCode, void *pvParameters )
6    {
7            *pxTopOfStack = ( StackType_t ) pvParameters;
8            pxTopOfStack--;
9
10           *pxTopOfStack = (StackType_t) 0xDEADBEEF;
11           pxTopOfStack--;
12
13           /* Exception stack frame starts with the return address. */
14           *pxTopOfStack = ( StackType_t ) pxCode;
15           pxTopOfStack--;
16
17           *pxTopOfStack = ( portINITIAL_FORMAT_VECTOR << 16UL )
18                           | ( portINITIAL_STATUS_REGISTER );
19           pxTopOfStack--;
20
21           *pxTopOfStack = ( StackType_t ) 0x0; /*FP*/
22           pxTopOfStack -= 14; /* A5 to D0. */
23
24       return pxTopOfStack;
25    }
```

The only aspect worth mentioning is the peculiar constant put on the stack at line 10. This is the stack position where the processor expects the return address to be when the main task function returns to the caller. Since this should never happen, according to the FREERTOS specifications, the stack initialization code shown above stores an invalid address there. In this way, any attempt to return from the main task function will trigger an illegal memory access and can be detected easily.

We have already examined the architecture-dependent functions that switch the processor from one task to another. Starting the very first task is somewhat an exception to this general behavior and, in the Cortex-M3, is performed by the private port-layer function prvPortStartFirstTask. The implementation, listed below, is based on generating and handling a software interrupt request.

```
1    static void prvPortStartFirstTask( void ) __attribute__ (( naked ));
2    void vPortSVCHandler( void ) __attribute__ (( naked ));
3
4    static void prvPortStartFirstTask( void )
5    {
6            __asm volatile(
7            " ldr r0, =0xE000ED08                    \n"
8            " ldr r0, [r0]                           \n"
9            " ldr r0, [r0]                           \n"
10           " msr msp, r0                            \n"
11           " cpsie i                                \n"
12           " dsb                                    \n"
13           " isb                                    \n"
14           " svc 0                                  \n"
15           " nop                                    \n"
16           );
17   }
18
```

```
19    void vPortSVCHandler( void )
20    {
21            __asm volatile (
22            "       ldr     r3, pxCurrentTCBConst2        \n"
23            "       ldr r1, [r3]                         \n"
24            "       ldr r0, [r1]                         \n"
25            "       ldmia r0!, {r4-r11}                  \n"
26            "       msr psp, r0                          \n"
27            "       isb                                  \n"
28            "       mov r0, #0                           \n"
29            "       msr     basepri, r0                  \n"
30            "       orr r14, #0xd                        \n"
31            "       bx r14                               \n"
32            "                                            \n"
33            "       .align 2                             \n"
34            "pxCurrentTCBConst2: .word pxCurrentTCB      \n"
35            );
36    }
```

The function `prvPortStartFirstTask` is called by FREERTOS to start the very first task after setting `pxCurrentTCB` to point to its TCB. It first fetches the operating system stack address from the first element of the exception vector table and stores it into MSP (lines 7–10).

In the Cortex-M3 architecture, the first 32-bit element of the exception vector table is not used as a real exception vector. It holds instead the initial value automatically loaded into the processor's stack pointer upon reset. FREERTOS picks it up as the top of its own stack. The actual assembly language code to retrieve this value consists of a double dereference at address 0xE000ED08. This is the address of the VTOR register that points to the base of the exception table [9].

It should be noted that the MSP (main stack pointer) register being discussed here is not the same as the PSP (process stack pointer) register we talked about earlier. The Cortex-M3 architecture, in fact, specifies two distinct stack pointers. With FREER-TOS the PSP is used when a *task* is running whereas the MSP is dedicated to exception handling. The processor switches between them automatically as its operating mode changes.

The initial context restoration is performed by means of a synchronous software interrupt request made by the `svc` instruction (line 14). This software interrupt request is handled by the exception handler `vPortSVCHandler`; its code is very similar to `xPortPendSVHandler`, but it only restores the context of the new task pointed by `pxCurrentTCB` without saving the context of the previous task beforehand. This is correct because there is no previous task at all. As before, the processor base priority mask `basepri` is reset to zero (lines 28–29) to enable all interrupt sources as soon as the exception handling function ends.

Before returning from the exception with a `bx` instruction, the contents of the link register LR (a synonym of R14) are modified (line 30) to ensure that the processor returns to the so-called *thread mode*, regardless of what its mode was. When handling an exception, the Cortex-M3 processor automatically enters *handler mode* and starts using the dedicated operating system stack mentioned earlier.

When the execution of a task is resumed, it is therefore necessary to restore the state from that task stack and keep using the same task stack to continue with execution. This is exactly what the exception return instruction does when it goes back to thread mode. A similar, automatic processor mode switch for exception handling is supported by most other modern processors, too, although the exact names given to the various execution modes may be different. The ColdFire-V2 implementation is quite similar and is not shown here for conciseness.

10.7   TICK TIMER

The next two functions manage the interval timer internal to Cortex-M3 processors, also known as SYSTICK. This timer is programmed to raise an interrupt request, or *tick*, at regular time intervals, thus providing the main reference used by FREERTOS to derive all its timing information.

It should be noted that recent versions of FREERTOS also support a *tickless* operating mode, in which timer interrupt requests are generated *on-demand* instead of *periodically*. However, this operating mode is beyond the scope of this book and will not be further discussed.

Namely, we assume that the configuration macro configUSE_TICKLESS_IDLE. which enables tickless mode, is not set to 1 and we omit the corresponding code in the listing.

```
 1   #define portNVIC_SYSTICK_CTRL_REG ( * ((volatile uint32_t *) 0xe000e010) )
 2   #define portNVIC_SYSTICK_LOAD_REG ( * ((volatile uint32_t *) 0xe000e014) )
 3
 4   #ifndef configSYSTICK_CLOCK_HZ
 5           #define configSYSTICK_CLOCK_HZ configCPU_CLOCK_HZ
 6           #define portNVIC_SYSTICK_CLK_BIT          ( 1UL << 2UL )
 7   #else
 8           #define portNVIC_SYSTICK_CLK_BIT          ( 0 )
 9   #endif
10
11   #define portNVIC_SYSTICK_INT_BIT                 ( 1UL << 1UL )
12   #define portNVIC_SYSTICK_ENABLE_BIT              ( 1UL << 0UL )
13
14   __attribute__(( weak )) void vPortSetupTimerInterrupt( void )
15   {
16           #if configUSE_TICKLESS_IDLE == 1
17               ...
18           #endif /* configUSE_TICKLESS_IDLE */
19
20           /* Configure SysTick to interrupt at the requested rate. */
21           portNVIC_SYSTICK_LOAD_REG =
22             ( configSYSTICK_CLOCK_HZ / configTICK_RATE_HZ ) - 1UL;
23           portNVIC_SYSTICK_CTRL_REG =
24             ( portNVIC_SYSTICK_CLK_BIT | portNVIC_SYSTICK_INT_BIT
25               | portNVIC_SYSTICK_ENABLE_BIT );
26   }
27
28   void xPortSysTickHandler( void )
29   {
30           ( void ) portSET_INTERRUPT_MASK_FROM_ISR();
31           {
32                   if( xTaskIncrementTick() != pdFALSE )
33                   {
34                           portNVIC_INT_CTRL_REG = portNVIC_PENDSVSET_BIT;
35                   }
36           }
37           portCLEAR_INTERRUPT_MASK_FROM_ISR( 0 );
38   }
```

In the listing above:

- The function `prvSetupTimerInterrupt` programs the timer to generate periodic interrupt requests at the rate specified by the `configTICK_RATE_HZ` configuration variable and starts it. The timer can be configured in two different ways depending on whether or not the configuration macro `configSYSTICK_CLOCK_HZ` is defined:
  1. If the macro is undefined, the SYSTICK timer is forced to run at the same clock speed as the CPU, that is, `configCPU_CLOCK_HZ`.
  2. If the macro is defined, it is assumed that its value is the SYSTICK timer frequency in Hertz, set externally to the porting layer. In this case, the timer frequency is not reprogrammed during initialization.

  In both cases, the tick interrupt frequency or rate, which also establishes the FREERTOS timer resolution, is taken from the configuration macro `configTICK_RATE_HZ`. The function has the weak attribute attached to it, in order to allow it to be overridden by another implementation, external to the porting layer.
- The function `xPortSysTickHandler` handles interrupt requests coming from the timer. In particular:
  1. It calls the FREERTOS function `xTaskIncrementTick`, within a critical region (lines 31–36). This function takes care of all aspects related to the tick timer, such as, for example, updating the current time, checking whether some task timeouts have expired, and so on.
  2. If the above-mentioned function did not return `pdFALSE`, it means that it woke up a task with a priority higher than the current one. As a consequence, it becomes necessary to request a context switch to be performed at the earliest opportunity.

     Even though the same result could be obtained by invoking the `portYIELD_FROM_ISR` macro, its body is included here directly, at line 34, still within the critical section.

Another aspect worth mentioning is the peculiar way the critical region that protects the body of `xPortSysTickHandler` is implemented. In general, this critical region should be implemented by saving the value of the interrupt mask before the critical region entry point—returned by `portSET_INTERRUPT_MASK_FROM_ISR`—into a variable, and then restoring it by passing that variable to `portCLEAR_INTERRUPT_MASK_FROM_ISR`.

However, the tick interrupt is configured so that it has the lowest possible priority. As a consequence, when its interrupt handler executes, all other interrupts must be unmasked. There is therefore no need to save and then restore the previous interrupt mask value, as its value is already known to be zero.

10.8 ARCHITECTURE-DEPENDENT SCHEDULER STARTUP

```
1   #define portNVIC_SYSPRI2_REG ( * ( (volatile uint32_t * ) 0xe000ed20 ) )
2   #define portNVIC_PENDSV_PRI \
3           ( ( ( uint32_t ) configKERNEL_INTERRUPT_PRIORITY ) << 16UL )
4   #define portNVIC_SYSTICK_PRI \
5           ( ( ( uint32_t ) configKERNEL_INTERRUPT_PRIORITY ) << 24UL )
6
7   BaseType_t xPortStartScheduler( void )
8   {
9           configASSERT( configMAX_SYSCALL_INTERRUPT_PRIORITY );
10
11          #if( configASSERT_DEFINED == 1 )
12              ...
13          #endif /* configASSERT_DEFINED */
14
15          portNVIC_SYSPRI2_REG |= portNVIC_PENDSV_PRI;
16          portNVIC_SYSPRI2_REG |= portNVIC_SYSTICK_PRI;
17
18          vPortSetupTimerInterrupt();
19
20          uxCriticalNesting = 0;
21
22          prvPortStartFirstTask();
23
24          prvTaskExitError();
25          return 0;
26  }
```

The very last function to be discussed here is xPortStartScheduler, whose listing is shown above. Like in the previous case, the listing has been simplified by omitting some debugging code that is conditionally compiled only when the configuration macro configASSERT_DEFINED is set to 1.

The function xPortStartScheduler is called during FREERTOS startup and, as its name suggests, must perform all architecture-dependent activities related to starting the scheduler. In particular,

- It sets the priority of the two interrupt sources used by FREERTOS (the PendSV software interrupt and the SYSTICK timer) to the value of the macro configKERNEL_INTERRUPT_PRIORITY taken from the FREER-TOS configuration (lines 15–16).
- It sets up and starts the tick timer interrupt, by calling the function vPortSetupTimerInterrupt (line 18).
- It initializes the uxCriticalNesting variable to zero. As previously discussed, this value indicates that no critical regions, based on disabling interrupts, are currently in effect.
- It starts the first stack, previously selected by the upper operating system layers, by calling prvPortStartFirstTask (line 22).

Under normal conditions, and as long as the operating system is running, prvPortStartFirstTask never returns to the caller, and xPortStartScheduler is not expected to return, either, unless FREER-TOS is stopped completely by calling vTaskEndScheduler. However, this capability is unsupported by the current version of the Cortex-M3 port.

Therefore, the role of the last two statements of `xPortStartScheduler` (lines 24–25) is merely to catch errors and prevent compiler warnings.

10.9   SUMMARY

This chapter went through the main components of the FREERTOS porting layer. Looking at how the porting layer of a real-time operating system is built is worthwhile for at least two reasons. First of all, it helps to refine concepts, like *context switch* (Section 10.4) and the implementation of critical regions by disabling interrupts (Section 10.5), because it fills the gap between their abstract definition and their concrete implementation.

Secondly, it better differentiates the general behavior of operating system primitives from the peculiarities and limitations of their implementation on a specific processor architecture. Due to lack of space, the presentation is far from being exhaustive but can be used as a starting point for readers willing to adapt an operating system to an architecture of their interest.

11 Performance and Footprint at the Toolchain Level

CONTENTS

Performance and footprint optimization are a very important aspect of embedded software development. A relevant example, discussed in Chapter 8, are real-time device drivers, in which performance is often critical to satisfy hardware-dictated timing requirements.

Optimization can be carried out at different levels of abstraction, and the choice affects code portability. For instance, optimizing the source code by means of assembly language inserts or other non-standard language features, as described in Chapter 9, is often beneficial to performance, and sometimes unavoidable, but is inherently detrimental to portability.

This chapter first introduces the reader to a set of compiler options that, directly or indirectly, affect performance and footprint. As a consequence, they are useful to start tackling the optimization problem, without requiring any modifications to the source code.

Another interesting way of enhancing code performance without impairing portability, is to work at the algorithmic and source code levels. The underlying idea is that, with modern compilers, it is often possible to reach a quality of machine code similar to the one attained by a proficient assembly language programmer.

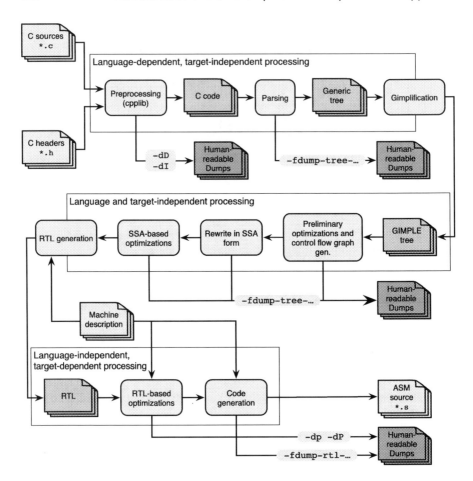

Figure 11.1 Overview of the GCC compiler workflow.

In fact, provided that the inner workings of the code generation algorithms are known, the algorithms and the corresponding source code can be reworked to drive the compiler to generate better code. The second part of the chapter briefly discusses a few basic techniques to achieve this goal, by means of a running real-world case study.

11.1 OVERVIEW OF THE GCC WORKFLOW AND OPTIMIZATIONS

This section provides a short overview of the GCC compiler workflow and optimization process, building on the more general concepts developed in Chapter 3. Interested readers are referred to specialized books, for instance [169], as well as the full-fledged documentation about GCC internals [159] for more detailed information.

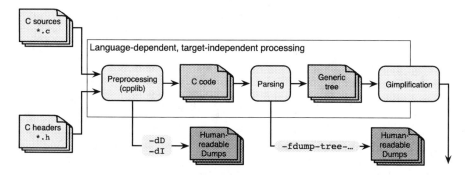

Figure 11.2 Language-dependent portion of the workflow, C language.

Figure 11.1 contains a summary of the compiler workflow, which will be used as a reference and starting point for the discussion. As shown in the three rows of the figure, the overall workflow can be divided into three parts, performed in sequence:

1. A *language-dependent* and *target-independent* part, which takes care of all the peculiarities of a specific input language, or language family, like its syntax and grammar.
2. A *language* and *target-independent* part, in which the bulk of code optimization is performed.
3. A *target-dependent* part, which mainly deals with code generation for the target processor architecture and architecture-dependent optimization.

11.1.1 LANGUAGE-DEPENDENT WORKFLOW

Usually, the first part of the compiler workflow—that is, the language-dependent and target-independent part depicted again in Figure 11.2 for better clarity—consists of two main phases. The figure depicts its internal organization for what concerns the C language, but the organization is very similar for other languages, too. For instance, the workflow is still the same for the C++ and Objective C languages, which belong to the same language family. The two phases are:

1. *Preprocessing* that, informally speaking, understands and implements all directives pertaining to the macro language used to transform C, C++, and Objective-C programs before they are compiled. For instance, macro definitions introduced by `#define` and their expansion in the source code are handled in this phase.
   The output of the preprocessing phase is still in the form of C source code, in which language constructs that have already been handled no longer appear. For instance, `#define` statements have been removed and macro invocations have been replaced by their respective body.
   Even though, strictly speaking, preprocessing is a separate step (or *pass*) in the workflow, for the C, C++, and Objective-C languages, it is performed by means

of a library, called `cpplib`. This library is common to both the standalone pre-processor, which can be invoked on its own, and the compiler.

2. *Parsing*, which transforms the source code into an internal, language-dependent representation that faithfully represents it, but is easier to manipulate than the source code text.

   The parsing process is driven by the language syntax and, even more importantly, by its grammar rules [1]. As a consequence, the general shape of the parsed output resembles the shape of the grammar rules themselves. This is the main reason why, even after parsing, language dependencies remain in the internal representation of the source code.

In most cases, regardless of the language, the internal language-dependent representations used by the compiler take the form of a *tree*, whose nodes are heavily decorated with information derived from the source code.

For instance, the internal representation of a data structure definition in the C language may take the form of a tree, in which:

- The root node represents the data structure definition as a whole and may be decorated with the name of the structure.
- Its children represent individual fields in the data structure. Each of them is decorated with the field name and its data type. In turn, the data type can be either a primitive data type (like `int`) that can be represented as a single node, or a more complex data type (like a pointer to another user-defined data structure). In the second case, the data type is described by its own subtree.

In the case of the C language, GCC makes use of the so-called GENERIC trees. Those trees are language independent in principle but, if deemed necessary, a front end for a certain language can extend the GENERIC trees it generates by introducing some custom, language-dependent node types.

The only requisite is that, in doing so, the front end must also provide an appropriate set of functions, often called *hooks*, associated to each of these custom node types. The functions are able to convert custom nodes into a different representation, called GIMPLE, to be used by the target-independent optimizer. For historical reasons, the intermediate representation used by the C and C++ front ends makes use of those extensions.

The third and last phase, depicted on the far right of Figure 11.2, is at the boundary between the language-dependent and the language-independent parts of the workflow. It is called *gimplification* in the official GCC documentation [159] because its purpose is to convert a GENERIC tree into a GIMPLE tree. For the C language, the translation takes place in a single pass and on a function-by-function basis.

The GIMPLE tree grammar is a simplified and more restrictive subset of the GENERIC tree grammar, which is more suitable for use in optimization. For instance, expressions can contain no more than three operands (except function calls, which can still have an arbitrary number of arguments as required by the language), and hence, more complex expressions are broken down into a 3-operand form by introducing temporary variables to hold intermediate results.

Moreover, side effects of expression evaluation are confined to be on the right-hand side of assignments, and cannot appear on the left-hand side. As before, temporary variables are introduced as needed to secure this property.

Last, GIMPLE trees keep no track of high-level control flow structures, such as loops. All those control structures are transformed and simplified—or *lowered* as is commonly said in compiler theory—into conditional statements (also called `if` statements in the following and in the compiler documentation) and jumps (also called `goto`).

For the history-oriented readers it is also interesting to remark that the features of GIMPLE trees were derived from the SIMPLE representation adopted by the McCAT compiler project [69], with several updates to make the representation, designed in a mainly research-oriented environment, more suitable for a production compiler.

As recalled previously, the conversion from GENERIC to GIMPLE trees is relatively simple and takes place in a single pass. However, for other languages, the internal representation produced by the parser in the first place may be quite far away from GIMPLE.

This is because, as also mentioned previously, the shape of the parser output heavily depends on the language grammar. As a consequence, the structure of the language-dependent part of the workflow becomes more involved, as additional conversion passes are needed.

In many cases, in order to keep compiler development effort to a minimum, these passes involve GENERIC trees, too, because a semi-automatic converter from GENERIC to GIMPLE trees is available.

11.1.2   LANGUAGE-INDEPENDENT WORKFLOW

As mentioned in the previous section, gimplification lies at the boundary between the first and the second part of the compiler workflow, that is, between the language-*dependent* and the language-*independent* part of the compiler.

Even though it is not the main topic of this book—and there is no way to provide thorough information about it here—it is still useful to briefly describe the internal representation of a source file, as it changes its shape while it progresses through the compiler.

There are two main reasons to do this:

- It helps to better understand how the source file is "seen" by the compiler and how code optimization options, to be described in Sections 11.2 and 11.3, work.
- A better grasp of how the compiler eventually transforms source code into machine code is also useful when it is necessary to modify the structure of the source code and help the compiler to produce better machine code. Source-level optimization will be the topic of Section 11.4.

Table 11.1

Compiler Options to Dump the Internal Code Representation

Option	Purpose
Language-dependent workflow	
-dD	Dump all macro definitions at the end of the preprocessing step.
-dI	Dump all #include directives at the end of the preprocessing step.
Language-independent workflow	
-fdump-tree-ssa	Dump the internal SSA representation. The output file name is formed by appending .ssa to the source file name.
-fdump-tree-copyrename	Dump the internal representation after applying the copy rename optimization. The output file name is generated by appending .copyrename to the source file name.
-fdump-tree-dse	Dump the internal representation after applying dead store elimination. The output file name is made by appending .dse to the source file name.
-fdump-tree-optimized	Dump the internal representation after all tree-based optimizations have been performed.
Target-dependent workflow	
-dp	Annotate the assembler output, enabled by using the -S compiler option, with a comment indicating which RTL code generation pattern was used.
-dP	Extended version of -dp, it also annotates the assembled output with the corresponding RTL.
-fdump-rtl-all	Dumps the RTL representation at various stages of the target-dependent workflow. Specific dumps can also be enabled selectively by replacing all with other options.

Some command-line options ask the compiler to dump into some output files a human-readable transcript of its internal representation at various stages of the compilation workflow. The real output produced by the compiler when it is invoked on some short and simple code excerpts will be used to guide the description that follows. The main options are summarized in Table 11.1.

Before going deeper into the description of the language-independent part of the workflow, let us refer back to Figure 11.2 and present some options of the same kind, but still related to the language-dependent part of the workflow pertaining to the C language. Namely:

- The option -dD (not to be confused with -dd, which is also valid but enables a totally different kind of dump) asks the compiler to dump all macro definitions (performed by means of #define directives) at the end of the preprocessing step.

- The option -dI has a similar meaning, but for #include directives. Also in this case, the output is produced at the end of the preprocessing step.

Beyond code optimization, both features are in some cases very useful to better realize what preprocessing actually did and how it affected the results of the compiler. This is especially true when dealing with complex code modules or—as sometimes happens with open-source software—when compiling a code module without having a thorough understanding of its internal structure, in particular for what concerns header files. For instance, at first glance:

- It may not be totally clear which macro definitions are in effect when compiling a certain source module, and in which header files those definitions are, in particular when conditional compilation is also in effect.
- Moreover, it is common that in a cross-compilation environment there are multiple versions of the same header file. For instance, there may be a stdio.h for the native compiler, plus one more stdio.h for each target architecture supported by the cross compilers. All these headers have the same name but reside at different places in the file system, and hence, it is hard to distinguish them from one another by just browsing the #include directives in the source code.

On the other hand, the option -fdump-tree asks the compiler to dump one or more tree-based internal representations. It applies to both the language-dependent and the language-independent parts of the workflow.

Its general syntax is -fdump-tree-<switch>-<options>, where

- <switch> selects one or more representations to be dumped. The special switch all directs the compiler to dump all its tree-based internal representations.
- <options> is a list of options, separated by hyphens (-), which change the way dumps are made. For instance, two useful options are: the details option that asks for more detailed dumps, and the lineno option that adds source code line numbers to the dumps.

For instance, the option -fdump-tree-all-details enables all dumps the compiler is able to produce, and they will contain more detailed information than normal. Reference [160] contains a thorough list of all available switches and options.

It is now time to refer to Figure 11.3, which shall be read from right to left, and consider the language-independent optimizations performed in the second portion of the compiler workflow.

Several preliminary optimizations are performed directly on GIMPLE trees. They correspond to the rightmost gray processing block shown in the figure. Two typical kinds of optimization performed at this stage are:

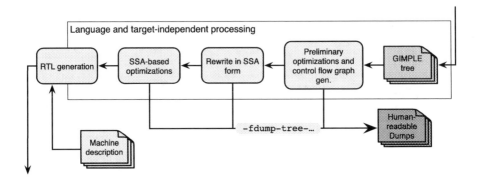

Figure 11.3   Language-independent portion of the workflow.

- *Useless statement removal.* This step removes obviously useless code, which can be detected without detailed control flow information.

  For instance, this optimization detects `if` statements with a constant predicate and simplifies them according to the value of the predicate, which is known at compile time.

  Another example is the removal of lexical environments, such as the ones that contain the local variables of a function, when they are completely empty, that is, they contain no variables.

- *Control flow simplification.* This step is performed in order to prepare to build the *control flow graph*, a key data structure for what concerns optimization, to be discussed immediately thereafter.

  In order to simplify control flow, `if` statements are rewritten so that, regardless of their original shape, they have exactly two `goto` statements, one in the `then` branch and the other in the `else` branch.

  Moreover, lexical binding information is explicitly attached to each statement instead of having to infer it from the statement position within the tree.

The *control flow graph* is a data structure built on top of an internal code representation—a tree-based representation in this case—that abstracts the control flow behavior of a unit of code. In GCC, control flow graph generation takes individual functions as units and works on them one by one. In the control flow graph *nodes* represent *basic blocks* of code, and *directed edges* link one node to another to represent possible *transfers of control flow* from one basic block to another.

A *basic block* is a sequence of statements with exactly one entry point at the beginning and one exit point at the end. To all purposes, it can therefore be considered as an atomic, indivisible unit by many optimization techniques. Accordingly, control flow graph generation scans the tree-based representation of a function, splits it into basic blocks and identifies all the directed edges that connect them.

At this point, the reason behind some of the preliminary optimizations discussed previously should become clearer. For instance, the two `goto` statements found in `if` statements after control flow simplification can be mapped into directed edges of the control flow graph in a straightforward way.

At a later time, the directed edges in the control flow graph are used as a base to establish *dependencies* among basic blocks that read and/or modify a given variable. In turn, this information is the starting point to build the *data flow graph* of a function, which drives further optimization techniques.

Referring back to Figure 11.3, the central step of the language-independent workflow is to translate, or *rewrite*, as it is called in the compiler documentation, the tree-based representation into static single assignment (SSA) form [43]. This form has been chosen because plenty of theoretical work has been performed to develop efficient, language-independent optimizations based on it.

One basic assumption the SSA representation is based upon is that program variables are assigned in *exactly one* place. Of course, this is an unrealistic assumption because in any real-world program variables are normally assigned multiple times. In order to address the issue the compiler attaches a *version number* to all program variables and manages it appropriately. In particular:

- Every assignment to a variable creates a *new version* of that variable, which makes it unique.
- When a variable is referenced, the compiler qualifies the reference with the version number that corresponds to the *most recent assignment* along the control flow path.

In some cases, depending on the program's control flow, it may be impossible to determine the exact version number for a variable reference. As can be seen in the bottom part of Figure 11.4, this typically happens after conditional statements in which the same variable is assigned in both branches, and is referenced thereafter.

In these cases, the compiler creates a fresh version of the variable—with its own unique version number—and inserts an artificial definition for it, which merges all possible incoming versions of that variable. The definition is called ϕ *function* or ϕ *node*.

In particular, the figure shows how the compiler qualifies variable a and inserts a required ϕ node when processing a very simple fragment of code depicted as a flowchart.

- Upon encountering the very first assignment to a, shown at the top of the figure, the compiler qualifies the variable with version number 1.
- When a is referenced on the right-hand side of the next statement, the compiler qualifies the reference with version number 1, according to the control flow graph (that, as a side note, corresponds to the arrows of the flowchart).
- The next statement is a new assignment to variable a, and hence, the compiler assigns a new version number to it. In the example, we assume that

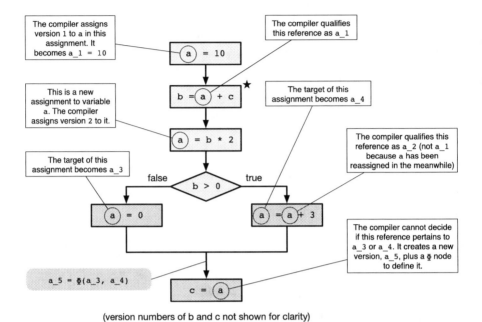

(version numbers of b and c not shown for clarity)

Figure 11.4 Version numbers and ϕ nodes in SSA generation.

version numbers are consecutive integers, although this is not strictly required, as long as they are all unique when they are generated. Following this assumption, the assignment refers to version 2 of variable a.

- In the right (true) branch of the conditional statement, a is referenced again. This time, the compiler qualifies the reference with version number 2 and not 1, because a has been reassigned and there is no way to "reach" version 1 any more.

- In both branches of the conditional statement, a is reassigned, too. The corresponding flowchart blocks have been highlighted in a darker shade of gray in Figure 11.4. Accordingly, the compiler generates two fresh version numbers (3 and 4) and qualifies the assignments appropriately.

The last statement of the flowchart, shown at the bottom of the figure and also highlighted by means of a darker shade of gray, is the most problematic from the point of view of version numbers. This is because the compiler cannot determine with certainty at *compile-time* whether the execution flow will follow the left or the right branch of the conditional statement at *run-time*.

In fact, this depends on the value of b and, in turn, it depends on the value of a and c through the assignment marked with ⋆ in the figure. The value of a is known at compile-time, but the value of c is not. For this reason, the version of a used in the last statement can be either 3 (if execution follows the left path) or 4 (if execution follows the right path).

Accordingly, the compiler creates version 5 of a and qualifies the reference to it accordingly. The definition of this new version of a is performed, as outlined previously, by inserting an artificial definition. It is shown as a rounded rectangle at the bottom left of Figure 11.4.

The artificial definition is carried out by means of a fictitious function ϕ that combines versions 3 and 4 of a in an unspecified way to produce version 5, thus remarking the dependency of version 5 from those other versions. It should also be noted that neither a_1 nor a_2 nor a_3 is mentioned in the ϕ node because none of these versions can possibly be used at this point.

Let us now consider a more specific and concrete example, by examining and commenting on the actual compiler output, corresponding to the C function definition listed below.

```
int f(int k) {
        int m;
        m = (k > 0 ? 3 : 5);
        return m;
}
```

The human-readable dump of the internal SSA representation, shown below, has been obtained by means of the -fdump-tree-ssa option, listed in Table 11.1. Concerning this output, the first thing worth noting is that, even though the internal representation is a tree, the dump takes the form of C-like code, to make it easier to read and understand.

```
1   ;; Function f (f)
2
3   f (k)
4   {
5       int m;
6       int D.1177;
7       int iftmp.0;
8
9   <bb 2>:
10      if (k_2(D) > 0)
11        goto <bb 3>;
12      else
13        goto <bb 4>;
14
15  <bb 3>:
16      iftmp.0_3 = 3;
17      goto <bb 5>;
18
19  <bb 4>:
20      iftmp.0_4 = 5;
21
22  <bb 5>:
```

```
23    # iftmp.0_1 = PHI <iftmp.0_3(3), iftmp.0_4(4)>
24    m_5 = iftmp.0_1;
25    D.1177_6 = m_5;
26    return D.1177_6;
27    }
```

By looking further into the output we can also notice that, even though the input function was extremely simple, the compiler performed all the activities mentioned so far. In particular:

- Besides m, two additional local variables have been added, that is, D.1177 to hold the function return value and iftmp.0 to store the result of the conditional expression before assigning it to m.
- Control flow has been simplified by converting the conditional expression into an if statement in which both branches consist of a single goto.
- The code has been split into 4 basic blocks, labeled from <bb 2> to <bb 5>. Basic blocks are connected by means of goto statements, which represent the directed edges of the control flow graph.
- Concerning goto statements, the one connecting <bb 4> to <bb 5> has been removed because it has been recognized as provably useless (the two blocks are in strict sequence).

Moreover, version numbers have been assigned to all variable assignments and references, as better detailed in the following.

- Argument k has a single version number, that is, 2. This version number is assigned to the variable at the beginning and does not change through the code because there are no additional assignments to it.
- Similarly, variables m and D.1177 have version number 5 and 6, respectively, all the way through the code. This is because they are assigned only once (at lines 24–25).
- On the contrary, temporary variable iftmp.0 gets two different version numbers, depending on the side of the conditional statement is it assigned to. Namely, version number 3 corresponds to the true branch of the statement (line 16), while version 4 corresponds to the false branch (line 20).
- When iftmp.0 is eventually assigned to m (line 24) a new version number becomes necessary because the compiler cannot determine which version of iftmp.0 will be used at runtime. Therefore:
  - The compiler introduces a new version of iftmp.0, version 1, to be used on the right-hand side of the assignment.
  - It also inserts a ϕ node (line 23) that highlights how version 1 of iftmp.0 depends on versions 3 and 4 of the same variable.

As a final remark, attentive readers have probably noticed that the compiler output contains additional annotations—for instance, the (D) annotation that follows the reference to k_2 at line 10. Unfortunately, they cannot be commented on in any detail here, and hence, readers are referred to the compiler documentation [159] for further information.

Most language-independent optimizations are performed on the SSA form. Among the main ones, we briefly recall:

1. *Dead code elimination.* Statements without side effects that may affect future computation and whose results are unused are removed. Since other optimizations may change the code and make more statements useless, this pass is repeated multiple times during the optimization process.
2. *Forward propagation of single-use variables.* This optimization identifies variables that are used exactly once. Then, it moves forward their definition to the point of use, in order to check whether or not the result leads to simpler code.
3. *Copy renaming.* In this pass, the compiler tries to rename variables involved in copy operations so that copies can be coalesced. This includes temporary compiler-generated variables that are copies of user variables. In this case, the compiler may rename compiler-generated variables to the corresponding user variables.
4. *Replace local aggregates with scalars.* Local aggregates without aliases are replaced with sets of scalar variables. By intuition, this makes subsequent optimizations easier and more effective because, after this pass, each component of the aggregate can be handled as an independent unit by the optimizer.
5. *Dead store elimination.* This optimization removes memory store operations that are useless because the same memory location is overwritten before being used.
6. *Forward movement of store operations.* Store operations are moved along the control flow graph, to bring them as close as possible to their point of use.
7. *Loop optimizations.* A complex group of related optimizations works on loops. Among them, the main ones are:
   - Loop-*invariant statements*—that is, statements that always produce the same result and no side effects at every iteration of the loop—are moved out of the loop in order to evaluate them only once.
   - Another optimization pass identifies *induction variables*, that is, variables that are incremented (or decremented) by a fixed amount upon each loop iteration, or depend linearly on other induction variables.
     Then, it tries to optimize them by reducing the strength of their computation, merging induction variables that consistently assume the same values through loop iterations, and eliminating induction variables that are used rarely and whose value can easily be derived from others.
   - When the loop contains conditional jumps whose result is loop-invariant, it is possible to evaluate them only once, out of the loop, and create a simplified version of the loop for each possible outcome of the conditional jumps themselves. This optimization is sometimes called loop *unswitching*.
   - Loops with a relatively small upper bound on the number of iterations can be *unrolled*, partially or completely. Informally speaking, complete loop unrolling replaces the loop by an equivalent sequence of instructions, thus deleting loop control instructions and possibly simplifying the code produced for

each loop iteration, according to the known value of induction variables on that iteration.

8. *Function call optimizations.* A further group of optimization is related to function calls and the management of their return value. Among them, we recall the following ones.

- When a function calls itself recursively but the call is the very last statement of the function (also called *tail recursion*), the recursion is transformed into a loop.
- If a function returns an aggregate and always returns the same local variable, then the local variable is replaced by the function return value. This saves a copy operation that, depending on the size of the aggregate, may introduce a significant performance overhead.
- If the return value of a function that returns an aggregate is stored into a variable immediately after the call, it is possible to use the variable itself to store the return value, thus saving a copy operation. This optimization and the previous ones are called *return slot* and *return value* optimization, respectively.

To continue the example, a dump of the internal representation of function f (defined on page 309) after *copy renaming*, is shown in the following.

```
 1   f (k)
 2   {
 3     int m;
 4
 5   <bb 2>:
 6     if (k_2(D) > 0)
 7       goto <bb 3>;
 8     else
 9       goto <bb 4>;
10
11   <bb 3>:
12     m_3 = 3;
13     goto <bb 5>;
14
15   <bb 4>:
16     m_4 = 5;
17
18   <bb 5>:
19     # m_1 = PHI <m_3(3), m_4(4)>
20     m_5 = m_1;
21     m_6 = m_5;
22     return m_6;
23   }
```

As can be seen, both temporary variables D.1177 and iftmp.0 have been removed because it was possible to rename them into local variable m, introducing several additional version numbers for it. Namely:

- iftmp.0_3 was renamed to m_3,

- `iftmp.0_4` was renamed to `m_4`,
- `iftmp.0_1` was renamed to `m_1`,
- `D.1177_6` was renamed to `m_6`.

The following listing shows instead the internal representation of function `f` after several other optimizations, namely *forward propagation of single-use variables* and *dead store elimination*, have been carried out.

```
f (k)
{
  int m;

<bb 2>:
  if (k_2(D) > 0)
    goto <bb 4>;
  else
    goto <bb 3>;

<bb 3>:

<bb 4>:
  # m_1 = PHI <3(2), 5(3)>
  return m_1;

}
```

In this case, both `m_3` and `m_4` are single-use variables. In fact, they are used only in the ϕ node that defines `m_1`. For this reason, they have been deleted and their definitions, that is, the constants 3 and 5, have been forward-propagated to their point of use. Moreover, the compiler determined that the sequence of stores into `m`, found at lines 20–21 of the listing on page 312, was useless and has removed it.

The very last step performed on the internal SSA representation before entering the target-dependent part of the compiler workflow is to remove version numbers attached to variables, leading to the final tree-based representation in *normal form*. It is shown in the following listing for our example function `f`.

```
f (k)
{
  int m;

<bb 2>:
  if (k > 0)
    goto <bb 4>;
  else
    goto <bb 3>;

<bb 3>:
  m = 5;
  goto <bb 5>;
```

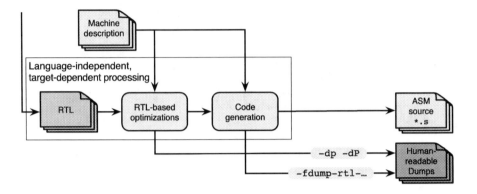

Figure 11.5   Target-dependent code generation and optimization.

```
<bb 4>:
  m = 3;

<bb 5>:
  return m;

}
```

11.1.3   TARGET-DEPENDENT WORKFLOW

As shown on the leftmost part of Figure 11.3, register transfer language (RTL) generation lies at the boundary between the language- and target-independent part and the *target-dependent* part of the workflow.

With respect to the tree-based representations discussed in the previous sections, the RTL representation is much lower-level and is close to a one-by-one representation of the machine instructions to be generated by the compiler. RTL generation is *pattern-based* and makes use of specific instruction patterns included in the target-dependent *machine description* built into the compiler.

Namely, named instruction patterns (with predefined, well-known names) are used to transform a portion of the parse tree (to which a certain pattern is applicable) into a sequence of RTL instructions. The process is repeated, one basic block after another, until the whole tree has been transformed into a sequence of RTL instructions.

As shown in Figure 11.5 all subsequent, target-dependent optimizations are carried out on the RTL representation. Many RTL-based optimizations have the same goal as their tree-based counterparts (for instance, dead code elimination). However, by means of the additional information available at the RTL level—mainly concerning the target architecture—further optimizations can now be done. For instance:

1. *Instruction combination* attempts to combine small groups of RTL instructions related by data flow. Namely, it combines the RTL expressions of the individual instructions into a single expression by substitution, simplifies the result algebraically, and checks the result against the instruction patterns in the machine description to see if a more efficient instruction sequence can be used.
2. *Register allocation* maps the abstract pseudo-registers used in RTL generation into real registers or stack positions.
3. *Instruction scheduling* reorders instructions according to their result availability latency, in order to avoid pipeline stalls and fill delay slots. This optimization is typically used for instructions with significant latency, for instance, memory load and floating-point instructions.
4. *Branch shortening* tries to use the most efficient machine instruction operation code for branching, depending on the predicted range of the branch.

At the very end, as also shown in Figure 11.5, the output templates found in the instruction patterns of the machine description are used for code generation.

It must also be remarked that, even in the target-dependent part of the workflow, it is still possible to dump the internal representation used by the compiler, for instance, by means of the $-dp$, $-dP$ and $-fdump-rtl-all$ compiler options listed in Table 11.1.

To conclude our running example, this is the bulk of the final RTL representation of function f. Additional RTL annotations have been omitted for conciseness.

```
(insn 54 50 55 r.c:7 (set (reg:CC 24 cc)
        (compare:CC (reg:SI 0 r0 [ k ])1
            (const_int 0 [0x0]))) 213 {*arm_cmpsi_insn} (nil))

(insn 55 54 56 r.c:7 (cond_exec (gt (reg:CC 24 cc)
            (const_int 0 [0x0]))
        (set (reg/i:SI 0 r0)
            (const_int 3 [0x3]))) 2322 {neon_vornv2di+72} (nil))

(insn 56 55 27 r.c:7 (cond_exec (le (reg:CC 24 cc)
            (const_int 0 [0x0]))
        (set (reg/i:SI 0 r0)
            (const_int 5 [0x5]))) 2322 {neon_vornv2di+72}
                (expr_list:REG_DEAD (reg:CC 24 cc)
        (nil)))

(insn 27 56 58 r.c:7 (use (reg/i:SI 0 r0)) -1 (nil))

(jump_insn 59 58 60 r.c:7 (return) 254 {return} (nil))
```

Even though, for the sake of simplicity, a detailed description of RTL syntax is outside the scope of this book, it is nevertheless useful to note the strong relationship between the RTL representation and the final machine code produced by the compiler, listed in the following.

```
00000000 <f>:
    0:   e3500000        cmp      r0, #0
    4:   c3a00003        movgt    r0, #3
    8:   d3a00005        movle    r0, #5
    c:   e12fff1e        bx       lr
```

As can be seen, the RTL representation contains 5 statements identified by `insn` or `jump_insn`:

- The first one is a comparison statement that sets the processor condition codes (`compare:CC`). It compares register r0 with the constant integer zero (`const_int 0`).
- The second and third statements are similar. Both represent conditionally executed instructions (`cond_exec`) depending on condition codes (`reg:CC`). They move different constants (`const_int 3` in the first case and `const_int 5` in the second) depending on whether register r0 was greater than (`gt`), or less than or equal to (`le`) zero, respectively, when the condition codes were set.
- The fourth statement indicates that register r0 is in use to hold the function's return value.
- The last statement denotes a return point of the function.

As can be seen from the machine code listing, all RTL statements except the fourth one indeed have a one-to-one correspondence with machine instructions.

11.2   OPTIMIZATION-RELATED COMPILER OPTIONS

When invoked without any optimization option, the compiler tries to compile source code as quickly as possible and make debugging as accurate as possible. For instance, source code statements are compiled independently from each other.

In this way, with the help of a debugger, it is possible to set a breakpoint between statements, change the value of any variable, continue execution, and get exactly the results that would be expected by looking at the source code.

On the other hand, as described in the previous section, when optimizations are turned on, the compiler transforms the code in various ways in the attempt to improve the speed and reduce the size of the machine code it eventually generates. For instance, those transformations include the removal of local variables and code reordering.

In this case, the concept of source code statement boundary becomes blurry at runtime, because code reordering does not take these boundaries into account. As a consequence, it may become impossible to set a breakpoint exactly between two statements and the debugger may be unable to access a local variable because it has been optimized away by the compiler. Both these aspects, and others, clearly limit the effectiveness of debugging optimized code.

As described in Table 11.2, the most straightforward way to turn on optimizations is to use one of the options starting with $-O$. The option $-O$ is valid by itself, and

Table 11.2

Compiler Options That Control Optimization

Option	Purpose
-O	This is the most generic option to turn on optimizations in the compiler. It corresponds to -O1 described below.
-O0	This is the default option. It turns off most optimizations in order to reduce compilation time and maximize the accuracy of debugging information.
-O1	This option turns on basic optimizations, for instance, copy renaming, with the goal of reducing both code execution time and size. Optimizations that increase the compilation time significantly are kept disabled.
-O2	With respect to -O1, this option turns on more optimizations that do not involve a trade-off between the execution speed and the size of the generated code.
-O3	Enables additional optimizations with respect to -O2, including optimizations that privilege execution speed at the expense of code size.
-Os	Enables a subset of -O2 optimizations, excluding the ones that may increase code size. As a result, the compiler attempts to optimize the code to reduce its size, even at the expense of speed.
-f...	A fairly large group of options, all starting with -f, can be used to turn on and off individual optimizations.
--param <name>=<value>	Allows the programmer to set various parameters that control optimization. Each option sets parameter <name> to the value <value> and multiple options can appear on the command line, one after another.

turns on a set of basic optimizations. As shown in the table, the option can also be followed by an integer number between 1 and 3 to set various optimization levels and turn on more and more optimization options.

In general, the higher the number, the "better" optimization becomes, at the expense of compilation time and memory requirements. However, especially in embedded software development, it is important to notice a difference between -O2 and lower optimization levels with respect to -O3.

In fact, up to optimization level -O2, the various optimization techniques that are enabled do not involve a tradeoff between execution speed and size of the generated code, because they generally improve both.

On the other hand, optimizations enabled by -O3 attempt to further improve execution speed even at the expense of code size, which may increase noticeably. Furthermore some of them, like loop unrolling, may affect execution time *predictability* in unexpected ways, as will be shown in Section 11.4.

Table 11.3
Additional Optimizations Enabled by Optimization Level 3

Option	Purpose
-finline-functions	Embed the body of simple functions into their callers.
-funswitch-loops	Move branches with loop invariant conditions out of the loop, inserting a simplified version of the loop in all branches.
-fpredictive-commoning	Enable predictive commoning optimization within loops.
-fgcse-after-reload	Perform an additional optimization pass in an attempt to remove redundant memory loads.
-ftree-vectorize	Perform loop vectorization, using the tree-based representation.

In particular, with respect to -O2, optimization level -O3 turns on the additional optimizations listed in Table 11.3. Each of them will be briefly discussed in the following.

- When the -finline-functions option is enabled, the compiler automatically tries to integrate the body of simple functions into the code that calls them. This works independently from functions explicitly marked with the inline keyword, as specified in the C89 and C99 standards.

    From the point of view of execution speed, this has two important advantages:

    1. The function call and return overheads are eliminated, along with the related need to save and restore at least some processor registers.
    2. Embedding the function body into the caller's code opens additional opportunities for optimization. For instance, when some function arguments are constant, the compiler can take advantage of it to specialize and simplify the code it embeds.

    Concerning code size, the effects of function inlining are harder to predict because, on one hand, if a certain function is called multiple times from different places, its body is replicated multiple times, too, thus obviously increasing total code size. On the other hand, the code that is embedded at a specific calling point is tailored to that specific call, and hence, it may often be simpler and shorter.

    Moreover, if a function is declared static, its address is never taken, and it was possible to inline it at every point it was called, the standalone function will not be put in the object code.

    The last point to be discussed is how the compiler decides whether or not a function is "simple enough" to be inlined when this option is enabled. It mainly depends on a compiler parameter that can be set by means of the --param option, also listed in Table 11.2.

Namely, the `max-inline-insns-auto` parameter sets the upper limit on the complexity of functions that the compiler will consider for inlining. The limit is expressed as a number of instructions, counted in the compiler's internal representation.

It should also be mentioned that a different way of obtaining a similar result is to use function-like macro expansion. With respect to automatic function inlining just described, it is more complex to use, because it is completely manual, but gives to the programmer better control on how expansion takes place and exactly which code is generated by it.

The use of function-like macro expansion to this purpose will be explained, by means of a practical example, in Chapter 15.

- The basic concepts related to loop unswitching, controlled by the `-fun-switch-loops` option, as well as its advantages and disadvantages with respect to execution speed and code size, have already been presented in the previous section and will not be repeated here.

- The idea behind predictive commoning optimization, enabled by the `-fpredictive-commoning` option, is quite simple. Informally speaking, the compiler tries to reuse part of the operations performed in previous iterations of the loop in the next iteration. Expensive operations, like memory loads and stores and considered first.

- The `-fgcse-after-reload` option enables an additional optimization pass that attempts to remove redundant memory loads that may be left over by other optimization methods.

- Loop vectorization, controlled by the `-ftree-vectorize` option, attempts to group operations performed in different iterations of a loop, in order to take advantage of vector operations commonly found in most modern processor architectures.

  For instance, let us consider a loop that adds two arrays of 8-bit characters element by element and stores the result in a third array. In the Cortex-M4 architecture [10], four iterations of such a loop can conveniently be replaced by two 32-bit word load instructions, to load 4+4 characters to be added into two registers, followed by a `SADD` instruction, which performs four 8-bit signed additions in parallel, and then by one 32-bit word store instruction to store the four results.

In addition, depending on the compiler version, optimization level `-O3` may also enable loop unrolling. This represents a typical case of "controversial" optimization that, on a case by case basis, may either improve or worsen execution speed and, even more, execution time predictability. As an example, a thorough description of the unforeseen side effects of loop unrolling on the execution time of a simple algorithm on the Cortex-M3 architecture will be given in Section 11.4.

Last, but not least, the optimization option `-Os` directs the compiler to reduce code size. Even though the selection of this optimization level may adversely affect execution speed, it is indeed quite useful in embedded software development when, as often happens, it becomes necessary to pack an ever-increasing amount of software in the internal flash memory of a microcontroller.

In those cases, all source modules containing code that is not critical for performance may be compiled with this options set. On the other hand, speed-oriented optimizations like $-\mathtt{O3}$, along with their code size penalties, may be reserved for code that must execute as fast as possible.

This is an alternative to using non-standard function attributes to the same purpose, as described in Chapter 9.

11.3   ARCHITECTURE-DEPENDENT COMPILER OPTIONS

Another quite important group of optimization-related compiler options is related to the target architecture. In the following, we will take the ARM architecture as a reference, because we are using it as a running example through the whole book, but the same observations are valid, at least to some extent, for all other architectures supported by GCC.

A first group of options, summarized at the top of Table 11.4, determines the assembly-level instruction set the compiler can use and how it will generate code. Within a certain processor family, like the ARM family, there are usually several *architectural variants* with different instruction sets and various degrees of hardware assistance in some areas, for instance, floating point instructions.

In this case, the same toolchain is able to generate code for any variant in the family and the kind of processor to be targeted can be specified by means of the $-\mathtt{mcpu}$ or $-\mathtt{march}$ options. These options are used to indicate the specific CPU model or, more generically, the architectural variant, respectively. The allowed models and variants depend on the base processor architecture and, to some extent, on whether or not the compiler has been configured to support them during toolchain generation.

When multiple architectural variants are present within a family, a subset of the instruction set is usually common to all of them, to ensure binary code compatibility across the family. The special CPU name $\mathtt{base}$ is always available for use, and indicates that the compiler should generate code that only uses instructions belonging to this subset. This code can be executed on any processor of the family, of course, at the expense of performance.

Then, some members of the family likely provide additional, more powerful instructions, which the compiler is allowed to take advantage of only if an appropriate $-\mathtt{mcpu}$ or $-\mathtt{march}$ option is specified. The additional option $-\mathtt{mtune}$ makes it possible to specify the target variant for optimizations and keep it separate from the variant used to determine the output instruction set, set by the above-mentioned options.

The purpose of this slightly obscure option is better explained by means of an example. In some cases it is important to retain binary code compatibility across the whole family, because it is not completely known in advance on which kinds of processor the code will run, but still gain better performance on a specific variant, on which it is indeed known that the code will run in most cases. In this scenario, we must still use $-\mathtt{mcpu=base}$ for compatibility, but we may also indicate the specific variant to be considered during optimization by using $-\mathtt{mtune}$.

Table 11.4
Main Architecture-Dependent Compiler Options, ARM Architecture

Option	Purpose
Instruction set and code generation	
`-mcpu=<name>`	Compile code for architectural variant `<name>`. The `base` variant, which is the default, is always available and corresponds to the simplest variant.
`-march=<name>`	Directs the compiler to generate code for architecture `<name>`, namely, it determines which assembly instructions the compiler can use.
`-mtune=<name>`	Specifies the `<name>` of the target variant for optimizations and keeps it separate from the variant used to determine the output instruction set, set by the `-mcpu` and `-march` options.
`-mthumb`	Generate code for the Thumb instruction set, rather than the ARM instruction set. The specific Thumb instruction set variant to be used is determined according to the `-mcpu` and `-march` options.
`-mthumb-interwork`	Generate code that supports calls from and to functions compiled with different instruction sets—namely, ARM and Thumb—at the expense of an increase of code size.
Floating-point support	
`-mhard-float`	Generate output containing floating point instructions, suitable for use when the processor includes a hardware floating point unit (FPU) or suitable FPU emulation software is available.
`-msoft-float`	Generate library function calls instead of floating point instructions. In order for this option to work, the language runtime libraries must define the requisite functions.
`-mfpu=<name>`	This option specifies which kind of FPU (or FPU emulation software) is available on the target and directs the compiler to generate the corresponding kinds of floating point instructions.
Code generation	
`-mstructure-size-boundary=<n>`	Instructs the compiler to round the size of all structures and unions to a multiple of `<n>`. A larger number may produce faster code, but may obviously increase the size of data structures. Software modules compiled with different values of `<n>` are potentially incompatible, and hence, this option must be used consistently across all modules to be linked together.
`-mlong-calls`	Specifies that the compiler must generate function call code that works regardless of how distant the target function is, unless it can prove that the function is reachable by means of a "short" function call.

In the ARM family, the matter is even more complex due to the fact that some members of the family simultaneously support two distinct instruction sets, namely:

- the 32-bit *ARM* instruction set, and
- the 16-bit *Thumb* instruction set.

The designs of the two instruction sets differ significantly because the first one is mainly aimed at obtaining top execution speed, whereas the second one favors code compactness. For the sake of completeness, it should also be mentioned that the Thumb instruction set design was later extended to include some 32-bit instructions, and hence, it became a variable-length instruction set. In these cases, the `-mcpu` or `-march` options may not be sufficient to uniquely indicate which instruction set the compiler must use. This is done by means of an additional option, that is, `-mthumb`.

A related problem is whether or not it should be possible to mix functions compiled using different instruction sets in the same executable program, and hence, whether or not these functions should be able to call each other. In the ARM family this possibility comes at a code size and performance cost, because the instruction sequence required for a function call becomes more complex. Hence, it is not enabled automatically and must be turned on manually, by means of the `-mthumb-interwork` option.

A second group of options, still related to the instruction set of the target architecture, concerns floating point calculations and is more closely related to embedded systems development. In fact, in many processors designed for embedded applications (and unlike most general-purpose processors) the hardware floating-point unit (FPU) is an *optional* component, which is often not included on the chip due to cost and power consumption considerations.

On systems without hardware support for floating-point instructions, floating-point calculations can still be performed in software, by means of two distinct approaches:

1. The compiler still generates floating-point instructions, as it would if the FPU were available. At *runtime*, the processor will trap them as undefined instructions as soon as it attempts to execute them, because there is no FPU. In turn, the trap handler will invoke an appropriate software module that is responsible for emulating the floating-point instruction, and then resume execution.
2. At *compile time*, the compiler generates calls to a floating-point library instead of floating-point instructions. The resulting object code is then linked against the floating-point library, which is responsible for performing floating-point operations in software, to build the executable image.

These two approaches are chosen by means of the options `-mhard-float` and `-msoft-float`, respectively. The main difference between them is that the first is binary-code compatible with systems actually equipped with a FPU. On the other hand, the second is usually faster on systems without a FPU because there is no extra overhead due to trap handling, which is slower than a regular function call on most processor architectures.

When $-$mhard$-$float is active, it is also necessary to specify the kind of FPU the compiler should target, by means of the $-$mfpu option. This is important because, especially in embedded microcontrollers, the same main processor may be coupled with different kinds of FPU, implementing different trade-off points between performance, complexity, power consumption, and cost. In turn, one FPU differs from another for what concerns the internal design and, often, the floating-point instruction set it supports.

The last group of architecture-dependent compiler options to be discussed in this section has to do with code generation. As an example, two important ones for the ARM family are listed at the bottom of Table 11.4.

- The first one, $-$mstructure$-$size$-$boundary controls how the compiler rounds the size of all structures and unions. Namely, if <n> is the argument of the option, the compiler rounds all sizes to a multiple of <n>.

  A larger number may produce faster code, especially on systems equipped with a cache, because structures and unions are better aligned with cache lines, but it may obviously increase their size, too.

  Moreover, software modules compiled with different values of <n> are potentially incompatible, and hence, this option must be used consistently across all modules that are linked together in the same executable image.

- The second option, $-$mlong$-$calls, specifies that the compiler must generate function call code that works regardless of how distant the target function is, unless it can prove that the function is reachable by means of the default "short" function call sequence.

  The option has a rather coarse level of granularity because it applies to all source code modules mentioned on the compiler command line, and hence, to all functions defined in them. It is also possible to choose the calling method to be used by the compiler more precisely, on a function-by-function basis, by means of a non-standard language extension—that is, a function attribute—described in Chapter 9.

  This is yet another example of how the same goal, in this case changing the way the compiler generates the function call code, can be achieved in two different ways. The first one does not affect source code portability, because it consists of a compiler option and does not even require any change to the source code itself.

  The second one is more sophisticated and powerful, but it does require updates to the source code and, what is worse, those updates make use of non-standard language features that might not be available on a different toolchain or architecture.

11.4   SOURCE-LEVEL OPTIMIZATION: A CASE STUDY

In this section, we examine how it was possible to optimize the C-language implementation of an algorithm of practical interest, mainly with respect to *execution time* and *jitter*, by working exclusively at the source code level.

The results are even more interesting because a satisfactory level of optimization was indeed achieved across different processor architectures with the same updates to the source code and without leveraging any detailed knowledge of the processor machine language. Namely, neither hand-written nor hand-optimized assembly code has been developed.

The two processor architectures being considered are the same ones already used as examples throughout the book. Namely, the first one is represented by the NXP LPC2468 microcontroller [126], which is a low-end component based on an ARM7TDMI-S processor core and running at 72 MHz. The second one is represented by NXP LPC1768 microcontroller [128], based on the contemporary ARM Cortex-M3 processor core running at 100 MHz.

11.4.1   SUMMARY OF THE ALGORITHM

The specific case study considered in this chapter concerns the implementation of the 8B9B encoding algorithm, fully described in Reference [31]. The basic idea of this algorithm is to appropriately encode the payload of controller area network (CAN) [91] frames before transmitting them on the network, in order to mitigate transmission time jitter due to *bit stuffing* and improve the error-detection capability of the CAN cyclic redundancy check (CRC). The encoding process is then reverted by receivers, to reconstruct the original payload.

In fact, CAN relies on a non-return to zero (NRZ) transmission scheme that does not send an explicit clock signal along with data. Instead, receivers rely on transitions, or *edges*, between data bits transmitted at opposite levels on the bus to synchronize a digital phase locked loop (DPLL) circuit on the input signal. The phase-corrected DPLL output is then used to sample the input signal and retrieve the incoming bit stream.

In order to ensure that a sufficient number of edges appear on the bus, every time the CAN controller in the transmitting node detects that 5 consecutive bits with the same value have just been sent, it inserts a dummy bit at the opposite level, called *stuff bit*. Symmetrically, CAN controllers in the receiving nodes check that a stuff bit actually appears after 5 consecutive bits at the same value and transparently delete it, before feeding the incoming data to the upper layers of the protocol stack.

Moreover, violations of the bit stuffing rule just mentioned—in particular, the transmission of 6 consecutive bits at the same value, without intervening stuff bits—are used to broadcast error indications to all nodes on the bus.

Unfortunately, besides the main, useful purpose just described, the bit stuffing mechanism has two important negative side effects, too, better described in the following.

- The exact duration of frame transmission depends not only on the size of the payload, as is natural, but also on its *content*. This is because, according to the definition given previously, the CAN controller may or may not insert a stuff bit at a certain position depending on the specific sequence of payload bits found there.

As a consequence, the reception time of a given frame with respect to its transmission time cannot be determined by simply considering the *nominal* duration of frame transmission, calculated according to the payload size plus the size of the frame header and trailer, which are both fixed and known, because the *actual* duration of frame transmission may be longer.

In turn, this gives rise to a reception time *jitter* that may be as high as 24 bit times in the worst case. This also means that if, as often happens, the system relies on the message reception time to trigger critical activities—for instance, actuation and sampling—its timing accuracy is negatively affected to the same extent.

- Another interesting consequence of bit stuffing is the detrimental way it interacts with the CAN CRC mechanism. Introducing a CRC is a very popular way to detect whether or not a frame has been corrupted while it was being transmitted on a communication network.

  Furthermore, as will be described in Chapter 15, besides being adopted in virtually all contemporary networks and buses, the same technique is also in common use in the software domain, to protect critical memory-resident data structures against corruption.

  The mathematical properties of the CRC calculation algorithm foreseen in CAN—and, specifically, of its generator polynomial—determine that it should be able to detect up to 5 erroneous bits, irrespective of where they occur in the frame, as well as error bursts up to 15 bits in length.

  However, due to bit stuffing, error detection capabilities are much worse. As was pointed out in several research works, for instance [36], only two bit errors sufficiently spaced apart in the frame may suffice to defeat the CAN CRC and make it unable to detect those errors.

For both reasons, avoiding the insertion of stuff bits when transmitting the payload assumes practical relevance, especially when dealing with demanding applications characterized by tight timing constraints and/or heavy dependability requirements. As was shown in other research work, it is also possible to avoid the insertion of stuff bits in all the other parts of the frame besides the payload, thus reaching completely jitterless CAN communication [32].

As shown in Figure 11.6, the basic principle of 8B9B encoding is fairly simple and revolves around the following steps.

- The encoder retrieves the data it operates on from an *input buffer* (shown on the left of the figure) and stores its result into an *output buffer* (on the right of the figure). Both buffers are defined as an array of bytes.
- The input buffer contains k payload bytes, with $0 \leq k \leq 7$, to be encoded. As shown in the figure, they are naturally aligned on byte boundaries.
- The encoder translates, by means of a suitable *forward lookup table* each individual payload byte into a *9-bit* codeword. In abstract, the table consists of $2^8 = 256$ entries, and each entry is 9 bits wide.

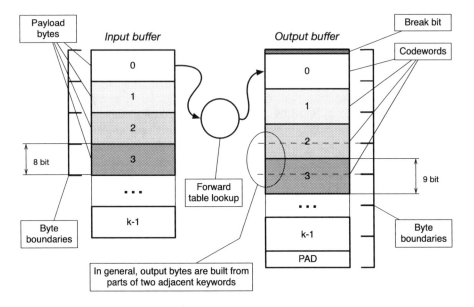

Figure 11.6   Basic principle of 8B9B encoding.

- Codewords are packed into the output buffer one after another and are preceded by a *break bit*. Because of the presence of the break bit itself and due to their size, codewords are generally *not* aligned on byte boundaries.
- After the last codeword (codeword $k-1$ in the figure) has been stored into the output buffer, it is also necessary to *pad* the buffer with a suitable bit pattern to completely fill the last byte.
- It can be proved that, except when $k = 0$, the number of bytes needed in the output buffer is always equal to $k+1$, when encoding k payload bytes. The case $k = 0$ is handled trivially, since no actual encoding is done in this case.

A thorough explanation of the design of the 8B9B algorithm and its lookup table in order to prevent bit stuffing from occurring is beyond the scope of this book. Interested readers are referred to [31], where all the technical details can be found. Informally speaking, 8B9B encoding relies on the following properties:

1. No codewords contain 5 or more consecutive bits at the same value, and hence, they do not trigger the bit stuffing mechanism by themselves.
2. No codewords start or end with 3 or more consecutive bits at the same value. For this reason, the bit pattern appearing across two consecutive codewords may not trigger the bit stuffing mechanism.
3. The break bit prevents 5 or more consecutive bits from appearing across the boundary between the encoded payload and the field that precedes it in the CAN frame, namely, the data length code (DLC) that is equal to the size of the en-

coded payload, expressed in bytes. For this reason, the break bit is chosen as the complement of the least significant bits of the DLC.

4. The pad is made of an alternating bit pattern of suitable length. Besides making the encoded payload size an integral number of bytes—and hence, making it suitable for transmission—it also has the same purpose as the break bit. Namely, it prevents 5 or more consecutive bits at the same level from appearing across the boundary between the encoded payload and the next field of the CAN frame.

In this book, we are instead more concerned about two general characteristics of the 8B9B encoding process, because they are common to many other algorithms. In particular:

1. Individual payload bytes are translated into items of a different size (9 bits in this case) by means of a *lookup table*. Since the table resides in main memory, it must be implemented and consulted in the most efficient way, for what concerns both storage space and access time.
2. Codewords are not an integral number of bytes in size and must be tightly *packed* in the output buffer. Therefore, codewords cross byte boundaries in the output buffer and each output byte, in general, consists of two adjacent sections. In turn, these sections consist of parts of two adjacent codewords.

   For instance, as shown in Figure 11.6, the fourth output byte consists of the trailing part of codeword 2 (light gray) as well as the leading part of codeword 3 (darker gray). As a consequence, output bytes must be built by *shifting* and *masking* two adjacent keywords into place.

For the sake of completeness, it must be mentioned that the 8B9B decoder proceeds symmetrically with respect to the encoder and, from the algorithmic point of view, it is quite similar to it. For this reason, only the encoder will be further analyzed in the following.

11.4.2   BASE SOURCE CODE

A straightforward implementation of the 8B9B encoding algorithm, to be used as the starting point for optimization, is shown below and commented on in the following.

```
1   typedef struct {
2       int dlc;
3       uint8_t data[8];
4   } payload_t;
5
6   #define PAD 0x166 /* Pad on 9 bits */
7
8   void encode(payload_t *in, payload_t *out)
9   {
10      int i;
11
12      int os;
13      int w, l, r;
14
15      assert(in->dlc <= 7);
16
```

```
17        if(in->dlc == 0)
18            out->dlc = 0;
19
20        else
21        {
22            /* If in->dlc is even <-> out->dlc is odd, except when in->dlc
23               is zero, but this case has already been handled.
24            */
25            out->dlc = in->dlc + 1;
26            out->data[0] = (out->dlc & 1) ? 0 : 128; /* Break Bit */
27
28            os = 7; /* Available bits in output byte i */
29            for(i=0; i<in->dlc; i++)
30            {
31                w = flt[in->data[i]]; /* 9 lsb */
32                l = w & lm[os];        /* os msb, msb @8 */
33                r = w & rm[os];        /* os lsb, lsb @0 */
34
35                /* Due to the 9/8 relationship, the presence of the break
36                   bit, and the limited maximum length of the payload, the
37                   9 output bits are always split between two output
38                   bytes.  In other words, os cannot be zero and the code
39                   becomes simpler.
40                */
41                out->data[i]   |= (l >> (9-os));
42                out->data[i+1] = (r << (os-1));
43                os--;
44            }
45
46            /* Pad */
47            out->data[i] |= (PAD >> (9-os));
48        }
49    }
```

- The encoding algorithm is implemented by the function `encode`, whose definition starts at line 8 of the listing. The function has two parameters, the input buffer `in` and the output buffer `out`.
- Both parameters are pointers to a data structure of type `payload_t`. As specified in lines 1–4 of the listing, this structure consists of
  - a fixed-size data buffer of 8 bytes, `data[]`, and
  - a length field `dlc` to indicate how many bytes of `data[]` are significant.

  Both the input and the output data buffers are always filled from the beginning, that is, from element zero.
- The `encode` function first ensures that the input buffer size is acceptable, by means of an assertion (line 15) and handles the special case of an empty input buffer appropriately (lines 17–18). In this case, in fact, no encoding is necessary and the function simply returns an empty output buffer.
- The actual encoding process starts at line 25, where the output buffer size is calculated. As described above, it can be proved that, if the input buffer is not empty, the output buffer size is always one byte more than the input buffer size.
- The next statement (line 26) computes the break bit as the complement of the output buffer size and stores it in the most significant bit of the first output data byte, that is, `out->data[0]`.

- In order to handle codeword packing appropriately, the variable `os` keeps track of how many bits are free in the current output byte during encoding. This variable is initialized to 7 at line 28, before entering the encoding loop, because the first output data byte has 7 free bits (in fact, one is taken by the break bit).
- The encoding loop (lines 29–44) is executed once for each input byte. Therefore, the `i`-th iteration of the loop handles input byte `in->data[i]` (line 31), storing the corresponding codeword partly into output byte `out->data[i]` and partly into `out->data[i+1]` (lines 41–42).
- The body of the encoding loop first consults the lookup table `flt[]` to obtain the codeword corresponding to `in->data[i]` and stores it into `w` (line 31). Since codewords are 9 bit wide, the most straightforward way of defining the lookup table is as an array of 256 read-only 16-bit integers, that is,

```
const uint16_t flt[] = { ... };
```

where the ... within braces is a placeholder for the array contents, which are not shown for conciseness.
- Then, the codeword must be split into two parts:
  1. the most significant part of the codeword, consisting of `os` bits, which must be stored in the least significant part of the current output byte `out->data[i]`;
  2. the least significant part of the codeword, consisting of `9-os` bits, which must be stored in the most significant bits of the next output byte `out->data[i+1]`.

In order to isolate and extract, or *mask*, the two parts the code makes use of two auxiliary lookup tables, `lm[]` and `rm[]`, respectively, defined as:

```
const uint16_t lm[8] =
  {0x000, 0x100, 0x180, 0x1C0, 0x1E0, 0x1F0, 0x1F8, 0x1FC};
const uint16_t rm[8] =
  {0x000, 0x0FF, 0x07F, 0x03F, 0x01F, 0x00F, 0x007, 0x003};
```

As can be seen from the listing above, element `lm[os]` contains a 9-bit mask with the `os` most significant bits set to 1 and the remaining ones set to 0. Symmetrically, `rm[os]` contains a 9-bit mask with `9-os` least significant bits set to 1 and the others to 0. In both arrays, the element with index 0 is unused.

In this way, as is done at lines 32–33 of the listing, the two parts of the codeword can be masked by means of a bit-by-bit and operation. The results are stored in local variables `l` and `r` for the left and right part, respectively.
- As discussed in the algorithm description, before storing them in the output bytes, `l` and `r` must be appropriately *shifted*, too. This is done directly in the assignments to the output bytes at lines 41 and 42 of the listing. Determining that the direction and amount of shift specified in the listing are correct is left as an exercise to readers.

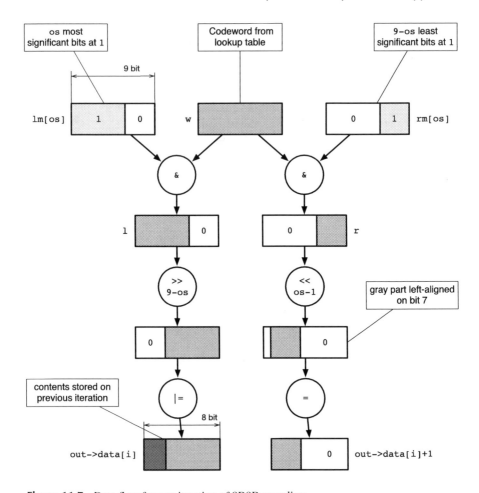

Figure 11.7   Data flow for one iteration of 8B9B encoding.

An additional point worth remarking is that, for what concerns the current output byte out->data[i], the assignment is done through a bit-by-bit or operation. This is because part of the current output byte has already been filled in the previous iteration of the loop (or before entering the loop for the first iteration), and hence, its contents must not be overwritten.

Instead, the next output byte out->data[i+1] has not been used so far, and hence, a direct assignment is appropriate. This way brings the additional benefit of setting the rest of out->data[i+1] to zero.

- The overall data flow pertaining to one iteration of the 8B9B encoding loop is summarized in Figure 11.7. In the figure, rectangular boxes represent data items and circles denote operations. When a data item is stored in a named variable, rather than being just the value of an expression, the variable name is shown beside the box, too.

- The last instruction within the loop body, at line 43, updates the value of `os` to prepare for the next iteration. It is in fact easy to prove that, if `os` bits were available for use in the current output byte, `os-1` bits are left for use in the next output byte, after storing a 9-bit codeword across them.
- The very last operation performed by the `encode` function is to pad the last output byte with a portion of the alternating bit pattern assigned to the `PAD` macro. This is done at line 47 that, quite unsurprisingly, bears a strong similarity with line 41.

As can be seen from the listing, in the initial implementation most effort has been spent to ensure that the code is *correct*—that is, it matches the definition of the 8B9B algorithm outlined in Section 11.4.1—rather than *efficient*.

This is a rather widespread approach in software design and development, in which an initial version of the code is produced first, in order to verify algorithm correctness. Then, possibly based on code inspection, benchmarking and profiling (when supported by the underlying target architecture) critical parts of the code are identified and optimized.

This also helps to avoid the so-called "premature optimization" issues, in which effort is spent optimizing parts of code that turn out to be unimportant for overall performance at a later stage.

11.4.3 OPTIMIZATIONS

In order to assess the encoder performance and gather information regarding possible optimizations, the software module discussed in the previous section was evaluated by encoding a large set of uniformly distributed, non-empty pseudo-random messages of varying lengths.

Namely, the software was embedded in a software test framework and cross-compiled for the NXP LPC2468 microcontroller [126]. Then, the amount of time spent in the encoding function, as a function of the payload size, was measured by means of a free-running, 32-bit counter clocked at the same clock frequency as the CPU. In this way, it was possible to collect a cycle-accurate encoding delay measurement for each message.

Working on a large number of pseudo-random messages was convenient to single out any data dependency in the delay itself. External sources of measurement noise were avoided by running the experiments in a very controlled environment, with interrupts disabled and unused peripheral devices powered down, to avoid unexpected activities on the internal microcontroller buses.

As can be expected, the overall delay of the encoding function has two distinct components.

1. A constant part b, which can be attributed to the prologue and epilogue code of the function, as well as local variables initialization.
2. A part linearly dependent on the payload length, caused by the encoding and decoding loops. Denoting with k the payload length, in bytes, this part can be expressed as $q \cdot k$, where q is a constant.

Table 11.5

Interpolated Encoder Delay, ARM7 Architecture

Code version	Constant part b (μs/byte)	Linear part q (μs/byte)
Base version (before optimization)	5.06	2.48
Code, data, and stack placement	1.18	0.50
Computed masks	1.09	0.43
Load and store reduction	1.03	0.36
Folded lookup table	1.08	0.43

Overall, the encoding time $t_e(k)$ can be modeled as a linear function of k as

$$t_e(k) = b + q \cdot k \; . \tag{11.1}$$

The two contributions to the delay, given by b and k, have been highlighted in Table 11.5, which shows the results of a linear fitting on the experimental data. Namely, the first row of the table pertains to the performance evaluation of the unoptimized, base code described in Section 11.4.2.

The experimental data clearly show that the encoder is very slow, namely, the amount of time needed to encode a payload of 7 bytes exceeds $22\,\mu$s according to (11.1) and the values of b and q shown in the first row of Table 11.5. Considering that the microcontroller runs at 72 MHz, the delay therefore exceeds 1584 clock cycles.

Even more importantly, the experimental results also show that the delay is affected by a significant amount of variability, or *jitter*, which increases as k grows, and is of the order of $1\,\mu$s in the worst case.

This becomes particularly disturbing when we contemplate one of the two main goals of the encoder itself, that is, reduce CAN communication jitter. To this purpose, it is of utmost importance to achieve full software execution time determinism, that is, ensure that the execution time of the encoding/decoding algorithms only depends on payload size, and not on any other factors.

After all, if full determinism were impossible to achieve, the original source of jitter (bit stuffing) would be simply replaced by another one (encoder jitter). On the other hand, a linear dependency on the payload size k, as expressed in (11.1), is both expected and acceptable.

Code, data, and stack placement

After analysis, and recalling the concepts outlines in Chapters 2 and 3, it was possible to determine that the slowness was due to the *kind of memory* where code, data, and stack were placed.

In fact, by default, the toolchain used for the evaluation (as configured by its default linker script) places all these components into a bank of DRAM external to the microcontroller, which is accessed by means of a 16-bit data bus, to reduce costs.

On the other hand, jitter is due to DRAM refresh cycles, which stall any instruction or data access operation issued by the CPU while they are in progress. For this architecture, both issues were solved by directing the toolchain to move code, data, and stack into the on-chip SRAM. The goal was achieved by means of two different techniques:

1. A non-standard extension to the C language, namely *object attributes*, were used to place critical code and data in their own sections. Then, the linker script was modified to allocate these sections in SRAM rather than DRAM. The whole procedure has been discussed in detail in Chapter 9.
2. Moving the stack into a non-default memory area requires a slightly different approach because, unlike code and data, it is not an object explicitly defined and visible at the C language level.

   Within the toolchain used in the case study, the C runtime library makes use of a symbol, to be defined in the linker script and called `__stack`, to locate the top of the stack. Therefore, moving the stack to a different memory bank implies examining the linker script and replacing the default definition with a different one.

The optimization just described satisfactorily achieved two important goals.

- As can readily be seen by looking at the second row of Table 11.5, coefficient q, which represents the slope of the linear part of the encoding delay, was reduced by a factor close to 5. The constant part of the encoding delay b improved similarly, too.
- Furthermore, all of the subsequent measurements revealed that jitter was *completely removed* from encoding delay.

It must also be remarked that this kind of behavior is not at all peculiar to the ARM7 processor architecture or the use of DRAM. On the contrary, it is just a special case of a more widespread phenomenon that should be taken into account in any case during embedded software development.

For instance, the processor exhibited a very similar behavior when the lookup table was stored in flash memory. In this case, the jitter was due to a component whose purpose is to mitigate the relatively large access time of flash memory (up to several CPU clock cycles in the worst case), by predicting future flash memory accesses and performing them in advance.

In this way, if the prediction is successful, the processor is never stalled waiting for flash memory access, since processor activities and flash memory accesses proceed concurrently. As was described in Chapter 2, this component is present in both the NXP LPC24xx and LPC17xx microcontroller families and is called memory accelerator module (MAM).

Unfortunately, although the prediction algorithm is quite effective with instruction fetch, because it proceeds sequentially in most cases, it does not work equally well for accesses to the lookup table—which are inherently random because they depend on payload content—thus introducing a non-negligible variability in table access time.

Overall, this first optimization result underlines the paramount importance of code optimization when developing critical code for an embedded system. In this particular case, it was possible to increase the performance by a factor of about 5 by simply choosing the right kind of memory to be used for code, data, and task storage. This is a relatively straightforward decision that, nonetheless, the toolchain was unable to make autonomously.

Computed masks

In the base implementation shown and commented on in Section 11.4.2, two auxiliary lookup tables, `lm[]` and `rm[]`, hold the bit masks that, as shown in Figure 11.7, are used on every iteration of the encoding loop to split the 9-bit value w obtained from the forward lookup table into two parts `l` and `r`, respectively.

In the optimization step to be discussed next, the bit masks have been computed directly, and hence, `lm[]` and `rm[]` have been deleted. The decision has been driven by a couple of observations, detailed in the following.

- Regardless of the kind of memory being used, memory accesses are, in general, slower than calculations that are carried out only within processor registers.
- In this case, the computation of the bit masks can be carried out in a straightforward way, by observing that:
  - For any `i`, `lm[i]` and `rm[i]` are the bitwise complement of each other. Namely, `rm[i]` can readily be generated by complementing `lm[i]`.
  - Upon each iteration of the encoding loop, exactly two bit masks are used and the two lookup tables are scanned in sequence. Namely, the first iteration of the look makes use of `lm[7]` and `rm[7]`, the second one makes use of `lm[6]` and `rm[6]`, and so on.
  - By definition, for any valid value of `i`, it is

$$lm[i-1] = lm[i] << 1 . \tag{11.2}$$

The listing of the optimized `encode` function is shown in the following.

```
 1    void encode(payload_t *in, payload_t *out)
 2    {
 3        int i;
 4
 5        /* int os: suppressed */
 6        int w, l, r;
 7        int lm;
 8
 9        assert(in->dlc <= 7);
10
11        if(in->dlc == 0)
12            out->dlc = 0;
13
14        else
15        {
```

```
16          /* If in->dlc is even <-> out->dlc is odd, except when in->dlc
17             is zero, but this cannot happen.
18          */
19          out->dlc = in->dlc + 1;
20          out->data[0] = (out->dlc & 1) ? 0 : 128; /* Break Bit */
21
22          lm = 0xFFFFFFFC;
23          for(i=0; i<in->dlc; i++)
24          {
25              w = flt[in->data[i]]; /* 9 bits, lsb @0 */
26              l = w & lm; /* 7-i bits, msb @8 */
27              r = w & ~lm; /* 2+i bits, lsb @0 */
28
29              /* Due to the 9/8 relationship, the presence of the break
30                 bit, and the limited maximum length of the payload, the
31                 9 output bits are always split between two output
32                 bytes.
33              */
34              out->data[i] |= (l >> (2+i));
35              out->data[i+1] = (r << (6-i));
36              lm <<= 1;
37          }
38
39          /* Pad */
40          out->data[i] |= (PAD >> (2+i));
41      }
42  }
```

With respect to the base version of the same function, shown previously, the following changes to the code are worth noting.

- The bit mask lm to be used in the first iteration of the loop is set before entering the loop, at line 22. Then, it is updated as specified in (11.2) at the end of each iteration, that is, at line 36.
- Then, lm and its complement are used at lines 26–27, instead of the lookup table entries lm[os] and rm[os].

Furthermore, in an effort to free one processor register and make it available to carry out the above-mentioned masks computation more efficiently, it was observed that the two variables i (the loop index) and os (the number of available bits in the current output byte) are strictly related. Namely, it is

$$os = 7-i \ . \tag{11.3}$$

Hence, all expressions involving variable os have been rewritten to make use of i, according to (11.3), and os was suppressed. Namely:

- the two occurrences of 9−os (lines 34 and 40) have been replaced by 2+i;
- the expression os−1 (line 35) has been replaced by 6−i.

As can be seen by comparing the third row of Table 11.5 with the second, replacing the two auxiliary lookup tables with arithmetic calculations improved the linear part of the encoding delay q by about 14% (namely, from 0.50 to 0.43μs/byte). On the other hand, the constant part of the delay b did not change significantly and went from 1.18 to 1.09μs. This was expected, as the optimization mainly concerned the body of the encoding loop.

Load and store reduction

The base version of the encoding algorithm, on each iteration, loads one input byte and, after using it as an index in the lookup table and performing some computations, as shown in Figure 11.7, stores some data into two adjacent output bytes.

In general, one of the output bytes used in a certain iteration is the same as one output byte already used in the previous iteration, and hence, each output byte is accessed twice.

Bearing in mind again that memory store operations are relatively expensive with respect to register manipulation, a local buffer has been introduced to carry forward the common output byte from one iteration to the next, and hence, store it only once.

In order for this optimization to be effective, it is necessary to ensure, beforehand, that the variable used as local buffer can indeed be allocated in a processor register by the optimizer. Otherwise, the compiler would allocate that variable on the stack and extra accesses to the output bytes would merely be replaced by extra accesses to the stack, with no benefit.

The check can readily be carried out by looking up how many registers are available on the target architecture, and then assessing how many registers the machine code of the `encode` function is using at the moment. The first item of information can be gathered from the processor documentation, while the second one can readily be obtained by using the compiler flags outlined in Section 11.1.3.

The following listing shows how the code has been modified to implement the optimization just described. Since all steps are independent from each other, this optimization step started from the source code produced by the previous one.

```
 1    void encode(payload_t *in, payload_t *out)
 2    {
 3        int i;
 4
 5        /* int os: suppressed */
 6        int w, l, r;
 7        int lm;
 8        int od;
 9
10        assert(in->dlc <= 7);
11
12        if(in->dlc == 0)
13            out->dlc = 0;
14
15        else
16        {
17            /* If in->dlc is even <-> out->dlc is odd, except when in->dlc
18               is zero, but this cannot happen.
19            */
20            out->dlc = in->dlc + 1;
21            od = (out->dlc & 1) ? 0 : 128; /* Break Bit */
22
23            lm = 0xFFFFFFFC;
24            for(i=0; i<in->dlc; i++)
25            {
26                w = flt[in->data[i]]; /* 9 bits, lsb @0 */
27                l = w &  lm; /* 7-i bits, msb @8 */
28                r = w & ~lm; /* 2+i bits, lsb @0 */
29
30                /* Due to the 9/8 relationship, the presence of the break
31                   bit, and the limited maximum length of the payload, the
```

```
32                    9 output bits are always split between two output
33                    bytes.
34               */
35               out->data[i] = od | (l >> (2+i));
36               od = (r << (6-i));
37               lm <<= 1;
38          }
39
40          /* Pad */
41          out->data[i] = od | (PAD >> (2+i));
42     }
43  }
```

Referring back to the listing, the following parts of the encode function have been modified:

- The newly-defined local variable od (line 8) is the local buffer used to carry forward the common output byte from one iteration to the next.
- The first output byte out->data[0] is in common between the break bit store operation and the first iteration of the loop. For this reason, the break bit store operation at line 21 has been modified to store the break bit into od instead of directly into memory.
- Upon each iteration of the loop, the operation at line 35 operates on an output byte partially filled by a previous iteration of the loop (or the break bit store operation, upon the first iteration) and fills it completely. Accordingly, it operates on od and stores the result into memory, out->data[i].
- On the contrary, the operation at line 36 fills the next output byte only partially. Therefore, the result must not be stored into out->data[i+1] yet, because it must be completely filled in the next iteration of the loop (or in the padding operation) beforehand. For this reason, the partial result is stored into od and carried forward.
- Finally, the padding operation at line 41 fills the last output byte completely. Therefore, it takes od as input and stores the result into memory.

Attentive readers may have noticed that both this optimization and the previous one mimic some automatic compiler optimizations discussed in Section 11.1. This is indeed true and further highlights the interest of knowing how compiler optimizations work, at least to some level of detail.

This knowledge is, in fact, useful to "assist" the compiler by manual intervention on the source code if, for any reason, the compiler is unable to automatically apply a certain optimization technique by itself.

Folded lookup table

The final optimization to be discussed is quite different from the previous ones. In fact, it is related to the encoder's *memory footprint* rather than execution time. Moreover, it highlights the fact that, sometimes, effective optimization relies on a thorough understanding of the algorithm to be implemented and its properties. In other words, it is sometimes better to work at an higher level of abstraction rather than focusing exclusively on low-level aspects of the architecture.

In this case, the starting point of the optimization is a basic symmetry property of the lookup table used by the encoder. In fact, it can be proved that, due to the way the table has been generated, it is

$$w = \text{flt}[d] \iff \sim w = \text{flt}[\sim d] \tag{11.4}$$

where, as usual, $\sim$ denotes the bitwise complement operator.

Informally speaking, this is because the properties of the patterns which were selected as table entries—summarized in Section 11.4.1—imply that if w is a valid pattern, also $\sim w$ is necessarily valid.

Moreover, since patterns were stored into the table in increasing numerical order, the most significant bit of the lookup table is always 0 in the first half of the table, whereas it is always 1 in the second half.

With the help of these properties, it is possible to reduce the size of the lookup table to one quarter of its original size by "folding" it twice.

1. Namely, using property (11.4), it is possible to store only half of the table and reduce its size from 256 to 128 9-bit entries. For efficient access, the size of each entry must nevertheless be an integral number of bytes, and hence, each entry actually occupies 16 bits in memory.
2. However, due to the second property stated previously, it is not necessary to store the most-significant bit of table entries explicitly and the lookup table can be further shrunk down to 128 8-bit entries.

The following listing shows the source code of the *encode* function after optimization.

```
1   const uint8_t flt[] = { ... };
2
3   void encode(payload_t *in, payload_t *out)
4   {
5       int i;
6
7       /* int os: suppressed */
8       int w, l, r;
9       int lm;
10      int od;
11      int x, m;
12
13      assert(in->dlc <= 7);
14
15      if(in->dlc == 0)
16          out->dlc = 0;
17
18      else
19      {
20          /* If in->dlc is even <-> out->dlc is odd, except when in->dlc
21             is zero, but this cannot happen.
22          */
23          out->dlc = in->dlc + 1;
24          od = (out->dlc & 1) ? 0 : 128; /* Break Bit */
25
26          lm = 0xFFFFFFFC;
27          for(i=0; i<in->dlc; i++)
28          {
29
```

```
30              x = in->data[i];
31              m = (x >= 0x80) ? 0x1FF : 0x000;
32              w = flt[(x ^ m) & 0x7F] ^ m; /* 9 bits, lsb @0 */
33
34              l = w &  lm; /* 7-i bits, msb @8 */
35              r = w & ~lm; /* 2+i bits, lsb @0 */
36
37              /* Due to the 9/8 relationship, the presence of the break
38                 bit, and the limited maximum length of the payload, the
39                 9 output bits are always split between two output
40                 bytes.
41              */
42              out->data[i] = od | (l >> (2+i));
43              od = (r << (6-i));
44              lm <<= 1;
45          }
46
47          /* Pad */
48          out->data[i] = od | (PAD >> (2+i));
49      }
50  }
```

With respect to the previous version of the code, the following updates concerning lookup table definition and access have been performed.

- According to the folding process, the lookup table definition (line 1) defines lookup table elements to be of type uint8_t (8-bit unsigned integer) instead of uint16_t (16-bit unsigned integer). The initializer part of the definition (shown as . . . in the listing for conciseness) now contains 128 entries instead of 256.
- Two additional local variables x and m (defined at line 11) hold, upon each iteration of the encoding loop, the current input byte as retrieved from the input buffer and an appropriate bit mask to complement it, if necessary, by means of an on-the-fly exclusive or operation upon lookup table access.
- Namely, at line 31, m is set to either 0x1ff (9 bits at 1) or 0 depending on whether the current input byte x was negative or positive.
- The mask m is used twice at line 32:
  1. It is used to conditionally complement the input byte x before using it as an index in the lookup table, by means of the exclusive or operation x^m. Since the mask is set to ones if and only if x was negative, the overall result of the exclusive or is to complement x if it was negative, or return it directly if it was either zero or positive.
  In any case, since the mask is 9-bit wide whereas the table has $2^7 = 128$ entries, the result of the exclusive or must be truncated to 7 bits before use, by means of the bitwise and operation & 0x7f.
  2. After lookup table access, the same mask is used again to conditionally complement the lookup table entry before storing it into w.

It is useful to remark why this computation has been carried out by means of conditional *expressions* and *exclusive or* operations instead of using a more straightforward conditional *statement*.

This was done in order to make execution time independent from predicate truth. For instance, the execution time of a conditional statement in the form

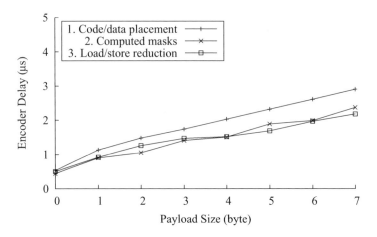

Figure 11.8 Encoder delay on Cortex-M3 architecture, loop unrolling enabled.

`if (x > 0x80) x = ~x` depends on the sign of x, because the bitwise complement and the assignment are performed only if x is negative.
On the contrary, the execution time of the conditional expression `m = (x >= 0x80) ? 0x1FF : 0x000` is constant, because both sides of the conditional expression have got the same structure. The expression `x^m` is executed in constant time as well.

The performance of this version of the code, determined in the usual way, is shown in the bottom row of Table 11.5. As can be seen from the table, unlike the previous optimizations, table folding is not "one-way," because it entails a trade-off between footprint reduction and loss of performance.

In fact, the price to be paid for the previously mentioned fourfold reduction in lookup table size is a performance penalty that, basically, brings the execution time of the encoder back to almost the same values measured before load and store reduction. The linear part of the delay is the same ($k = 0.43\mu s$/byte in both cases) and the difference in the constant part of the delay is minimal (1.09 versus $1.08\mu s$).

Unexpected optimization outcomes

In the last part of this section, we will highlight how sometimes optimization outcomes may be surprising and, at first glance, hard to explain. Previously, we described how the dependency of the encoding delay from the payload size could be expressed, as it looks natural, as a linear function (11.1) on the ARM7 processor architecture.

On the contrary, when the same experiments were repeated on the more modern Cortex-M3 architecture, the results were quite different, even with the same toolchain configuration and options. The results are shown in Figure 11.8 and show that, albeit

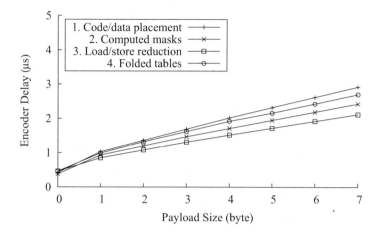

Figure 11.9 Encoder delay on Cortex-M3 architecture, loop unrolling disabled.

the sample variance was still zero in all cases—meaning that the architecture nevertheless behaved in a fully deterministic way—the clean, linear relationship between the payload size k and the encoding time found with the ARM7 architecture was lost after code optimization.

After some further analysis of the optimizations automatically put in place by the compiler, the reason for the peculiar behavior was identified in a specific optimization, that is, *loop unrolling*. According to the general description given in Section 11.2, in this particular scenario the compiler partially unrolled the encoding loop, by a factor of two.

In other words, in an effort to achieve a better execution efficiency, the compiler duplicated the loop body so that a single loop iteration handles up to two input bytes and this gave rise to the up/down pattern in the delays, depending on whether k is odd or even.

When this optimization was turned off, by means of the -fno-unroll-loops option, linearity was almost completely restored, as shown in Figure 11.9, at a minor performance cost. The comparison between Figures 11.8 and 11.9 confirms that the compiler took the right choice when applying the optimization, because the *average* performance of the unrolled encoding loop, across the whole range of values of k, is indeed better.

However, in the case being considered here, the optimization also leads to other detrimental effects that the compiler's optimization algorithms do not consider at all, namely, encoding time linearity. This is one of the reasons why the compiler offers the possibility of turning off optimizations on an individual basis. Interestingly enough, loop unrolling did not take place on the ARM7 architecture. This difference is most probably due to the different compiler versions in use and to the dissimilarities between the ARM7 and Cortex-M3 architectures.

To conclude this section, we observe that, as shown in Figure 11.9, the very same optimizations that were effective on the ARM7 architecture brought similar performance improvements on the Cortex-M3 architecture, too.

11.5 SUMMARY

The main topic of this chapter was the improvement of code performance and footprint by working at the toolchain and source code levels. In order to do this the first section of the chapter, Section 11.1, contains an overview of the compiler workflow and optimizations.

The next two sections build on this knowledge to describe how some compiler options affect code generation and allow programmers to achieve different trade-offs between performance and footprint. In particular, Section 11.2 is about architecture-independent options, while Section 11.3 discusses architecture-dependent options.

On the other hand, Section 11.4 describes source-level code optimization. To enhance the practical relevance and usefulness of the discussion, the whole section makes use of a real-world embedded algorithm as a case study, rather than focusing exclusively on theoretical concepts.

12 Example: A MODBUS TCP Device

CONTENTS

This chapter concludes the first part of the book. It complements the smaller examples contained in the previous chapter with a more comprehensive case study. In it, readers can follow the whole development process of an embedded system from the beginning to the end.

At the same time, they will get more practical information on how to apply the techniques learned in the previous chapters and how to leverage and bind together existing open-source software modules to effectively build a working embedded system of industrial interest. Namely, the case study involves the design and development of a simple MODBUS TCP data logger using a low-cost ARM-based microcontroller, the LPC2468 [126].

12.1 TOOLCHAIN AND OPERATING SYSTEM

The most natural choice for an open-source software development toolchain to be used in the case study was to adopt the GNU Compiler Collection GCC and related components, already described in detail in Chapter 3.

Table 12.1

Software Development Toolchain Components

Component	Purpose	Version
BINUTILS [64]	Binary utilities: assembler, linker, ...	2.20
GCC [62]	C Compiler Collection	4.3.4
NEWLIB [145]	C Runtime Libraries: `libc`, `libm`, ...	1.17
GDB [63]	Source-level debugger	6.6

As a reference, Table 12.1 lists the exact version numbers of those components. It should be noted that, although it is quite possible to build all toolchain components starting from their source code, as in the case study, it is often more convenient to acquire them directly in binary form, especially if the target architecture is in widespread use.

As an example, at the time of this writing, Mentor Graphics offers a free, lite edition of their GCC-based Sourcery CodeBench toolchain [116], previously known as CodeSourcery, which is able to generate code for a variety of contemporary processor architectures, including ARM.

Another aspect already noted in Chapter 3 is that, more often than not, component version numbers to be used shall not be chosen "at random." Similarly, always choosing the latest version of a component is often not a good idea, either.

This is due to several main factors, which must all be taken into account in the decision.

- Toolchain components work together in strict concert. However, they are maintained by different people and the coordination between the respective software development groups is loose.

  Even though some effort is made to ensure backward compatibility whenever a component is updated, there is absolutely no guarantee that, for instance, *any* version of the compiler works with *any other* version of the assembler.
- Merely downloading the latest version of all components roughly at the same time does not provide additional guarantees as well, for a similar reason. This is because component release dates are not tightly related to each other.
- There may be adverse interactions between the version numbers of the *native* toolchain used to build the cross-compiling toolchain and the version numbers of the *cross-compiling* toolchain to be built.

Hence, the best approach in selecting toolchain component version numbers is to make use of all the available information. Often, the documentation of the major software components to be included in the project is very valuable from this point of view.

For instance, most real-time operating system developers give some suggestions about which version numbers are the best ones. They are usually the version numbers the developers themselves used when they compiled the operating system on the target architecture in order to test it.

Since the operating system—and, in particular, its porting layer, as discussed in Chapter 10—is a very complex component, it is likely to leverage even obscure or little-used parts of the toolchain. It is therefore sensible to assume that, if the toolchain was able to successfully build it, it will likely also be able to build other, simpler software modules.

In addition, operating system developers may also offer several useful *patches*— that is, corrections to the source code, to be applied to a specific version of a toolchain component—that fix toolchain problems they have encountered during development and testing.

Last, but not least, the *release notes* of each individual toolchain component give more information about known component (in)compatibilities and additional dependencies among components. For instance, release notes often include constraints to specify that, in order for a component to work correctly, another component should be more recent than a certain version number.

For what concerns real-time operating systems for embedded applications, a relatively wide choice of open-source products is nowadays available, for instance [17, 131, 146]. They represent different trade-offs between the number of functions they make available through their application programming interface (API) on one hand, with respect to their memory requirement and processing power overhead on the other.

For this case study, the choice fell on the FREERTOS real-time operating system [18, 19], which was thoroughly described in Chapter 5. The main reason for this choice was to minimize memory occupation, because memory is often a scarce resource in small embedded systems. At the same time, as will be better detailed in the following, this operating system still provides all the functions needed to effectively support all the other firmware components.

The port of a small real-time operating system like FREERTOS to a new embedded system platform of choice is usually not an issue. This is because, on one hand, these operating systems are designed to be extremely portable—also due to the limited size and complexity of their porting layer, an example of which was presented in Chapter 10.

On the other hand, their source code package is likely to already support the selected architecture with no modifications required. In this case, operating system deployment becomes even easier because a working C compiler is all that is needed to build and use it.

12.2   GENERAL FIRMWARE STRUCTURE

The overall structure of the firmware that realizes the MODBUS TCP data logger is summarized in Figure 12.1. In the figure, major components are represented by white blocks. In addition, light gray blocks represent the blocks' APIs and dark gray

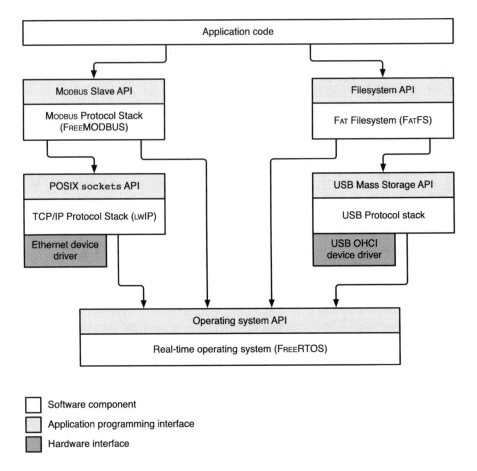

Figure 12.1 General structure of the MODBUS device firmware.

blocks represent the hardware interface of blocks that interact with some hardware device. Hierarchical relationships among components that make use of each other are represented by arrows.

Furthermore, the C language runtime libraries have been omitted from the figure for clarity, because all the other modules make use of, and depend on, it.

As can be seen, some firmware components have already been thoroughly discussed in the previous chapter and will only be briefly recalled here. Namely:

- The FREERTOS *operating system*, presented in Chapter 5 and shown at the bottom of the figure, implements scheduling, basic timing, and inter-task communication and synchronization services, and makes them available to all the other components through its native API.
- The *TCP/IP protocol stack* used for the case study is LWIP [51], depicted on the center left of Figure 12.1 and fully discussed in Chapter 7. With

respect to other open-source products, it was chosen because it represents a good trade-off between tiny protocol stacks like UIP [53] (by the same author), and feature-rich protocol stacks, usually derived from the Berkeley BSD general-purpose operating system [115] and characterized by a much bigger memory footprint.

In addition, LWIP has already been used successfully for distributed computing with embedded systems, for instance in [137], and its UDP protocol performance has been proven satisfactory for an embedded computing environment [38].

Last, but not least, another advantage of choosing a simple, streamlined protocol stack is that it is relatively easy to use and its adaptation to new processor architectures and network devices is faster, easier, and produces more reliable code. For what concerns this aspect the two main interfaces of LWIP, besides its user APIs, are:

1. An adaptation layer toward the operating system API, presented in Chapter 7, along with LWIP iself.

2. A *network device driver*, in this case toward the embedded Ethernet controller of the LPC2468. The design and implementation of this driver was considered as a case study in Chapter 8.

Other components are more typical of the application considered as a case study and will be better described in the following. Namely, they are:

- An open-source MODBUS *slave protocol stack*, FREEMODBUS [170], which implements the well-known MODBUS TCP industrial communication protocol [119, 120] and will be more thoroughly described in Section 12.3. As shown on the top left of the figure, it is layered on top of LWIP for what concerns network communication, and makes use of several operating system services for internal communication and synchronization.

- A file allocation table (FAT) *filesystem* module, a universal serial bus (USB) *protocol stack* and its associated *device driver*, which interfaces with the USB open host controller interface (OHCI) [41] device embedded in the LPC2468.

All these modules will be presented in Section 12.4. Together, they give to the system the ability to access flash drives and other kinds of removable USB mass storage devices, for which the FAT filesystem [118] is currently the most popular choice.

Accordingly, there are several open-source, portable implementations currently available, for instance [35, 123, 172]. All of them have similar characteristics, for what concerns footprint and performance, and similar capabilities. Among them, FATFS [35] was chosen for the case study.

The USB-related software modules were developed starting from the ones provided by the microcontroller vendor [124], suitably improved as described in the following.

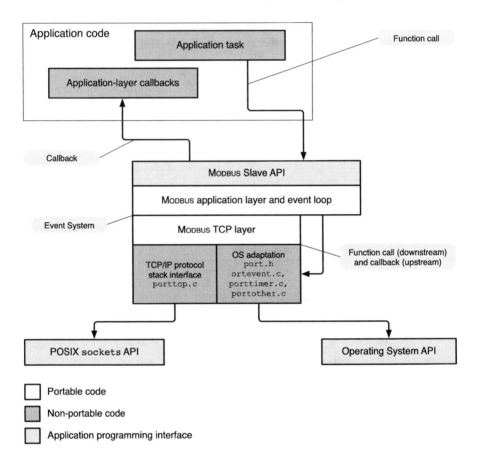

Figure 12.2   Inner structure and interfaces of the MODBUS slave protocol stack.

- Finally, the *application code*, shown at the top of Figure 12.1, glues all components together and performs *data logging*. Namely, it stores on a USB flash-disk resident file information about the incoming MODBUS TCP traffic, which can be useful to debug MODBUS master protocol stacks and gather performance data about them.

12.3   MODBUS **SLAVE PROTOCOL STACK**

The internal organization of the FREEMODBUS protocol stack is based on a layered design, depicted in Figure 12.2. In the following, individual layers will be discussed in bottom-up order. In the figure, layers are color-coded based on their role and characteristics. In particular:

- *White* blocks indicate portable layers, which can be moved from one architecture/project to another by simply recompiling them.

Table 12.2

FREEMODBUS **TCP/IP Protocol Stack Interface Functions**

Function	Description
Initialization and shutdown	
xMBTCPPortInit	Initialize interface and listen for client connections
vMBTCPPortDisable	Forcibly close all client connections
vMBTCPPortClose	Shut down communication completely
Connection management	
xMBPortTCPPoll	Accept incoming client connections, receive requests
Data transfer	
xMBTCPPortGetRequest	Retrieve a request previously received
xMBTCPPortSendResponse	Send a response to the last request retrieved

- *Light gray* blocks indicate APIs that a certain module offers, or makes use of. More specifically, an arrow going from a layer toward a standalone API indicates that the layer makes use of that API, belonging to another firmware component. On the other hand, an API positioned above a layer indicates that the layer makes the services it implements available to other modules through that API.
- *Dark gray* blocks correspond to code modules that are partially or completely *non-portable*. A non-portable code module may require updates when it is moved from one architecture/project to another, and hence, software development effort is likely to be more focused on them.

The same convention has been followed in the next figures of this chapter, too.

12.3.1   TCP/IP PROTOCOL STACK INTERFACE

As shown in Table 12.2, the FREEMODBUS TCP/IP protocol stack interface functions can informally be divided into three groups. Each group will be discussed individually in the following, taking the implementation template provided with FREEMODBUS itself as a reference, which was used as a starting point for the port toward LWIP.

The template makes use of the POSIX sockets API [86], thus privileging maximum portability versus performance, as outlined in Chapters 7 and 9. Another restriction of the template implementation is that it supports only *one* active client connection from MODBUS masters at a time. However, this restriction can easily be lifted.

Concerning this aspect, the number of connections a MODBUS slave supports also determines whether or not the master shall close the connection to a specific slave before talking to the next one, when there are multiple masters wanting to communicate with the same group of slaves. This not only affects the design and implementation of the MODBUS master firmware, but may also have significant side effects at the TCP/IP protocol stack level, which must be carefully scrutinized.

For instance, the number of TCP connections that are simultaneously open has important effects on LWIP memory consumption. In turn, this affects how LWIP memory pools have to be configured and, as a whole, the memory management strategy of the embedded system.

Before proceeding further, it is important to informally introduce some connection-related terminology that will be used in the following and may seem confusing at first. When considering MODBUS TCP connection, a distinction must be made between the nomenclature used at the application layer, and the one used at the TCP layer. Namely:

- At the application layer, connections are initiated by the MODBUS *master* by issuing a connection request to the target MODBUS *slave*.
- At the TCP layer, slave nodes listen for incoming connection requests—and are therefore called *servers*—while master nodes connect to a server, and hence, they become *clients* of that slave.

As a consequence, a master node at the application layer acts as a client at the TCP layer. Similarly, a slave behaves as a server.

Initialization and shutdown

The first group of functions is responsible for *initializing and shutting down* the TCP/IP interface layer. Namely:

- The function `xMBTCPPortInit` initializes the TCP/IP protocol stack interface layer and opens a TCP socket to listen for incoming connection requests from a MODBUS master. The two main LWIP protocol stack functions invoked to this purpose are `socket` (to open the socket), `bind` (to assign it the required IP address and port number), and `listen` (to listen for incoming connection requests).
  The only argument of `xMBTCPPortInit` allows the upper layers to specify a non-default TCP port, instead of port 502, to be used to listen for incoming connection requests.
- The function `vMBTCPPortDisable` may be called by the upper protocol layers at any time to forcibly close all currently open connections. The listening socket shall not be closed, though, and the interface layer shall still accept new, incoming connection requests.
- The function `vMBTCPPortClose` may also be called by the upper protocol layers at any time. Besides closing all currently open connections, like `vMBTCPPortDisable` does, it shall also close the listening socket, thus shutting down communication completely. A fresh call to `xMBTCPPortInit` is then required to resume operations.

Connection management

The function `xMBPortTCPPoll` is extremely important and belongs to a group by itself because it is concerned with *connection management*. In particular, it must perform the following activities when called:

- Monitor the *listening socket* for incoming connection requests, and monitor open *client sockets* for incoming MODBUS requests. Monitoring includes passive wait until an event of interest occurs within the TCP/IP protocol stack and is implemented by invoking the `select` function on the set of sockets to be monitored.
- Accept *incoming connection requests* and establish new connections with MODBUS masters, by means of the function `accept`.
- Retrieve and gather chunks of *incoming data* from client sockets—by means of the LWIP `recv` function—in order to build full MODBUS requests and pass them to the upper protocol layers. This goal is achieved by parsing the MODBUS TCP application protocol (MBAP) header [120], found at the very beginning of each MODBUS request and that contains, among other information, the overall length of the request itself. Once the expected, total length of the request is known, `xMBPortTCPPoll` keeps gathering and buffering incoming data until a whole request becomes available and can be passed to the upper layers of the protocol stack.
- Close client connections, by means of the LWIP `close` function, upon normal shutdown or when an error is detected. Normal shutdown occurs when the master closes the connection. In this case, the `recv` function returns zero as the number of bytes transferred when called on the corresponding socket on the slave side. Errors are detected when either `recv` itself or any other LWIP function returns an error indication when invoked on a socket.

In order to properly synchronize with the upper protocol layer, besides carrying out all the previously mentioned activities, `xMBPortTCPPoll` must also:

- Return a Boolean flag to the caller, true denoting successful completion and false denoting an error.
- Post event `EV_FRAME_RECEIVED` through the event system to be discussed in Section 12.3.2 when a MODBUS request has been received successfully.

Data transfer

The third and last group of functions allows the upper protocol layers to perform *data transfer*, handling one MODBUS TCP request and response transaction at a time. The two functions in this group are:

- `xMBTCPPortGetRequest`, which retrieves a MODBUS TCP request previously gathered by `xMBPortTCPPoll` and returns a pointer to it, along with its length, to the caller.

- `xMBTCPPortSendResponse`, which sends a MODBUS TCP response to the master that sent the last request retrieved by `xMBTCPPortGetRequest`.

Along with the event system, these two functions also implicitly serialize request and response handling by upper protocol layers. This is because requests are retrieved and handed to the upper layers one at a time, and the upper layer must provide a response to the current request before retrieving a new one.

12.3.2 OPERATING SYSTEM ADAPTATION LAYER

The operating system adaptation layer consists of three modules of code, all layered on top of the FREERTOS API, and implementing different functionality.

Event system

The two portable layers of the protocol stack, to be discussed in Sections 12.3.3 and 12.3.4 are synchronized by means of a simple *event system*, implemented in the module `portevent.c`. The main purpose of the event system, in its most general form, is to allow the upper layer to wait for an event, which is generated by the lower layer.

The event system specification supports either an *active* or a *passive* wait. When appropriate operating system support is available, like in this case, passive wait is preferred because it avoids trapping the processor in a waiting loop—with a detrimental effect on overhead and processor power consumption—until an event occurs.

It should also be noted that the event system defines a boundary between two kinds of code within the protocol stack:

1. Higher-level code that *consumes* events, and possibly waits in the attempt until events become available.
2. Lower-level code that *produces* events and never waits.

This boundary must be considered, for instance, when the goal is not just to make use of the protocol stack but also to *extend it*, for instance, to add a new protocol module. In this case, it is important to remember that lower-level code is not allowed to block internally unless a separate task, with its own execution context, is introduced to this purpose.

The event system must implement a single event *stream* and make available four functions, summarized at the top of Table 12.3, to interact with it. In particular:

- The two functions `xMBPortEventInit` and `xMBPortEventClose` are invoked by the upper layers to initialize the event system before use, and shut it down when it is no longer needed, respectively. They are responsible for allocating and releasing any internal data structure needed by the event system itself.
- The function `xMBPortEventPost` is called to insert, or *post* an event into the event stream. Each event has an identifier of type `eMBEventType`

Table 12.3

FREEMODBUS Operating System Adaptation Functions

Function	Description
Event System	
xMBPortEventInit	Initialize the event system
vMBPortEventClose	Close the event system
xMBPortEventPost	Post an event
xMBPortEventGet	Get an event
Timer	
xMBPortTimersInit	Initialize the timer module
vMBPortTimerClose	Close timer
vMBPortTimersEnable	Start timer
vMBPortTimersDisable	Stop timer
pxMBPortCBTimerExpired	Pointer to the timer expiration callback
Critical Regions	
ENTER_CRITICAL_SECTION	Macro to open a critical region
EXIT_CRITICAL_SECTION	Macro to close a critical region

attached to it, which distinguishes one class of event from another. Accordingly, this function accepts one argument of that type.

- The function xMBPortEventGet looks for and possibly consumes an event. When successful, it returns the corresponding event identifier, of type eMBEventType, to the caller.

The implementation of xMBPortEventGet can be carried out in two different ways, depending on whether or not operating system support for passive wait is available for use:

1. If passive wait is *not* available, xMBPortEventGet simply *polls* the event stream and checks whether or not there is a pending event. If there is one, it consumes it and returns the event identifier to the caller. Otherwise, it notifies the caller that the event stream is empty, without delay. In the last case, it is the caller's responsibility to retry the operation, if appropriate, at a later time.
2. Otherwise, the function must block the caller, by means of a suitable operating system-provided task synchronization mechanism—for example a message queue—until a message is available, and then return the event identifier to the caller. The function shall also return immediately if an event is already available at the time of the call.

The event system must be capable of storing just *one* posted event, before it is consumed. However, trying to post a new event when another one is still pending must *not* be flagged as an error.

The MODBUS slave protocol stack has been designed to be quite permissive and flexible with respect to the event system, and hence, it is permissible to "lose" events in this case. More specifically, when a new event is posted by the lower layer while there is already a pending event in the event system, it can be handled in two different ways:

- keep the pending event and ignore the new one, or
- replace the pending event with the new one.

Due to the rather large degree of freedom in how the event system may be implemented, the upper layer considers at least the following two possible event handling scenarios and handles them properly.

As a side note, this provides evidence of how designing and implementing portable code, which can be easily layered on top of other software components without posing overly restrictive requirements on them, may introduce additional complexity into it.

1. The `xMBPortEventGet` function may return to the caller (with or without waiting) and signal that *no events* are available. This must not be considered an error condition and the operation shall be retried.
2. It is also possible that the same function returns *fewer* events than were generated. In fact, this may happen when events appear close to each other, and the event system drops some of them. In this case, it is not specified which events are kept and which are lost with respect to their generation time. Then, the upper layer needs to rely on extra information, usually provided by the lower layer through a side channel, to keep memory and properly handle the events.

In this case study, FREERTOS synchronization support is available, and hence, the event system can be implemented in a straightforward way by means of a message queue with a buffer capacity of one message of type `eMBEventType`. The queue provides both the required synchronization and the ability to transfer event identifiers (as message contents) between `xMBPortEventPost` and `xMBPortEventGet`.

Within `xMBPortEventPost`, the `xQueueSend` function that operates on the message queue (see Section 5.4) is invoked with a timeout of *zero*, thus asking the operating system to return immediately to the caller if the message queue is full. As a consequence, the event system always drops the *most recent* events when necessary.

In other cases—namely, when it is invoked from an interrupt handling rather than from a task context—`xMBPortEventPost` has to use `xQueueSendFromISR` to operate on the queue. However, its behavior is still consistent with respect to the previous scenario, because `xQueueSendFromISR` implicitly operates as `xQueueSend` with zero timeout does.

Timer

The purpose of `porttimer.c` is to implement a single timer, exporting the four interfaces listed in the middle part of Table 12.3 and described in the following.

Namely:

- The functions xMBPortTimersInit and vMBPortTimerClose have the same purpose as xMBPortEventInit and xMBPortEventClose, respectively. Moreover, the argument of xMBPortTimersInit indicates the timer's *timeout* value, that is, the amount of time that must pass after the timer is started until it *expires*, expressed with a granularity of $50 \mu s$.
- The function vMBPortTimersEnable starts the timer. When the timer is started, it counts down from its timeout value. As specified previously, the timeout value is specified upon initialization and cannot be changed afterward. If the timer is already counting down when vMBPortTimersEnable is called—because it was already started in the past and it has not expired yet—the count restarts from the original timeout value rather than keep counting.
- The function vMBPortTimersDisable can be called to prematurely stop a timer before its expires. Calling vMBPortTimersDisable on a timer that is not running—either because of a previous call to the same function or because it already expired—has no effect and must not be reported as an error.

Upon expiration, the timer module invokes the callback function specified by the function pointer pxMBPortCBTimerExpired, possibly from an interrupt context. The timer is one-shot and stops after expiration, until it is possibly started again.

It should be noted that, as in most other timeout notification mechanisms, there is an inherent race condition between the invocation of the premature stop function and timer expiration, which must be resolved externally from the timer module. In fact, the timer module cannot guarantee by itself that the timer will not expire (and the callback function will not be invoked) *during* the execution of vMBPortTimersDisable.

Within the FREEMODBUS protocol stack, the timer is used for a very specific purpose, that is, to detect the end of the MODBUS frame being received and corresponds to the $t_{3.5}$ timer defined in the MODBUS RTU specification [121]. If the protocol stack is used exclusively for MODBUS TCP communication, as happens in this case, the timer module is *not required* and its implementation can indeed be omitted.

In any case, it is worth noting—as an example of how operating system support, albeit present, may not be fully adequate to support all possible upper-layer software requirements—that the timer, if required, could not be implemented using the FREERTOS timer facility when it operates in *tick-based* mode (see Section 10.7 for further details).

This is because, in order to support the required resolution, the tick frequency would need to be increased from the default of 1 kHz to 20 kHz and tick interrupt handling overheads would thus become inordinate. In this case, a direct implementation on top of one of the hardware timers embedded in the LPC2468 would be more appropriate.

Critical regions

The third and last aspect to be discussed in this section concerns the implementation of *critical region* entry and exit code. Namely, in the FREEMODBUS code, critical regions are surrounded by the ENTER_CRITICAL_SECTION and EXIT_CRITICAL_SECTION macros, which are responsible for mutual exclusion.

The operating system adaptation layer shall provide a definition for these macros in the header file port.h. Depending on its complexity, the actual code of critical region entry and exit—invoked by the above-mentioned macros—can be provided either directly in port.h if it is simple and short enough, or in the source module portother.c.

Only a single mutual exclusion domain must be provided for the whole protocol stack, because FREEMODBUS critical regions have been kept very short, and hence, unnecessary blocking has been reduced to a minimum. In the FREERTOS port, the two macros have been mapped to the portENTER_CRITICAL and portEXIT_CRITICAL system services, respectively, as has been done for the fast critical regions of LWIP, as described in Chapter 7.

A notable special case that could not happen in LWIP is that, according to its specification, FREERTOS may invoke the mutual exclusion macros not only from a task context, but also from an *interrupt handling* context. The two previously mentioned system services are not supported in the last case, so they must not be invoked.

The issue can be solved by checking whether or not ENTER_CRITICAL_SECTION and EXIT_CRITICAL_SECTION have been invoked from an interrupt handling context, by querying the CPU status register through a short assembly language insert, according to the general technique presented in Section 9.4. If this is the case, the macros simply do nothing.

Since mutual exclusion is implemented by disabling interrupts, mutual exclusion with respect to both regular tasks and interrupt handlers (as well as any callback function invoked by them) is guaranteed anyway, on a single-processor system. Mutual exclusion between callback functions invoked from an interrupt handler is also implicitly enforced in the same way, as long as interrupt handlers are not themselves interruptible.

It is worth mentioning that, as most other low-level mechanisms, this way of protecting critical regions may have some unforeseen interactions with other parts of the protocol stack that must be carefully considered in a case-by-case basis.

In particular, passive wait (for instance, within an event system function) within a critical region may not be supported in some operating systems. This is the case, for example, of FREERTOS on Cortex-M3 processors, due to the way rescheduling is implemented.

12.3.3 MODBUS **TCP LAYER**

As shown in Figure 12.2, the MODBUS TCP layer resides below the MODBUS application layer and event loop, to be described in Section 12.3.4, and the TCP/IP

Table 12.4

FREEMODBUS MODBUS **TCP Layer Functions**

Function	Description
eMBTCPDoInit	Prepares this layer and the ones below it for use
eMBTCPStart	Enable protocol stack layer
eMBTCPStop	Disable protocol stack layer
eMBTCPSend	Send a MODBUS response
eMBTCPReceive	Retrieve a MODBUS request

protocol stack interface, presented in Section 12.3.1. To this purpose, it exports the functions listed in Table 12.4 and described in the following.

- The function eMBTCPDoInit is responsible for initializing and preparing for using the MODBUS TCP layer and the layers below it. Since the MODBUS TCP layer is fairly undemanding, no specific initialization procedure is necessary for it. Therefore, this function simply invokes the function xMBTCPPortInit to initialize the TCP/IP protocol stack interface layer.
- The two functions eMBTCPStart and eMBTCPStop enable and disable the protocol stack layer, respectively. The implementation of the first function is empty, because TCP/IP communication is automatically enabled upon initialization. The second function is mapped directly onto the lower-level function vMBTCPPortDisable.
- The main purpose of eMBTCPSend is to encapsulate MODBUS protocol data units coming from the application layer into MODBUS TCP protocol data units, by setting up a proper MBAP header, and then convey them to the TCP/IP protocol stack through the corresponding interface.

  Doing this takes advantage of the fact that MODBUS is a master-slave protocol, and hence, the slave never generates a reply without having first received a request from the master. For this reason, when preparing a reply, the slave protocol stack can take advantage of the MBAP header that was received with the request, which has already been thoroughly checked for validity.

  Indeed, some fields of the two MBAP headers have to be related, according to the protocol specification. For instance, the *transaction identifier* field of the reply must match the one found in the request, so that the master can use it to correctly associate the reply with the corresponding request.
- Along the receive path, processing is even simpler because the MBAP header has already been partially parsed within the TCP/IP protocol stack interface, in order to determine how long the incoming MODBUS protocol data unit is. For this reason, the only duty of the function eMBTCPReceive—besides forwarding the protocol data unit from one

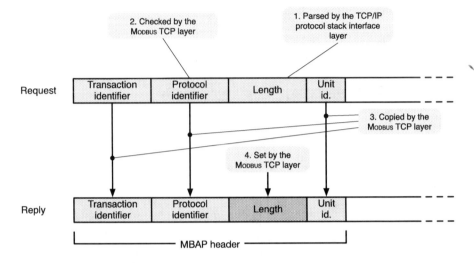

Figure 12.3   MBAP header processing in the MODBUS TCP layer and below.

layer to another—is to check that the *protocol identifier* field of the MBAP header really indicates that the request belongs to the MODBUS protocol.

In summary, Figure 12.3 depicts how the MBAP header is managed and transformed by the MODBUS TCP layer when going from a request to the corresponding reply.

12.3.4   MODBUS **APPLICATION LAYER AND EVENT LOOP**

This layer contains two important high-level elements. The corresponding functions are listed in Table 12.5.

1. *Initialization* code, to configure the MODBUS protocol stack for a certain combination of data link and physical layers, and prepare it for use. Symmetrically, the *shutdown* code turns off the protocol stack when it is no longer needed.
2. The main *event handling function*, a function that—when it is invoked by the task responsible for handling MODBUS transactions after protocol stack initialization—looks for incoming requests and handles them.

The initialization function eMBTCPInit initializes the protocol stack as a whole, when it is configured to work on top of TCP/IP connections, that is, for MODBUS TCP. As part of the process, it directly or indirectly invokes the appropriate initialization functions of the lower protocol layers. Namely, it calls eMBTCPDoInit within the MODBUS TCP layer and xMBPortEventInit to initialize the event system.

A different function also defined in the same layer, eMBInit, initializes the protocol stack when it is configured for a TIA/EIA–485 interface bus at the physical level.

Table 12.5
FREEMODBUS **Application Layer Functions**

Function	Description
Initialization and shutdown	
eMBTCPInit	Initializes FREEMODBUS for TCP communication
eMBInit	Initializes FREEMODBUS for RTU or ASCII communication
eMBClose	Shuts down the protocol stack
eMBEnable	Enable the protocol stack for use, after initialization
eMBDisable	Disable the protocol stack temporarily, without shutting it down
Event handling	
eMBPoll	Main event handling function
Application-layer callbacks	
eMBRegisterCB	Register an application-layer callback

Table 12.6
Links between the Application Layer and the MODBUS **TCP Layer**

Application layer function pointer	MODBUS TCP layer function	Purpose
pvMBFrameStartCur	eMBTCPStart	Enable lower protocol stack layer
pvMBFrameStopCur	eMBTCPStop	Disable lower protocol stack layer
peMBFrameSendCur	eMBTCPSend	Send a MODBUS response
peMBFrameReceiveCur	eMBTCPReceive	Retrieve a MODBUS request
pvMBFrameCloseCur	vMBTCPPortClose	Shut down communication

In this case, the communication protocol can be either MODBUS ASCII or MODBUS RTU [121]. However, this operating mode is beyond the scope of this discussion.

The initialization function also sets up the interface between the application layer and the MODBUS TCP layer. Table 12.6 lists the names of the function pointers used at the application layer and gives a terse description of the function they correspond to in the MODBUS TCP layer.

Using function pointers instead of hardwired function names for most of the interface helps to accommodate different communication protocols, implemented by distinct lower layers, transparently and efficiently. As a side note, referring back to the table, the vMBTCPPortClose function technically belongs to the TCP/IP interface layer, and not to the MODBUS TCP layer.

Conversely, the function eMBClose shuts down the protocol stack, by simply invoking the corresponding MODBUS TCP layer function through its function pointer.

Similarly, the `eMBEnable` and `eMBDisable` functions enable and disable the protocol stack, respectively. As in the previous case, these functions do very little by themselves, except invoking the corresponding MODBUSTCP layer functions through their function pointers.

The `eMBPoll` function is responsible for event handling and contains the application layer protocol state machine. Accordingly, the protocol state machine is driven by events taken from the event system. Events are generated either by the state machine itself or by the lower protocol layers.

It should be noted that `eMBPoll` does not necessarily *wait* for events (whether or not that happens depends on the event system implementation), and hence, it can return to the caller without doing anything. This is not considered an error condition, and the function will return `MB_ENOERR` in this case. It is expected that calls to `eMBPoll` will be enclosed within a polling loop in an even higher layer of code, above the protocol stack.

Since part of MODBUS requests handling is application-dependent, the event handling function just mentioned works in concert with appropriate *callback functions* that must be implemented at the application layer. As will be discussed in Section 12.5, the function `eMBRegisterCB` can be invoked from the application layer to register a callback function for a certain category of MODBUS requests.

12.4   USB-BASED FILESYSTEM

The USB-based filesystem is based on two distinct software modules, which will be discussed in the following.

12.4.1   FAT FILESYSTEM

The FATFS [35] filesystem, whose inner structure is depicted in Figure 12.4, makes available through its API a rather general set of functions to open, close, read from, and write to files resident on a mass storage device formatted according to the FAT filesystem specification [118].

Configuration options

Besides the main interfaces toward the mass storage device and the operating system, which will be discussed later, two more non-portable headers are required. Namely:

- The header `integer.h` provides appropriate type definitions for signed and unsigned 8-, 16-, and 32-bit integers.
- The header `ffconf.h` configures the filesystem features and capabilities.

The FATFS configuration options can be divided into three groups listed in Table 12.7.

1. The options in the *features/capabilities* group make available different trade-offs between the capabilities of the filesystem module and its speed, with respect to code/data footprint and memory requirements.

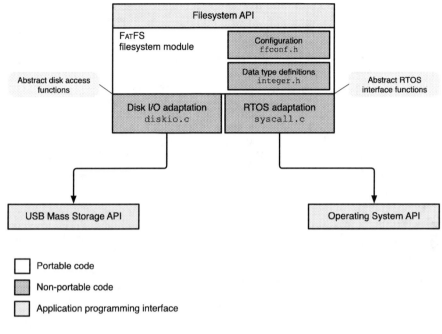

Figure 12.4   Inner structure and interfaces of the FATFS filesystem.

Table 12.7

FATFS Main Configuration Options

Option	Description
Features and capabilities	
_FS_TINY	Reduce filesystem footprint to a minimum
_FS_READONLY	Do not provide filesystem write functions
_FS_MINIMIZE	Select a minimal subset of the API to reduce footprint
_FS_RPATH	Support path names relative to the current directory
_USE_LFN	Support long file names (LFNs)
_LFN_UNICODE	Use Unicode character set in long file names
Device characteristics	
_VOLUMES	Maximum number of logical drives per physical device
_MAX_SS	Maximum sector size, in bytes
_MULTI_PARTITION	Support multiple partitions on a logical drive
Operating system interface	
_FS_REENTRANT	Enable concurrent filesystem access by multiple tasks
_FS_SHARE	Enable file locking upon concurrent access
_WORD_ACCESS	Allow unaligned word access

Table 12.8

Filesystem Requirements Concerning Disk I/O

Function	Mapping	Description
disk_initialize	MS_Init	Initialize disk I/O subsystem
disk_status	MS_TestUnitReady	Get disk status
disk_read	MS_BulkRecv	Read contiguous disk blocks
disk_write	MS_BulkSend	Write contiguous disk blocks

2. The option group concerned about *device characteristics* determines how many logical drives at a time the filesystem module can handle, the maximum sector size to be supported, and whether or not it is possible to access multiple partitions on the same device.
3. The *operating system interface* option group controls whether or not the filesystem modules shall be reentrant and optionally enables file locking. Moreover, on processor architectures that support them, it is possible to enable unaligned word accesses in the filesystem code, thus achieving a higher speed.

The last option group deserves special attention because it affects not only the footprint of FATFS itself, but also the complexity of its adaptation layer—the operating system interface in particular—and the operating system memory requirements.

This is because, when reentrancy support is enabled, FATFS relies on the underlying operating system for proper synchronization among concurrent filesystem requests. Accordingly, several additional operating system interfaces are required to declare, create, delete, and use mutual exclusion locks—called *synchronization objects* by FATFS.

For this specific case study, the main purpose of the mass storage subsystem is data logging, performed by a single application task after collecting information from other components of the real-time system. For this reason, FATFS reentrancy has been disabled to speed up the development of the adaptation layer. On the other hand, the full set of FATFS features/capabilities has been enabled, at a small footprint cost, to make data logger development more convenient.

Disk I/O adaptation layer

The purpose of the disk I/O adaptation layer, implemented in the diskio.c source file, is to act as a bridge between the filesystem module and the disk input/output interfaces exported by the USB protocol stack to be presented in Section 12.4.2.

Table 12.8 lists the requirements of the filesystem module for what concerns disk I/O and outlines how they have been mapped onto the underlying layer. As shown in the table, the mapping is fairly straightforward in this case study, because there is a

Table 12.9

Filesystem Requirements upon the Operating System and Runtime Library

Function	Mapping	Description
ff_memalloc	malloc	Dynamic memory allocation
ff_memfree	free	Dynamic memory release
get_fattime	*custom implementation*	Get current time of day (wall-clock time)

one-to-one relationship between the function required by the filesystem and what is offered by the underlying layer.

The main duty of the adaptation layer is, therefore, limited to perform argument and return code conversion.

Operating system adaptation layer

The mapping between the filesystem module onto the underlying operating system and runtime library is carried out within the syscall.c source file and is summarized in Table 12.9.

The two dynamic memory allocation functions required by FATFS have been mapped directly upon the corresponding facilities provided by the C runtime library, malloc and free.

On the other hand, the function get_fattime, also required by FATFS, cannot be implemented in terms of the FREERTOS API. This is because the filesystem code invokes this function whenever it needs to know the current, absolute time of the day, in order to update some filesystem information, such as the last modification time of a file.

Although FREERTOS provides a quite accurate time base *relative* to the boot-up time, it does not directly support the notion of *absolute* time. Even more importantly, absolute time is not kept across reboots and relative time simply restarts from zero every time the system boots up.

As a consequence, the get_fattime has been implemented directly and based upon the real-time clock (RTC) hardware module embedded in the LPC2468. This module keeps the correct time of the day even when the main power to the processor is switched off, by means of a backup power source, without operating system intervention.

It should also be noted that the actual filesystem requirements with respect to the operating system depend on its configuration and may comprise additional interfaces. Table 12.9 lists the requirements corresponding to the minimal configuration described previously and, for instance, does not include synchronization object management.

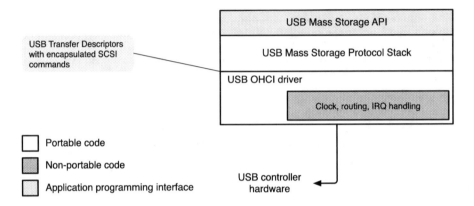

Figure 12.5 Inner structure and interfaces of the USB protocol stack.

12.4.2   USB PROTOCOL STACK AND DRIVER

The protocol stack for USB-based mass storage devices is composed of four layers, shown in Figure 12.5. As before, non-portable components—that is, components that must be either specifically developed for, or adapted to, every combination of toolchain, operating system, and mass storage device type—are highlighted in dark gray.

USB host controller interface driver

Unlike other kinds of controller—for instance the Ethernet controller discussed in Chapter 8—the USB host controller interface is mostly uniform across different kinds of systems and vendors. In particular, the vast majority of host controllers currently available adhere to the UHCI [81], OHCI [41], EHCI [82], or xHCI [84] standards.

Only several low-level aspects of the interface are not covered by the standards and are therefore platform-dependent. They correspond to the only dark gray box visible in Figure 12.5. In particular, the standards do not cover the details of:

- the exact location of the bank of host controller operational registers in the microcontroller address space;
- the way USB controller signals are routed to the microcontroller's I/O ports, to the USB transceiver and, ultimately, to the physical USB port;
- the clock source to be used by the USB controller and its frequency;
- how USB controller interrupts are routed to the processor and how they are associated with the corresponding interrupt handler.

As a consequence, the development of a USB device driver is a relatively easy task for what concerns the *amount* of code to be developed, even though it is time consuming because it requires a very accurate and detailed *knowledge* of the target microcontroller's inner workings.

In order to assist software developers, many microcontroller vendors provide the source code of an exemplar driver that works with their products and can be further extended by system developers according to their specific needs.

In our case, the LPC2468 microcontroller [125] includes an OHCI-compliant [41] USB 2.0 [40] host controller. The vendor itself provides an extremely simplified, *polling*-based driver as part of a larger software package fully described in [124].

In this kind of controller, all interactions between the device driver and the controller itself, after initialization, take place by reading from and writing into the controller's operational registers and by exchanging data through a shared memory area called host controller communication area (HCCA). Both the register and memory layouts are fixed and completely defined in [41].

With respect to a full-fledged host controller driver, the main limitation of the prototype provided in [124] is related to the device enumeration algorithm, which lacks the capability of crossing USB hubs. As a consequence, as it is, the driver is capable of handling only one device, directly connected to the host controller port. On the contrary, dynamic device insertion and removal are supported, so that the driver is, for instance, more than adequate to support a USB mass-storage device for data-logging purposes.

The main improvement introduced to make the driver work efficiently in an operating system-based environment was to replace the polling-based, busy waiting loops provided by the original driver with a passive, *interrupt*-based synchronization. As a consequence, two synchronization semaphores managed by means of appropriate FREERTOS primitives were introduced for the two sources of events of the controller, outlined in the following.

1. The *Root Hub Status Change (RHSC)* event, which is generated when a USB device is connected to, or disconnected from, the USB controller. When a device connection is detected, the device driver reacts by issuing, according to the USB specification, a port reset command. After that, device initialization is completed by assigning an address to it and retrieving its configuration descriptor. After confirming that the device is indeed a mass storage device—an operation accomplished by the USB mass storage interface module—it is configured for use.
2. The *Writeback Done Head (WDH)* event. This event is triggered when the USB controller has finished processing one or more transmit descriptors (TDs), has linked them to the *done queue* data structure, and has successfully updated the pointer to the head of the done queue accessible to the device driver. This event therefore provides a convenient way for the device driver to be notified of the completion of the transfer described by a certain TD and reuse the TD for further transfers.

The main interfaces exported by the device driver are summarized in Table 12.10. The two initialization functions listed at the top of the table prepare the host

Table 12.10
USB Host Controller Driver Interfaces

Function	Description
Initialization	
Host_Init	Host controller initialization
Host_EnumDev	Device enumeration
Control information and data transfer	
Host_ProcessTD	Bulk and control endpoints I/O
Host_CtrlSend	Send control information
Host_CtrlRecv	Receive control information

controller for use and wait for the insertion of a mass storage device, respectively. After device initialization, the other interfaces allow the caller to:

- send and receive control information through the device's control endpoint descriptor;
- send data to the device's bulk output endpoint descriptor;
- receive data from the device's bulk input endpoint descriptor.

This minimal set of capabilities corresponds to what is required by the USB mass storage interface module, to be discussed next.

USB mass storage interface

The specifications a device belonging to the USB mass storage class must obey are fully enumerated in [166]. Although, in principle, multiple mechanisms are specified to transport command, data, and status information to/from the device, software development focused on the so-called *bulk-only transport* [165]. This is because, at least for USB flash drives, this is the simplest and most commonly supported transport.

Similarly, even if a given transport mechanism can be used with multiple command sets, depending on the kind of device and its capabilities, the most widespread choice for USB flash drives is to implement a subset of the *SCSI command set* [5], and hence, only this option is currently supported by the USB mass storage interface layer being described. As a side note, this is also the transport/command set combination specified for use with USB floppy drives [164] and other removable USB mass storage devices, such as hard disks.

Table 12.11 lists the functions made available by the mass storage interface module. The first two functions are used during device initialization. Namely:

- The function MS_ParseConfiguration parses the configuration information retrieved from the device by the host controller interface driver, to

Table 12.11

Main USB Mass Storage Interface Entry Points

Function	Description
MS_ParseConfiguration	Parse device configuration and confirm it is supported
MS_Init	Prepare the device for use, retrieve block size and capacity
MS_TestUnitReady	Check if the device is ready for use
MS_BulkRecv	Read contiguous sectors from the device
MS_BulkSend	Write contiguous sectors to the device

check whether or not the device is supported. The main checks performed at this stage are aimed at ensuring that the device is indeed a mass storage device that understands the transport mechanism and command set just discussed.

- The function MS_Init makes sure that the device is ready for use by issuing the SCSI command *test unit ready* and waiting if necessary. The same capability is also made available to the upper software layers for later use, by means of the function MS_TestUnitReady. In this way, the same check can be performed at that level, too, for instance, before issuing additional commands to the device. During initialization, the block size and the total storage capacity of the device are also retrieved and returned to the caller, since this information is required by the filesystem module.

After device initialization, the main purpose of the USB mass storage interface is to encapsulate the SCSI commands *Read 10* and *Write 10* into bulk-only transport messages and perform data transfer to the device by means of the Host_ProcessTD function of the host controller driver.

As their names imply, these two commands are used to read and write, respectively, a number of contiguous storage blocks. These crucial functions are made available to the upper software layer by means of the last two functions listed in Table 12.11, that is, MS_BulkRecv and MS_BulkSend.

12.5   APPLICATION CODE

As shown in Figure 12.2, the most important elements within application code, besides the application task that is mainly responsible for system initialization, are the MODBUS application-layer *callbacks*.

Application-layer callbacks are functions that are responsible for implementing MODBUS requests coming from the master, obey them, and provide an appropriate reply. They inherently depend on the application, because their contents, behavior, and interface with the other parts of the system depend on what the embedded system is supposed to do.

These functions are called callbacks because they are invoked in a peculiar way, that is, they are *called back* from the MODBUS application layer, discussed in Section 12.3.4, when an incoming MODBUS request is received. They must carry out the transaction on the slave and then return to the caller with its result.

All application-layer callbacks have a uniform interface, involving two arguments. For better efficiency, those arguments are used both as inputs to the function and outputs from it.

1. The first argument, `pucFrame`, is a pointer to a message buffer. The callback function finds the incoming request there, and it must use the same buffer to store the reply.
2. The second argument, `pusLength`, points to an integer variable. Upon callback invocation this variable contains the length of the incoming message, and the callback function shall store into it the length of the reply before returning.

To avoid multiple copies of message contents through the protocol stack, callback functions are trusted to fill the message buffer starting at a fixed index. The buffer elements before that index are reserved for data link headers and should not be touched.

The link between the application layer and the callback functions is configurable, by means of the `xFuncHandlers` array. Each element of this array, of type `xMBFunctionHandler`, associates a MODBUS function code with the callback function to be called when there is an incoming request bearing that function code.

Besides static initialization, the array can also be modified by means of the `eMBRegisterCB` application layer function (see Table 12.5). This function is able to remove an existing association between a function code and a callback function or to establish a new one.

In order to facilitate callback development, the FREEMODBUS protocol stack provides callbacks for the most common MODBUS requests. They take charge of the low-level aspects of callback processing and, in turn, invoke an application-layer function with a higher-level interface.

For instance, the function `eMBFuncReadInputRegister` handles *read input registers* MODBUS requests [119]. It accepts the two low-level arguments just described and invokes the application-layer function `eMBRegInputCB` with 3 higher-level arguments:

1. a pointer to a buffer where the application-layer function must store the values of the registers requested by the master, `pucFrameCur`;
2. the initial register number requested by the master, `usRegAddress`;
3. the number or registers to be stored into the buffer, `usRegCount`.

12.6   PERFORMANCE AND FOOTPRINT

The experimental evaluation took place on a prototype of the embedded system, implemented on an Embedded Artists LPC2468 OEM board [56]. In particular, we evaluated the overall memory footprint of the system and assessed the performance it can achieve in MODBUS TCP communication.

12.6.1 MEMORY FOOTPRINT

Memory requirements represent an important design constraint in small-scale embedded systems, often even more than performance requirements. This is because, as outlined in Chapter 2, many low-cost microcontrollers lack an external memory interface, and hence, the whole software load must necessarily fit in the on-chip flash and RAM memory.

Even when an external memory interface is indeed available—as is the case for the microcontroller considered in this case study—the cost, size, and complexity of the additional components needed to implement an external memory, as well as their interconnection through the printed circuit board, are likely to be a significant drawback.

The memory footprint of the main firmware components discussed so far in this chapter has been evaluated by building the components from their source code, by means of the toolchain described in Section 12.1. The results are shown in Table 12.12.

The footprint has been evaluated by means of the `size` command of BINUTILS. The resulting data have been aggregated into two categories, namely, *read-only* (executable code) and *read-write* (comprising initialized and uninitialized data).

This distinction is especially important in an embedded system because the two categories likely correspond to two different kinds of memory. Namely, read-only data must be stored in (nonvolatile) flash memory, while read-write data takes space in both flash and RAM.

This is because flash memory holds the initial values of the read-write data, which is copied into RAM during system initialization. The RAM memory image of the segment is used as read-write storage thereafter.

As can be seen from the table, the memory requirements are quite acceptable and well within the amount of on-chip memory of the LPC2468. Overall, the components discussed in this paper take about 150 Kbyte of flash out of 512 Kbyte (less than 30%) and about 37 Kbyte of RAM out of 96 Kbyte (less than 39%). In this way, plenty of space is still available to the application code without resorting to external memory.

The rather large amount of RAM required by the portable modules of LWIP and FREERTOS can be easily explained by observing that:

- The portable modules of LWIP contain memory pools used by both LWIP itself and network applications for network buffers and other data structures, like connection descriptors and protocol state vectors.
- The portable modules of FREERTOS allocate the memory pool from which the operating system draws most of its internal data structures, for instance, task descriptors. The required heap size mainly depends on the number of

Table 12.12

Firmware Memory Footprint

Component	Memory footprint (B)	
	Read-only (ROM)	**Read-write (RAM)**
LWIP **protocol stack**		
Portable modules	87024	11737
Operating system adaptation	4200	4
Ethernet device driver	4928	12
USB-based filesystem		
FATFS filesystem	27728	520
Disk I/O adaptation	288	0
Operating system adaptation	100	0
USB protocol stack and driver	7548	8
MODBUS **TCP connectivity**		
Application layer callbacks	1796	34
Application layer and event loop	2620	172
MODBUS TCP layer	372	0
TCP/IP adaptation	2228	284
Operating system adaptation	1052	16
FREERTOS **operating system**		
Portable modules	10436	25056
Porting layer	2432	24
Total	152752	37867

tasks, because the largest object allocated there is the per-task stack. The configuration used in the case study supports a minimum of 6 tasks.

When necessary, it is often possible to further reduce the footprint of these components by means of an appropriate configuration. For instance, the LWIP configuration used in the case study includes support for both TCP and UDP for generality. If one of these protocols is not used by the application, it can be removed from the configuration and the memory required by it can be put to other uses, as discussed in Chapter 7.

It should also be noted that the amount of memory required by the application code, except the MODBUS application-layer callbacks, and by the C language runtime libraries has not been quantified, because it strongly depends on the application code. In fact, the library modules are included in the executable image only on-demand and the system components by themselves require only a few of them, whereas the library usage by the application code is usually much heavier.

Table 12.13

MODBUS **Communication Performance, Write Multiple Registers**

Number of 16-bit registers	Average round trip time (ms)
2	6.6
32	6.7
62	6.8
92	7.0
123	7.1

As a consequence, just listing the memory footprint of the library modules needed by the system components would have been overly optimistic, whereas listing the memory footprint of all library modules included in the firmware under study would have been misleading, because these figures would not be a reliable representative of any other application.

For similar reasons, the footprint data listed in Table 12.12 do not include the amount of RAM used as buffer by the Ethernet and USB controllers, which has been drawn from their own dedicated banks of RAM.

12.6.2 SYSTEM PERFORMANCE

The basic performance of two major components of the embedded system considered in the case study—namely, the MODBUS *slave protocol stack* and the *filesystem*—has been evaluated by means of synthetic test programs representing realistic workloads.

MODBUS **slave protocol stack**

The performance of the MODBUS slave protocol stack has been evaluated by exercising it using a matching MODBUS master implementation based on [57], also running on another identical embedded board.

The average round-trip time $RTT(n)$ obtained when exercising a typical MODBUS application-layer primitive (*write multiple registers*) was evaluated for a varying number of registers n. The results are shown in Table 12.13.

As shown in the table, the RTT exhibits a linear dependency on the number of registers being written. The overall communication delay is around 7 ms for all legitimate values of n. This is considered to be adequate for networked embedded systems [61].

The RTT is dominated by software processing time, as the Ethernet data transfer time never exceeds $30\,\mu s$, even for the longest MODBUS PDU. This is two orders of magnitude smaller than the measured RTT. The software processing time can be easily justified by the complexity of the communication software which includes not

Table 12.14

Filesystem Performance, Read and Append Rates

Data chunk size (B)	Read rate (B/s)	Append rate (B/s)
16	7200	7100
64	24870	23690
512	88600	75340
521	87600	74630

only the MODBUS TCP protocol stacks themselves (both the master and the slave sides), but also another complex component, LWIP.

Since, for each *write multiple registers* command issued by the MODBUS master, a reply from the slave is expected, it indicates that the RTT comprises two receive-path delays and two transmit-path delays through the whole protocol stacks.

Filesystem

Since the primary intended use of the filesystem is data logging, its performance was evaluated by considering sequential read and append operations with different data chunk sizes. Table 12.14 shows the read and append rates (expressed in bytes per second) as a function of the data chunk size (in bytes). The rates have been calculated by averaging the time needed to transfer 100 MB of data.

The first three data chunk sizes are a sub-multiple of the sector size used by the filesystem, that is, 512 B. This ensures that read and append operations never overlap with a sector boundary and makes the measurement adequate to determine the relative contribution of software overheads, which are inversely proportional to the data chunk size, with respect to the total number of USB I/O operations, which are constant across the experiments.

The experimental results in this group exhibit a strong dependency on the data chunk size for what concerns both read and append rates. Namely, the ratio between whole-sector operations and operations typical of fine-grained data logging (and hence, operating with much smaller data chunk sizes) is about one order of magnitude and is likely to affect application-level software design.

The lower performance of append with respect to read (from -2% to -15% depending on the data chunk size) can easily be justified by considering that appending to a file incurs extra overhead because it is necessary to update the file allocation table information on the filesystem as new blocks are linked to the file being appended to. In addition, write operations on a flash drive may be slower than reads due to the significant amount of time needed to erase flash memory blocks before writing.

The fourth data chunk size value used in the experiments, 521 bytes, is the smallest prime number greater than the sector size. The value has been chosen in this way to keep software overheads related to I/O requests from the application program as close as possible to the whole-section read/append case and, at the same time, maximize the probability that read and append operations cross a sector boundary.

In this way, it was possible to evaluate the extra overhead introduced in this scenario. The experimental results show that the additional overhead does not exceed 2% and confirm that the filesystem module behaves satisfactorily.

12.7   SUMMARY

The main goal of this chapter was to provide a complete example of how a complete and fully working embedded system can be assembled exclusively from open-source components, including the C-language software development toolchain used to build it.

Indeed, the discussion began by providing, in Section 12.1, some summary information about how the main toolchain components (binary utilities, compiler, C runtime library, and debugger) should be chosen so that they work correctly together. The same section also provides information about the rationale behind the choice of using the FREERTOS operating system instead of other open-source products, and about the consequences of this choice on the software development process.

Then, Section 12.2 provided an overall picture of the firmware structure, outlining its main components and how they are connected to each other through their APIs—by means of appropriate adaptation or porting layers. More information about individual components and their internal structure, also useful to conveniently port them to new architectures and projects, was given in Sections 12.3 through 12.5.

Last, but not least, Section 12.6 provided some insights into the memory footprint and performance of the main firmware components.

Part II

Advanced Topics

13 Model Checking of Distributed and Concurrent Systems

CONTENTS

This chapter discusses the first advanced topic of the book. It briefly presents how the well-known limits of software testing techniques can be addressed by means of formal verification. As embedded software complexity grows, the correct use of this technique is becoming more and more important, in order to ensure that it still works in a reliable way. From the practical point of view, very often software verification is carried out through model checking. This chapter is devoted to this topic using the well-known model checker SPIN and its input language PROMELA to draw some introductory examples. The discussion of a full-fledged case study is left to Chapter 14.

13.1 INTRODUCTION

When a distributed concurrent system has been designed and implemented, software testing techniques can be applied to check whether it works as expected or not. However, software testing techniques can just be used to show the presence of bugs, but they lack the ability to prove the absence of them. Instead, this goal can be achieved with *formal verification*.

Model checking [15] is one popular technique of formal verification. It has been widely used in the hardware industry. More recently, its adoption in the software industry keeps growing as well, especially for what concerns safety critical applications. This book will focus on its application in the verification of distributed concurrent software systems.

More specifically, model checking is an *automated* technique that, given a (finite-state) model M of a system and a logical property ϕ, *systematically* checks whether or not this property holds on the model, that is, $M \models \phi$. The logic property should also be specified with some formal notation, for instance, linear temporal logic (LTL), which will be discussed in Section 13.4.

Model checking aims at proving that the property of interest holds by *exhaustively* exploring every possible state the system under analysis can reach during its execution, starting from its initial state. Collectively, these states are called the *state space* of the system.

The system transitions from one state to another by means of elementary computation steps. The possible computation steps originating from a certain state are specified in the model itself. Therefore, a path from the initial state up to one possible final state (in which no more transitions are possible) represents one of the possible *computations* the system can perform.

Based on the (initial) design or implementation of a system, it is possible to build the corresponding model, upon which verification can be carried out by suitable model checkers. Depending on the verification results, possible refinement techniques can be applied to achieve better implementation.

PROMELA, which is a shorthand for *PRO*tocol *ME*ta *LA*nguage, is a language which can be used to write models for concurrent systems. With respect to ordinary programming languages like C, PROMELA offers *formal semantics*. Moreover, it is also more powerful in expressing concepts like nondeterminism, passive wait for an event, concurrency and so on, as will be better discussed in Section 13.3. On the other hand, PROMELA suffers from some weak points. For example, it just supports a limited set of data types, for example pointers are not available in PROMELA.

Since the model is written in a language different than the one used for the implementation of the original system, it is essential to ensure that the model is a *faithful* representation of the real system. In other words, the discrepancy between the model and the corresponding system should be kept to a minimum. Otherwise, the results of model checking are just meaningless.

Recently, it has become possible to abstract a PROMELA model automatically from the system implementation. However, for the current time, it only supports systems developed with the C programming language [97]. Moreover, automatic translation may include more details than necessary for what concerns the verification of a property, and in turn this makes the verification inefficient. As a consequence, this topic will not be further discussed in the book.

SPIN is one of the most powerful model checkers. Actually, SPIN stands for *S*imple *P*romela *IN*terpreter, namely, SPIN performs model checking based on the formal model written in PROMELA. SPIN can work in two modes, *simulation* or *ver-*

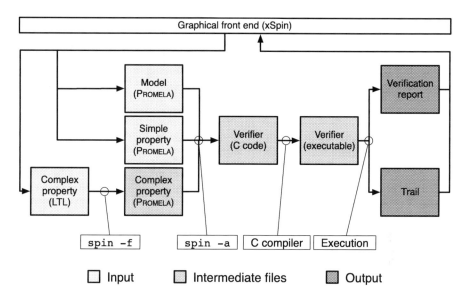

Figure 13.1 Simplified view of the SPIN verification flow.

ification. Simulation examines just *one* possible computation at a time, whereas verification explores them *all*. More information about this can be found in Section 13.2.

In order to help readers to build a more solid foundation, some simple examples are provided while introducing the basic concepts. A much more complex example, which shows how model checking can help to detect low probability issues of an industry level distributed multi-master election protocol, is presented in Chapter 14.

It is well known that model checking, or verification in general could be quite time and memory consuming, not only because it requires examining the whole state space but also because a state could become quite complex as system complexity grows. Some hints to improve the verification performance are discussed in Section 13.5.

As mentioned, SPIN requires writing a separate model for the system under analysis and it is also important to make sure that the model is consistent with respect to the real system, which is not a trivial job at all. Instead, unlike SPIN, SPLINT, a tool which can be used to detect code mistakes and vulnerabilities (especially for what concerns memory access errors) by means of static code analysis, can work directly on the C source code. And it will be discussed in Chapter 16.

13.2 THE SPIN MODEL CHECKER

Figure 13.1 depicts a simplified view of the SPIN verification flow. First of all, the model as well as simple correctness properties can be specified directly with PROMELA. Instead, more complex properties can be defined by means of linear temporal logic (LTL), which can be translated automatically by SPIN into PROMELA when it is invoked with either the −f or −F command-line options. LTL will be discussed in a more detailed way in Section 13.4.

When both the model and the property to be verified (shown in the light gray box of Figure 13.1) are available, SPIN is able to generate a verifier corresponding to them when invoked with the command-line option −a. Moreover, the verifier is written in the C programming language, hence it is possible to compile it with the native C compiler to get the executable file, which then can be run in order to carry out the verification. By convention, the executable file of the verifier is called *pan*, which is a shorthand for Protocol ANalyzer.

After the verification process is completed, SPIN produces two kinds of outputs: a verification report and optionally a trail if the verification result turns out to be negative. Both outputs are shown as dark gray boxes in Figure 13.1. The verification report summarizes the verification results, state space exploration information, memory usage, possible values of variables defined in the model, and so on. On the other hand, the trail provides a counterexample which leads to the violation of the target property. After that, users can follow a guided simulation driven by the trail to find out how it takes place exactly with all necessary details. This could be quite helpful to locate the issue and fix it.

In addition, several graphical front ends, which permit more convenient interaction with SPIN rather than through the command line, are available for a variety of operating systems. More information about this can be found in [73]. Besides generating the verification report and optionally the trail, these tools (for example xspin), are powerful also in the sense that they are able to perform post-processing of the SPIN textual output and present it in a more convenient way. For example, xspin can derive a message sequence chart (MSC) [94] from the trail which demonstrates inter-process communication of a model.

It is worth mentioning that examples shown in the following make use of SPIN version 5. Moreover, technical details internal to SPIN are out of the scope of this book, except that some hints to improve the verification performance are provided in Section 13.5. More interested readers could refer to References [74, 76].

Last but not least, from version 5 on, SPIN also supports model checking on multi-core systems [73]. Reference [77] presents the extensions to SPIN in order to make it support model checking on multi-core machines, as well as the related performance measurement.

13.3   THE PROMELA MODELING LANGUAGE

This section discusses in a general way the structure of the PROMELA modeling language. Some intuitive examples are provided to help readers better understand the concept. For a thorough explanation, interested readers could refer to specialized books like [76, 20].

13.3.1   PROCESSES

Processes are the main "actors" in a PROMELA model. They contain all the executable statements of the model. A process can be declared by means of the *proctype* keyword as follows:

```
proctype P(<param>) {
  <body>
}
```

in which, P indicates the process name, <param> is a list of parameters, and <body> represents the body of process P. Executable statements should be put in <body>. Moreover, statements are separated but *not* terminated by a semicolon (;). As a consequence, in the following sequence of statements belonging to process P,

```
proctype P(byte incr) {
  byte temp;
  temp = n + incr;
  n = temp;
  printf("Process P: n = %d\n", n)
}
```

it is not necessary to have a semicolon at the end of the last statement. By the way, as can be seen from the above listing, the PROMELA syntax still shares a lot in common with the C programming language, including comments. This makes the model written with it still intuitive to understand.

Processes declared in the above way must be instantiated *explicitly* one or more times, by means of the run operator. If a process has parameters in its declaration, then the same process can be instantiated multiple times, with different parameters. For instance,

```
run P(3);
run P(7)
```

As we can see, the run operator takes as arguments a previously defined proctype and, a possibly non-empty list of parameters which match the formal declaration of the proctype. It is worth mentioning that the processes created with run do not necessarily start their execution immediately after run completes.

Processes declarations can also be active,

```
active [<n>] proctype P () {
  <body>
}
```

The difference is that, during declaration, process P is also instantiated *implicitly* exactly n times. It is still legal, but not so useful to specify parameters in this way of declaration. This is simply because there is no way to specify different parameters for different instances of P within a single declaration. As a result, all instances of process P are identical.

```
1   byte n;
2
3   proctype P(byte incr) {
4      byte temp;
5      temp = n + incr;
6      n = temp;
7      printf("Process P: n = %d\n", n)
8   }
9
10  init {
11     n = 0;
12     atomic {
13        run P(3);
14        run P(7)
15     }
16  }
```

Figure 13.2 Example of init process.

By the way, when illustrating the PROMELA syntax, optional terms are specified with (non-quoted) square brackets ([...]), whereas a list of terms is enclosed in angle brackets (<...>). If [<n>] is omitted, the default value 1 will be used. This means that there is just one instance of process P.

In PROMELA, there is a special process, namely the init process,

```
init {
  <body>
}
```

The init process, if declared, is always the *first* process to be activated. It is typically used for:

- Instantiation of other processes.
- Complex initialization of some global variables. For instance, nondeterministic choice of a value. This can be useful to model different configurations/settings of a system, external inputs to a system, and so on.

Figure 13.2 shows a simple example of the init process. It initializes a global variable n and creates two copies of process P with different parameters. In the example, the global variable is initialized in quite a straightforward way. Better examples will be shown at a later time when control statements which permit nondeterministic execution are introduced in Section 13.3.4.

It is worth remarking that, by convention, instantiation of a group of processes is performed within an atomic sequence, as shown in the above example (line 12). It ensures that executions of those processes start at the same time. One advantage of doing it in this way is that all possible interleaving among processes can be examined. More information about this is provided in Section 13.3.5.

Table 13.1

Basic PROMELA Data Types on a Typical 32-Bit Architecture

Name	Size (bit)	Range
bit	1	$0, 1$
bool	1	false, true
byte	8	$0, \ldots, 255$
short	16	$-32768, \ldots, 32767$
int	32	$-2^{31}, \ldots, 2^{31} - 1$
unsigned	$n \leq 32$	$0, \ldots, 2^n - 1$

13.3.2 DATA TYPES

Table 13.1 shows the basic data types supported in PROMELA, together with the number of bits used to represent them as well as the range indicated by the corresponding type on a typical 32-bit architecture. Data type representation on a given architecture depends on the underlying C compiler, because the verifier is compiled with the native C compiler of the computer on which the verification will run as discussed in Section 13.2.

However, unlike in the C programming language, *short* and *unsigned* cannot be used as prefix of *int*. In particular, the unsigned keyword is used in quite a different way in PROMELA than C. It is used to specify variables holding unsigned values in a user-defined number of bits n, where $1 \leq n \leq 32$. For instance,

```
unsigned x: 5;
```

the above listing declares an unsigned variable x stored in 5 bits.

Variable declaration can be performed in a quite straightforward way, by indicating the data type followed by the variable name, and optionally the initial value given to the variable, as follows,

```
bit x;
bool done = false;
byte a = 78;
```

The above listing declares three variables, a bit variable x, a Boolean variable done, as well as a byte variable a. What's more, the Boolean variable and the byte variable are initialized to false and 78, respectively. In PROMELA, if no initial value is provided, a variable is initialized to 0 by default.

The syntax for variable assignment is similar to C. For instance,

```
byte x;
x = 4
```

declares a variable of type byte and then assigns a value of 4 to it. However we should pay attention to the fact that, in PROMELA, this is *not* exactly the same as byte x = 4. This is because, this may introduce additional, unnecessary, or even unexpected system states. For instance, in the above listing and in some states, the value of x is 0 instead of 4.

This is especially important when modeling a concurrent system, where other processes could observe the value of x, and may perform actions correspondingly.

It is worth noting that data types, like characters, strings, pointers, and floating-point numbers, which are commonly used in most programming languages, are not available in PROMELA.

Other data types supported in PROMELA include *array*, *mtype*, and *chan*. More specifically, only one-dimensional arrays are directly supported by PROMELA. Arrays can be declared in a similar way like in the C language. For instance,

```
byte a[10] = 8;
```

the above listing declares an array called a with 10 elements, each of which contains a byte value initialized to 8. If no initial value is provided, all elements get a default value of 0. Array can be accessed with index and the first element of an array is at index zero. It is also possible to specify multi-dimensional arrays in PROMELA, by means of the structure definition, which will be shown later.

An *mtype* keyword is available in PROMELA, which can be used to define symbolic names for numeric constant values. For instance,

```
mtype = {mon, tue, wed, thu, fri, sat, sun};
```

declares a list of symbolic names. Then, it can be used to specify variables of mtype type and assign values to the variable, as follows,

```
mtype x = fri;
```

Symbolic names improve readability of PROMELA models. Internally, they are represented with positive *byte* values. Equivalently, variables of mtype type are stored in a variable of type *byte*.

There can be one or more mtype declarations in a PROMELA model. However, if multiple declarations are provided, they will be *implicitly* merged into a single mtype declaration by concatenating all lists of symbolic names. This data type can also be used in channel (*chan*) declarations, which will be introduced in the following, to specify the type of message fields.

Channels are used to exchange *messages* among processes. They can be declared as follows,

```
chan ch = [capacity] of {type, ..., type}
```

where ch represents the name of a channel variable that can be used by processes to send or receive messages from a channel. type, ..., type defines the structure of messages held by a channel. capacity indicates the number of messages that can be stored in a channel. It can be either zero or a positive value, which represents a *rendezvous unbuffered* channel or a *buffered* channel, respectively. The difference between the two types of channel together with possible operations over channels will be discussed in Section 13.3.6.

```
chan c = [4] of {byte, mtype}
```

The above listing declares a buffered channel c that can hold 4 messages, which are made up of a byte value and a field of mtype type.

Channels are commonly declared as global variables so that they can be accessed by different processes. Actually, any process can send/receive messages to/from any channel it has access to. As a result, contrary to what their name indicates, channels are not constrained to be point-to-point. Local channels are also possible [20], but their usage is beyond the scope of this book.

If the aforementioned data types are not expressive enough, PROMELA also supports user-defined structured data types by means of the *typedef* keyword, followed by the name uname chosen for the new type and a list of fields that make up the structure, as follows,

```
typedef uname {
  type f_name1;
  ...
  type f_nameN
}
```

Individual fields can be referred in the same manner as in the C language, for instance uname.f_name1. It is worth mentioning that a typedef definition must always be global, whereas it can be used to declare both global and local variables of the newly defined type. The following listing defines a two-dimensional array with each of its elements of *byte* type. Moreover, one element of it is assigned with value 10.

```
typedef arr{
  byte b[5];
}

arr a[3];
a[1].b[4] = 10
```

For what concerns variable declaration, there are exactly two kinds of *scope*[1] in PROMELA, namely:

- One single *global* scope. Global variables are visible to all processes in the system and can be manipulated by all of them
- One *local* scope for each process. The scope of a local variable is the *whole* process, *regardless* of where it is declared. And there are no nested scopes. All local variable declarations are silently moved to the very beginning of the process where they are declared. This may have unexpected side effects, for instance, in the following listing

```
1  active proctype P() {
2    byte a = 1;
3    a = 2;
4    byte b = a
5  }
```

---

[1] It is worth noting that an additional *block* scope has been added to SPIN version 6 [73].

Table 13.2

Summary of PROMELA **Operators in Decreasing Precedence Order**

Operators	Assoc.	Meaning
() [] .	L	Parentheses, array indexing, field selection
! ~ ++ --	R	Negation, bitwise complement, postfix increment/decrement
* / %	L	Multiplication, division, modulus
+ -	L	Addition, subtraction
<< >>	L	Left/right bitwise shift
< <= >= >	L	Relational
== !=	L	Equality/inequality
&	L	Bitwise and
^	L	Bitwise exclusive or
|	L	Bitwise or
&&	L	Logical and
||	L	Logical or
-> :	R	Conditional expression
=	R	Assignment

the value of b is 1, instead of 2. This is because the declaration of b (line 4) is implicitly moved above the assignment (line 3).

Last but not least, it is recommended that programmers should pick up the most appropriate data types for variables in a model. This is because the data type size directly affects the *state vector* size during verification. A system *state* of a PROMELA model consists of a set of values of its variables, both global and local, as well as location counters, which indicate where each process currently is in its execution.

Generally speaking, during verification, the model checker needs to store the state in memory and examine every possible value of a variable. As a result, if variables are defined with data types which are unnecessarily large, it could easily lead to state space explosion and make the verification lengthy or even unfeasible. For instance, the capacity of a channel should be selected carefully. Otherwise, due to the rather complex structure of channels with respect to other data types, the state vector size may increase in a significant way with inappropriate channel capacity values, which in turn impairs verification efficiency. Detailed information about this topic is provided in Section 13.5.

13.3.3 EXPRESSIONS

The way a PROMELA expression is constructed is, for the most part, the same as in the C language and also most C-language operators are supported. They are listed in Table 13.2 in decreasing precedence order; operators with the same precedence are listed together. The most important difference of expressions in PROMELA with respect to C relates to the fact that it must be possible to repeatedly evaluate a

PROMELA expression to determine whether or not a process is executable. Expressions are therefore constrained to be *side-effect free*, that is, their evaluation must not modify the system state in any way. Otherwise, while evaluating whether the system can move from the current state to the next one, the state changes in between. There is a single exception to this rule and it will be discussed at the end of this section.

For the above reason, unlike in C, PROMELA assignments are not expressions because they change the value of the variable on their left side. And the postfix increment and decrement operators (++ and −−) can be used only in a standalone assignment, like a++, but not in an expression, for instance on the right side of an assignment. It is worth noting that prefix increment and decrement are not available in PROMELA. Even if they existed, they would share the same problem as their postfix counterparts.

The syntax for conditional expressions is slightly different from the C language and it looks like:

```
(expression_1 -> expression_2 : expression_3)
```

As we can see, the ? operator in C is replaced by −>. Moreover, the conditional expression as a whole should always be surrounded by round braces, in order to avoid being misinterpreted as the arrow separator used in the selection and repetition statements, which will be discussed in Section 13.3.4. Since assignments are not expressions, they cannot be used in conditional expressions as well. For instance, the following fragment of code will cause a syntax error in PROMELA,

```
(a > b -> (k = 0) : (k = 1))
```

All other aspects of the expression syntax are the same as in C.

Both individual expressions and assignments are valid PROMELA *statements*. Besides these, there are five more *control statements* in PROMELA, to model how the computation flow proceeds in a process, namely: sequence, selection, repetition, jump, and the *unless* clause. They will be discussed in a more detailed way in Section 13.3.4.

Executability is an important concept in PROMELA and it is tied to how *passive wait* and *synchronization*—two essential aspects of any concurrent system—are formalized in a PROMELA model.

More specifically, the *executability* of PROMELA processes is evaluated at the *statement* level. Any PROMELA statement may or may not be executable in a given state. Whenever a process encounters a non-executable statement during its execution, it blocks (passively waits) without proceeding further, until the statement becomes executable again in a future state of the computation.

The executability rules that apply to a given statement depend on the specific type of statement:

- Assignments are always executable;
- For what concerns expressions, an expression is executable if it is evaluated to be not *false*. Otherwise, the process where it locates will be blocked until

it becomes true. This rule applies when an expression is used either standalone or at a guard position of a statement, for instance, in a selection statement or a repetition statement, which will be introduced in Section 13.3.4. For instance, a process that encounters a standalone expression i==3 blocks until i is set to 3 by other processes. As we can see, expressions can profitably be used to synchronize multiple processes.

- The executability rules for more complex types of statement, like those aforementioned, will be given along with their description.

It is worth noting that, besides being used in the init process to instantiate other processes, the *run* operator is the only operator allowed inside expressions that can have a side-effect. This is because run P() returns an identifier for the process just created. Upon successful creation, a positive process identifier is returned. Otherwise, the return value is 0. Most importantly, processes identifiers are also part of system state. As a consequence, the run operator should be used carefully inside expressions. For interested readers, more detailed information can be found in [76].

13.3.4 CONTROL STATEMENTS

There are five different types of control statement, namely sequence, selection (if), repetition (do), jump (goto), as well as the unless clause. They differentiate from each other in the sense that they model different computation flows within a process.

A sequence of statements is separated by semicolons (;) and they will be executed one after another in a sequential manner. Besides basic statements like assignment and expression, it could also include more complex control statements like those that will be introduced in the following.

Selection statement

The selection statement models the choice among a number of execution alternatives. And its general syntax is as follows:

```
if
:: guard_1 -> sequence_1
   ...
:: guard_n -> sequence_n
fi
```

It starts with the if keyword and ends with the fi keyword. It includes one or more branches, starting with double colons (::). Each branch is made up of a *guard* and the sequence of statements following it, if the guard is evaluated to be true. A guard is usually an expression which results in a Boolean value, either true or false.

By the way, the arrow separator is an alternative for the semicolon (;) separator. But semantically, they are the same. The arrow separator is more commonly used in selection statements and repetition statements, which will be introduced later, between a guard and the following sequence of statements to better visually identify those points in a model where execution could block. It is mainly for improvement of readability.

The selection statement is executable if *at least* one guard is true. Otherwise, the process is blocked. Instead, if more than one guard are evaluated to be true, a *nondeterministic* choice exists among them. When at this point, the verification can proceed with either branch. As a result, this corresponds to multiple execution paths in verification. There is a special guard *else*, which is true if and only if all the other guards are false.

For a concurrent program, it follows one possible execution path at a single run. Different runs may lead to different results. In order to verify whether a certain property holds on it or not, it should be checked for every possible execution path. As we can see, PROMELA provides exactly what is needed.

```
int i=3, j=0;
void P(void) {
    if (i<4) j=1;
    else if (i>2) j=2;
    else j=3;
}
```

```
int i=3, j=0;
active proctype P() {
    if
    :: (i<4) -> j=1
    :: (i>2) -> j=2
    :: else -> j=3
    fi
}
```

The above two listings show the conditional statement in C (left) and the selection statement in PROMELA (right). The two listings may not make any sense from the logic point of view. However, syntactically, they are not wrong. They are used here just to demonstrate the difference between the two languages. At first glance, the two if statements may look quite similar to each other. However, they are not equivalent.

This is because, in C, the if predicates are evaluated in a *sequential* manner. In other words, if predicate A comes before predicate B, then A is evaluated first. If A is true, the statements associated to it are executed and B will not be evaluated. After the execution of the statements associated to A, the if statement terminates. Only if A is false, will B be examined.

Since the predicate i<4 is true in this example, the value of j will always be 1.

Instead, for what concerns the PROMELA code, since the first and second guards are both true, there is a nondeterministic choice between their associated statements and both execution paths are possible. In this case, j gets the values 1 and 2 in different execution paths. This way of reasoning is also confirmed by the SPIN verification output.

```
unreached in proctype P
        line 4, "pan_in", state 6, "j = 3"
        (1 of 9 states)

Values assigned within interval [0..255]:
        j          : 1-2,
        i          : 3,
```

The first part reports that the statement `j=3` is unreachable during verification, whereas the second part indicates that `j` is assigned with value 1 and 2. By the way, `pan_in` is a conventional way of naming a model when it is supplied to SPIN. It is worth noting that some verification output is omitted for clarity.

Repetition statement

```
do
:: guard_1 -> sequence_1
   ...
:: guard_n -> sequence_n
od
```

The repetition statement is embraced with the pair of keywords `do` and `od`. It also contains one or more branches starting with the guard and followed by a sequence of statements, like in the selection statement.

It behaves as follows: after the execution of statements associated to a true guard, it goes back to the beginning of `do` and the guards are evaluated again. If any of the guards are true, the execution will continue with the statements following it. This process will be repeated until it encounters the `break` keyword. It terminates the loop and control passes to the statement following `od`.

It is worth remarking that the while loop implemented in the C code (shown in the left listing) cannot be translated to the PROMELA code shown on the right.

```
int i=0, j=0;
void P(void) {
  while (i<3) {
    i++
  }
  j=i;
}
```

```
int i=0, j=0;
active proctype P() {
  do
  :: (i<3) -> i++
  od;
  j=i
}
```

This is because the body of the C while loop, namely `i++`, is executed three times and then the assignment `j=i` is performed. As a result, both `i` and `j` get value of 3.

Instead, the PROMELA process never goes beyond the `do` loop. The reason is that, after being incremented for three times, `i` becomes 3. After that, the execution goes back to the beginning of the `do` loop. When it tries to evaluate the guard again, it is no longer true. And the repetition statement is no longer executable and process `P` will block at this statement indefinitely, if there are no other processes in the system operating on `i`.

The SPIN verification output is shown in the following,

```
pan: invalid end state (at depth 5)
pan: wrote pan_in.trail
```

```
unreached in proctype P
        line 7, "pan_in", state 6, "j = i"
        line 7, "pan_in", state 7, "-end-"
        (2 of 7 states)

Values assigned within interval [0..255]:
        i        : 0-3,
```

The above results show that statement j=i in the PROMELA model is unreachable because execution is blocked within the repetition statement. And, i obtains any value between 0 and 3.

In order to obtain the same results as the C code, a second branch should be added to the do loop of the PROMELA code, as follows,

```
int i=0, j=0;
active proctype P() {
  do
  :: (i<3) -> i++
  :: else -> break
  od;
  j=i
}
```

As discussed before, the else guard is true if and only if all other guards are false. Consequently, when i becomes 3, the first guard is false. Instead the else guard becomes true and the statement that comes after it, namely break, is executed, which terminates the loop. The execution proceeds to the assignment statement and j gets the value of 3 as well.

If the else guard in the above loop is replaced by the true keyword, as follows

```
int i=0, j=0;
active proctype P() {
  do
  :: (i<3) -> i++
  :: true -> break
  od;
  j=i
}
```

it will behave in yet another way. More specifically, according to the SPIN verification output shown in the following, variable j gets all possible values between 0 and 3. Besides, it also shows the state space exploration information, for instance, a system state is stored in a 16-byte state vector.

```
State-vector 16 byte, depth reached 9, errors: 0
        19 states, stored
         0 states, matched
        19 transitions (= stored+matched)
         0 atomic steps
```

```
Values assigned within interval [0..255]:
        j          : 0-3,
        i          : 0-3,
```

The aforementioned result can be easily justified by the fact that, unlike `else`, the `true` guard is always true. As a consequence, when i is less than 3, both guards within the repetition statement are true. This represents a nondeterministic choice. In other words, at each iteration, it is possible to either increment the value of i or terminate the loop. Verification examines all possible computation paths, from terminating the loop immediately as soon as entering the loop, to performing three times of increment and then terminating.

This variant of the `do` loop can be used to nondeterministically choose a value within a certain *range*, in order to use it in the subsequent code. It can be profitably used in the `init` process to perform complex initialization of global variables.

There are two other types of statement, namely jump (`go`) and the `unless` clause. The jump statement breaks the normal computation flow and causes the control to jump to a label, which is a user-defined identifier followed by a colon. For instance,

```
label_1:
    ...
```

```
if
:: (i > 6) -> goto label_1
:: else -> ...
fi
```

in the above listing, if the guard i>6 is true, it will jump to where `label_1` is and execute the code that follows that label. As we can see, if there is more than one jump statement, it makes the code less structured and may impair the readability of a model. Consequently, it is recommended to limit the use of the jump statement.

An exception to this rule is that when the original system can be well represented as a finite state automaton, where the system is made up of multiple states, the system takes different actions under different states, and the system moves from one state to another through a transition. In this case, a state can be modeled as a label followed by a group of statements, whereas the transition can be modeled by a `goto` statement.

The `unless` clause is used to model exception catch. An `unless` block can be associated with either a single statement or a sequence of statements. Moreover, it is executed when the first statement of the block becomes executable. For instance,

```
active proctype P() {
  byte i=4, j=2, k;

  do
  :: {
        k = i / j;
        i--;
        j--;
```

```
    } unless {
        j == 0 ->   break
    }
  od
}
```

in the above example, i is divided by j and the result is stored in variable k. After that, both i and j are decremented. This process is repeated until j becomes 0, which is an invalid divisor, an exception caught by the unless clause.

The unless clause is commonly used for the situation that an event occurs at an arbitrary point in a computation.

For more information about the jump statement and the unless clause, interested readers can refer to [20].

13.3.5 ATOMICITY

In PROMELA, only elementary statements (namely expressions and assignments) are *atomic*. During verification, an atomic statement is always executed as an *indivisible* unit. Consequently, variables will not change their value in between. If there is any side effect associated with an atomic statement (for instance the assignment statement), the effect is visible to other processes either *entirely* or *not at all*. For what concerns the following PROMELA code, where a variable a of the int type is first declared and initialized to 0 and then assigned with value 1,

```
int a = 0;
a = 1
```

in PROMELA, any process observing this variable will see either 0 or 1, but nothing else.

On the contrary, a sequence of statements is not executed atomically. This is also because processes' location counters are also part of system state and they work at the statement level.

It is possible to make a sequence of statements executed closer toward an atomic way with two individual specifiers, namely *atomic* and *d_step*, as follows.

```
atomic {
    sequence
}

d_step {
    sequence
}
```

The difference between them is that,

- The atomic sequence is *executable* if and only if its first statement is executable. As a consequence, the first statement works as a *guard* for the atomic sequence. Once the atomic sequence starts executing, it is executed *without* interleaving with other processes until it ends or it reaches a non-executable (in other words, *blocking*) statement. In the second case, atomic execution *resumes* when the statement becomes executable again, possibly because it has been unblocked by other processes.

- d_step applies even stricter rules for what concerns atomicity. It is assumed to be executed as if it were a single indivisible statement. More specifically, it is illegal to block within a d_step sequence, except for the guard. If it involves any nondeterministic choice, the nondeterminism is resolved by always choosing the *first* alternative with a true guard. Moreover, it is not allowed to jump into a d_step sequence or out of it. As a result, any goto statement is forbidden.

From the verification point of view, d_step sequence is more efficient than atomic sequence because there is only one possible execution path regarding a d_step sequence whereas atomic sequence at least can interleave with other processes at certain points.

As a result, d_step is recommended to be reserved for fragments of sequential code which can be executed as a whole. Instead, atomic is preferred for implementing *synchronization* primitives.

The following example shows a case where atomicity is needed and how an atomic sequence can help to address the issue. The following fragment of code (unfortunately not completely correct, as we will see later) is an attempt to force multiple processes to enter a critical region in a mutually exclusive way.

```
1   bool lock = false;
2   byte crit = 0;
3
4   active [2] proctype P () {
5       !lock;
6       lock=true;
7
8       crit++;
9       assert(crit == 1);
10      crit--;
11
12      lock=false
13  }
```

Within the critical region (lines 8–10), both processes will access the shared variable crit, by first increasing its value by 1 (line 8) and then bringing it back (line 10). If mutual exclusion is implemented successfully, there should be only one process in the critical region at a time. As a result, when the value of crit is evaluated at line 9 by any process, it should always be 1.

By the way, the above code also includes an assertion (line 9). An assertion can be specified by means of the assert keyword, followed by an expression. It can be placed anywhere between two statements. Assertion is mainly used to help SPIN trap violation of simple correctness properties during verification. SPIN will complain if the expression within an assertion is found false.

Mutual exclusion is implemented by means of code at lines 5–6 and line 12. The reasoning behind the code is that the value of the global variable lock tells whether or not any process is within the critical region at any given instant. At the beginning, no processes are in the critical region, and hence, its initial value is false. Thus, the expression !lock will not be executable when lock is true, that is, it will block any process trying to enter the critical region while another process is already inside it.

When a process successfully goes beyond that expression (line 5)—meaning that it is allowed to enter the critical region—it will execute the assignment lock=true (that is always executable) to prevent other processes from doing the same. The lock variable is reset to false when the process is about to exit from the critical region (line 12). This assignment will make the expression !lock executable, and hence, allow another process formerly blocked on it to proceed and enter the critical region.

However, as said before, this solution to the mutual exclusion problem is not fully correct, which is also confirmed by the SPIN verification output.

```
pan: assertion violated (crit==1) (at depth 7)
pan: wrote pan_in.trail

Values assigned within interval [0..255]:
        crit      : 0-2,
        lock      : 0-1,
```

As we can see, SPIN complains that the assertion is violated. In particular, the crit variable can obtain the value of 2. SPIN wrote a counterexample found during verification back into a trail file, namely pan_in.trail.

As will be shown in the following guided simulation output driven by the trail file, the issue is related to the *atomicity* of execution. More specifically, the sequence of statements (line 5–6) of the PROMELA code is not executed atomically and its execution by a process can *interleave* with other processes' activities.

```
 1   Starting P with pid 0
 2   Starting P with pid 1
 3     1:     proc  1 (P) line   5 [(!(lock))]
 4     2:     proc  0 (P) line   5 [(!(lock))]
 5     3:     proc  1 (P) line   6 [lock = 1]
 6     4:     proc  1 (P) line   8 [crit = (crit+1)]
 7     5:     proc  1 (P) line   9 [assert((crit==1))]
 8     6:     proc  0 (P) line   6 [lock = 1]
 9     7:     proc  0 (P) line   8 [crit = (crit+1)]
10   spin: line   9 "pan_in", Error: assertion violated
11   spin: text of failed assertion: assert((crit==1))
```

As shown in the guided simulation output, two instances of process P are created, with different process identifiers, namely pid 0 and pid 1 (lines 1–2). The guided simulation output also shows how the two processes interleave with each other and give rise to the error step by step. More specifically, process 1 first evaluates `!lock` (line 3) and proceeds beyond it, because `lock` is `false` at the moment. Then, before process 1 sets `lock=true` according to its next statement, process 0 also evaluates the expression `!lock` and it is still true (line 4). As a result, both processes are allowed to enter the critical region. After that, process 1 first keeps going for three steps within the critical region by setting the lock, incrementing the shared variable, and performing assertion (lines 5–7). And then process 0 tries to do the same and the assertion fails (lines 8–9).

A straightforward solution to this issue is to execute both the test expression and the subsequent assignment atomically, which can be achieved by means of the following construct:

```
atomic {
  !lock;
  lock=true
}
```

According to the rules of atomic sequence, since the assignment statement is always executable, when a process evaluates the guard and it is true, it will proceed to the end of atomic sequence without interleaving with other processes. This indicates that if other processes attempt to enter the critical region at the same time, the atomic sequence within them is *not* executable, as the guard is not executable.

As we can see, this models how a synchronization semaphore (which has been discussed in Section 5.3) behaves. Actually, d_step can also be used in this simple example and it makes no difference.

It is important to remark that introducing an atomic or d_step sequence in a model is reasonable if and only if it is guaranteed that the modeled system works in exactly the same way. Otherwise, verification results may be incorrect as some possible interleavings among processes are not considered.

Referring back to Figure 13.2, since the instantiation of two processes is embedded in an atomic sequence within the init process, interleaving of the init process with other processes is not allowed as long as the atomic sequence does not encounter any blocking statement. In this way, instantiation of the two processes will be completed before any of them can be executed. The execution of the two processes will start afterward so that all possible ways of interleaving between them can be modeled.

13.3.6 CHANNELS

As mentioned in Section 13.3.2, channels are used to exchange messages among processes. There are two types of operations that can be performed upon a channel, namely *send* and *receive*. The basic form of send and receive operations are as follows,

```
chan ch = [capacity] of {type, ..., type};
ch ! exp, ..., exp;
ch ? var, ..., var
```

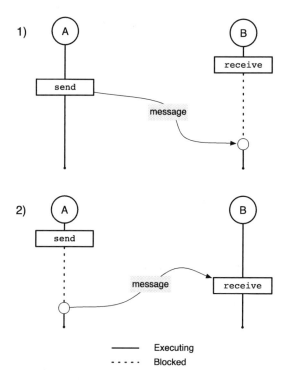

Figure 13.3   Message exchange through rendezvous channel.

The ! represents the basic form of the send operation, whereas ? represents the receive counterpart. exp, . . ., exp is a sequence of expressions whose values are sent through the channel. Instead, var, . . ., var is a sequence of variables that are used to store the values received from the channel. If one of them is specified with _ (underscore), it indicates that the corresponding field of the message is not interesting to the receiver and its value will simply be discarded.

For both the send and the receive operations, the number and types of the expressions and variables must *match* the channel declaration.

If the capacity of a channel is 0, it models a *rendezvous* channel. Since there is no room in this kind of channel to store the message, it is required that both the sender and receiver should be ready when the message is about to be exchanged. Moreover, the exchange of message will be performed in one atomic step.

More specifically, when two processes A and B would like to exchange messages through a rendezvous channel, as shown in the two parts of Figure 13.3,

1. If process B encounters the receive statement first, while at the same time process A is not yet at the point to send the message, process B will be blocked at the

receive statement. The receive statement becomes executable when process A is about to send the message.

2. The reverse is also true. That's to say, the send statement in Process A becomes executable when B is at its receive statement.

When the capacity of a channel is greater than 0, it represents a *buffered* channel. It can be specified as follows,

```
chan ch = [k] of {type, ..., type}
```

where k is used to indicate the capacity of a buffered channel and k>0. The basic form of send and receive statements can also be used for it. Regarding a buffered channel, the send statement is executable as long as the channel is *not* full. In other words, the number of messages held in the channel is less than k. Instead, the receive statement is executable as long as the channel is *not* empty, namely the number of messages stored in it is greater than 0.

It should be noted that ordinary send and receive statements operate on a channel in a first in first out (FIFO) order. Moreover, channel contents are also part of the system state.

There are four predefined Boolean functions available in PROMELA which can be used to check the state of a channel, whether it is empty or full. They are,

```
empty(ch)
nempty(ch)
full(ch)
nfull(ch)
```

As we can see, they all take the name of a channel as argument. There are four functions instead of two because the negation operator (!) cannot be applied to them for a reason related to how SPIN works internally. Interested readers could refer to Reference [20] for more information. Instead, nempty (not empty) and nfull (not full) are used to specify the counterparts of empty and full, respectively. These functions can be used as a guard to determine whether a certain operation, either send or receive, can be carried out or not.

For instance, the following code specifies one sender and two receivers, which access the same channel that can store only a single one-byte message. The two receivers intend to receive a message from the channel if it is available, otherwise they simply terminate without doing anything. The skip keyword simply concludes a selection statement and moves the execution to the next statement. It is always executable.

```
1  chan c = [1] of { byte };
2
3  active proctype S() {
4     c ! 1
5  }
6
7  active [2] proctype R() {
8     byte v;
```

```
9    if
10    :: nempty(c) -> c ? v
11    :: empty(c) -> skip
12    fi
13    }
```

However, it does not work as expected. Instead, the SPIN verification output reports there is invalid end state.

```
pan: invalid end state (at depth 4)
pan: wrote pan_in.trail
```

By following the trail, the guided simulation output shown in the following indicates that it is related to atomic execution. More specifically,

```
1    Starting S with pid 0
2    Starting R with pid 1
3    Starting R with pid 2
4     1:    proc  0 (S) line    4 [values: 1!1]
5     1:    proc  0 (S) line    4 [c!1]
6     2:    proc  2 (R) line   10 [(nempty(c))]
7     3:    proc  1 (R) line   10 [(nempty(c))]
8     4:    proc  2 (R) line   10 [values: 1?1]
9     4:    proc  2 (R) line   10 [c?v]
10    5: proc 2 terminates
11   spin: trail ends after 5 steps
```

the sender sends a message to the channel (lines 4–5). This leads to both receivers passing their guard since the channel is no longer empty (lines 6–7). One receiver receives the message (lines 8–9) and then terminates (line 10). Instead the other receiver is stuck in its receive operation and never terminates. SPIN reports this situation as an *invalid end state*.

The issue can be addressed by enclosing the check and the receive operation in an atomic sequence by means of the `atomic` keyword. Instead, `d_step` cannot be adopted here, because `c?v` could also block besides the guard.

Besides ordinary receive statement, several other types of receive statement are also available, including random receive, nondestructive receive, as well as polling.

Random receive can be specified with the double question mark (`??`) in the following way,

```
ch ?? exp_or_var, ..., exp_or_var
```

Unlike the basic form of receive statement, the list of variables can also contain *expressions*. It receives the first message in the channel that *matches* the given expressions. This message could be at any position in the target channel. As a result, it does not follow the FIFO order. Variables are handled as usual. If there is no match, the random receive is *not* executable.

The next type of receive can be built on top of both ordinary receive and random receive by means of the angle brackets (`<>`),

```
ch ? <var, ..., var>
ch ?? <exp_or_var, ..., exp_or_var>
```

The meaning is the same as the ordinary and random receive. The most important difference is that the (matched) message is *not* removed from the channel. Variables are still assigned with values of the corresponding message fields. Hence, this type of receive still has side effects and thus they cannot be used as expressions.

On the contrary, polling receive statements written as follows

```
ch ? [var, ..., var]
ch ?? [exp_or_var, ..., exp_or_var]
```

are *expressions*. They are true if and only if the corresponding receive statements are executable. Variables are *not* assigned. As a result, there is no side effect and they can be used as a guard.

13.4 PROPERTY SPECIFICATION

When the PROMELA model is ready, properties which express requirements on the behavior of the system should be specified. After that, the model checker can be used to check whether the properties hold on the corresponding model or not.

As mentioned in Section 13.3.5, simple correctness properties, for instance mutual exclusion, can be specified by means of an assertion. Since an assertion is checked only in a specific location of a specific process, it lacks the ability to evaluate global properties which should be satisfied regardless of what processes are executing, for example, absence of deadlocks. Complex properties like those can be specified by means of a higher-level formalism, namely linear temporal logic (LTL) [139] formulae.

In its simplest form, an LTL formula can be built from the following elements. By the way, the SPIN syntax for them is shown in parentheses. Interested readers could refer to [113] for a thorough discussion of this topic.

- *Atomic propositions.* Any Boolean PROMELA expression is a valid atomic proposition in LTL, for instance, x==3.
- *Propositional calculus operators*, such as *not* (!), *and* (&&), *or* (| |), *implication* (->), and *equivalence* (<->). Atomic propositions can be combined by means of propositional calculus operators to get propositional calculus formulae, for instance, (x==3) && (y<9).
  In particular, A->B means that if A is true then B must be true. On the contrary, if A is false, it does not matter what the value of B is and the overall result will be true. Logically, it is equivalent to !A| | B. Instead, A<->B requires that both A->B and B->A are true.
- *Temporal operators*, such as *always* ([]) and *eventually* (<>), can be applied to the previous two elements to specify temporal features.

The following LTL formula specifies that, for every possible computation, it will always be true that if F is true, then it implies that sooner or later G will eventually become true.

```
[] ( F -> <> G )
```

The above example highlights an important difference between the temporal operators and the other two. More specifically,

- Both atomic propositions and propositional calculus formulae can be evaluated by looking only at a *single* state in the computation.
- Instead, the evaluation of temporal operators involves looking at future states in the computation.

The above example can be read in an alternative way, that is, whenever there is a state within which F is true, it is necessary to examine the following states to see whether there is a state within which G is true.

Properties specified as LTL formulae can be automatically translated by SPIN into PROMELA code before verification, and become part of the model being verified.

13.5   PERFORMANCE HINTS

Time and memory resources to be used during verification are always a critical issue, common to all model checkers. This is because verification is assumed to be done over the whole state space and state space could easily get exploded when the system to be verified becomes complex. Even though it is impossible to fully describe how SPIN works internally within limited space, some high-level information can be useful to use it in a more effective way.

As defined in Section 13.3.2, a state is represented as a state vector. In particular, it consists of values of global and local variables, including channels, and location counters. On the one hand, nondeterminism and interleaving are important aspects to be modeled for concurrent and distributed systems. On the other hand, they can significantly affect the size of the state space. For example:

```
chan c = [N] of {byte};

active proctype fill() {
  do
  :: c ! 1
  :: c ! 0
  od
}
```

In the above listing, the `fill` process fills channel c with N byte-sized values, chosen in a nondeterministic way, and then blocks because the channel buffer is full. During an exhaustive state space exploration, the number of states x grows exponentially with respect to N, that is, $x \simeq 2^{N+1}$. This is because, just considering the channel variable c, the content of its buffer goes from being completely empty (at the beginning of verification) to being completely full (at the end) and, at any step, it can contain any sequence of values, from all 0s to all 1s. In order to perform an exhaustive search, it is necessary to consider and check the property of interest in all these states.

It is not the case that the whole state space is built all at once, and then verification is carried out on it. On the contrary, state vectors are calculated *on the fly* and stored into a hash table. More specifically, when storing a state vector, a hash function is applied to calculate an index which indicates the position where the state vector should be stored within the table. If there is already a state vector stored in that position, and it is not the same as the current one, a hash conflict occurs. In this case, the new state vector is stored elsewhere and linked to the existing state vector through a linked list. By intuition, as the number of hash conflicts increases, storing a new state becomes less and less efficient because the linked lists grow. As a consequence, more time is required to linearly scan the lists, just to check whether or not a certain state vector is already in the table.

Verification is the process of checking whether a given property holds in each state. As a consequence, if a state has already been checked, there is no need to check it again. Let us assume that the verification is currently considering a certain state, and it is about to execute a statement. After execution, the values of some variables in the model may change. This leads to another state, namely, a state *transition* takes place. What SPIN does next is to look into the hash table:

- If the new state has *not* been stored yet, then it checks whether the property holds on that state. If it does, SPIN stores the state vector into the hash table, otherwise SPIN just found a *counterexample*.
- If the state has already been stored in the table, it is a proof that the property holds on that state. In this case, there is no need to store it again and the program can move to the next step.

Overall, during verification, SPIN keeps storing state vectors into the hash table and looking up newly created state vectors to see whether or not the corresponding states have already been visited before. This process may be highly time and memory consuming, depending on a lot of different aspects. For instance, if the hash algorithm in use is not so effective, it is possible to have long linked lists somewhere in the hash table, whereas other parts of the table still remain empty.

By the way, the verification process ends when either a counterexample is found (with the default verification option of SPIN), or all processes reach their ends, or no more states can be generated (for example, all processes are blocked).

When trying to improve the performance of SPIN, the goal is to achieve the best trade-off between *speed* and *memory* requirements. More specifically, speed depends on many factors, such as how large the hash table is, how effective its management algorithms are, and how efficient SPIN is in updating and searching it. However, speed improvements in most cases have some impact on memory requirements.

This topic can be addressed in different ways and at different levels, starting from the model itself down to configuring the way SPIN performs verification bearing in mind how some SPIN algorithms work internally. Some of the most commonly used optimization methods are summarized in the following:

- **Writing efficient models.** More efficient models can be developed by carefully considering which aspects of a model have a bigger effect on *state vector size* and on the *number* of states.

  For instance, using as few concurrent processes as possible, enclosing sequences of statements into atomic sequences when possible, and introducing nondeterministic constructs only when necessary can considerably reduce the number of distinct states to be considered during verification.

  Similarly, in order to reduce the state vector size, it is useful to keep the amount of channel buffer as small as possible and declare numeric variables with the narrowest size compatible with their range requirements.

  The approaches mentioned above can improve the performance in terms of both memory and time, because less space is needed to store states and state comparison will be faster as well. However, special care is necessary to make sure that the simplified model still is a faithful representation of the real system, or at least, the side effects of the simplification are well understood.

- **Allocating more memory for the hash table.** As discussed above, the time spent walking through the linked list associated with a hash table entry is significantly higher than the time spent on calculating its index. If more memory can be allocated for the hash table, the probability of having hash conflicts will usually be reduced and the verification time will improve. However, in this case we need to *sacrifice* memory for efficiency.

  It is worth mentioning that, in any case, the hash table must fit in the *physical* memory available on the machine where SPIN is running on. Even though most operating systems do provide an amount of *virtual* memory which is much bigger than the physical memory, the overhead associated with disk input/output operations due to virtual memory paging would certainly degrade SPIN's performance more significantly than hash conflicts. This is especially true because paging algorithms rely on memory access predictability (or locality) to achieve good performance, whereas hash table accesses are practically pseudo-random, in other words, hard to predict.

- **Compressing state vectors.** State vectors of several hundred bytes are quite common. Instead of storing them as they are, it is possible to compress them. In this case, the memory requirement can be reduced. However, more time will be spent on compression whenever it is necessary to store a state vector. This option therefore represents a trade-off between memory and time requirements.

- **Partial order reduction.** The details of partial order reduction are quite complex and fully described in [78]. Informally speaking, it may happen that, starting from a certain state, it can be proved that the execution order of several computation steps does not affect the final state, and the different execution orders cannot be distinguished in any way. This may happen, for instance, when two concurrent processes are working independently on their own local variables.

In this case, instead of considering all possible execution orders, and generating the corresponding intermediate states, it is enough to follow just one execution order. In some cases, this is a quite effective method since it can reduce the size of the state space sharply.

- **Bitstate hashing and hashing compact.** Instead of allocating more memory to accommodate a large hash table, it is also possible to reduce its memory requirements.

  For what concerns bitstate hashing, which is also known as supertrace, a state is identified by its index in the hash table. As a result, a single bit is sufficient to indicate whether a state has already been visited or not. However, it is possible that two different states may correspond to the same index, due to hash conflict. If the "visited" bit is already set for a certain state, when coming to verify whether the property of interest holds on a different but conflicting state, the result is positive, even if this may not be the case in reality. In other words, some parts of the state space may not be searched and it is possible to miss some counterexamples. However, if a counterexample is found, it does represent a true error.

  For what concerns hashing compact, instead of storing a large hash table, the indexes of visited states in that table are stored in another, much smaller, hash table. It has the same issue as bitstate hashing, because two different states may have the same index in the large hash table, and thus collide.

  Both methods are quite effective to reduce the memory requirements of state storage. However, they are *lossy* and entail a certain probability of having a *false positive* in the verification results, that is, a property may be considered true although some counterexamples do exist. The false positive probability can be estimated and often brought down to an acceptable level by tuning some of the algorithm parameters [75].

Except for the first one, the other optimization methods can be enabled in SPIN by configuration, through either command-line options or graphical front ends. For instance, state vector compression can be specified with the -DCOLLAPSE option when compiling the verifier to obtain the executable file.

13.6 SUMMARY

This chapter provides an overview of formal verification of distributed concurrent systems through model checking, which evaluates whether or not a property holds on the model corresponding to the practical system by exhaustively exploring the state space.

SPIN is one of the most widespread model checkers. The general verification flow of SPIN is introduced in Section 13.2. Besides being able to carry out verification in an efficient way, it can also provide a trail if the property to be verified is violated and a counterexample is found. In this case, users can follow the trail to better understand the issue and then may fix the problem with other methods. Model checking is most effective when it is used together with debugging techniques.

PROMELA, which is the input language for SPIN, can be used to write formal models for concurrent programs. With respect to ordinary programming languages, PROMELA is powerful in specifying concepts like nondeterministic execution, concurrency, passive wait as well as atomicity. These concepts are illustrated in Section 13.3 by means of some intuitive examples.

When the model is ready, property to be verified on the model should also be specified in a formal way. Simple properties like assertions can be indicated directly with PROMELA, whereas more complex properties can be defined by means of linear temporal logic, which is explained in Section 13.4.

Last but not least, some hints to improve the verification performance are provided in Section 13.5, from writing an efficient model down to tuning the way SPIN works, taking into account internal technical details of SPIN.

In order to help readers better understand the concept of model checking, a more complex example can be found in the next chapter. It considers a multi-master election protocol which has already been used in practice and uses model checking techniques introduced in this chapter to identify low probability issues and optimize the implementation.

14 Model Checking: An Example

The general concept of model checking is introduced in the previous chapter. Topics covered there include the PROMELA modeling language which can be used to write a formal model for a practical concurrent system, the linear temporal logic formalism that can be adopted to specify properties to be verified over the model, as well as the SPIN model checker together with possible techniques to improve the verification performance.

This chapter will show a case study based on a distributed multi-master election protocol that has been deployed in industry for real-time communication. It demonstrates how model checking can be used to identify low-probability issues which may never occur in practice and how counterexamples found by the SPIN model checker can be helpful to correct them.

14.1 INTRODUCTION

Fieldbus networks are still popular nowadays, especially for low-cost distributed embedded systems, because their underlying technology is mature and well-known to most designers. MODBUS [119, 121] is one of the most popular fieldbuses. It is based on the master-slave communication paradigm and can be deployed over the TIA/EIA–485 [163] (formerly called RS485) physical level channel.

The basic MODBUS protocol was originally conceived as a low-cost and low-complexity solution for single-master systems only. It is possible to extend it to support multiple masters on the same fieldbus segment, as shown in Reference [29],

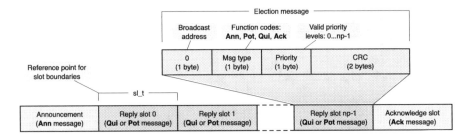

Figure 14.1   Simplified structure of a master election.

while maintaining backward compatibility with existing MODBUS slaves. In this extension, the *master election* protocol—a protocol executed periodically to decide which is the highest-priority master connected to the fieldbus segment and give it permission to operate on the bus until the next election—plays a key role.

The design of this type of protocol is seemingly straightforward and the designer may believe that its correctness can be assessed satisfactorily by intuition and testing. However, as will be shown in the following, formal verification can help to identify and fix subtle and low-probability issues, which seldom occur in practice, and therefore, may be extremely difficult to detect during pre-production testing.

At the same time, the case study is also useful as an example of how a typical real-time protocol—not necessarily MODBUS-related—can be modeled and analyzed in an effective way by means of a contemporary model checker, SPIN [74]. In particular, it shows how to handle the concept of *time* and how to model a multi-drop, inherently broadcast physical *channel*, like TIA/EIA–485, in an accurate but efficient way, that is, without introducing any undue complexity in the analysis, which is often one of the most important issues of practical model checking.

The design of the protocol to be analyzed is briefly recalled in Section 14.2, then Section 14.3 describes how the corresponding formal model has been built. The results of the analysis and the ensuing corrective actions are presented in Section 14.4.

14.2   DISTRIBUTED MASTER ELECTION PROTOCOL

The master election protocol is initiated periodically by the currently *active* master of the MODBUS segment. The main purpose of the protocol is to elect the highest-priority master connected to the segment at the moment, make it the new active master, and allow it to operate on the bus until the next election round. An auxiliary function of the protocol is to make all masters aware of the presence, state, and priority of the others. Each master is characterized by:

- A *state*. The *active* master has full control of the MODBUS segment. A *potential* master takes part in the elections proclaimed by the active master, and is willing to become the active master. A *quiescent* master takes part in the election as a potential master does, but does not want to become the active master.

- An 8-bit, integer *priority*, which must be unique for, and preassigned to, each master. To simplify the protocol, the priority is also used to identify each master. Moreover, it is assumed that priorities lie in the range $[0, np - 1]$, and np is known in advance.

Unlike the states, which may change over time as a result of an election, the masters' priorities are constant. As shown in Figure 14.1, the election comprises three distinct phases:

1. After *ann_t* ms (about 500 ms in a typical implementation) since the end of the previous election, the active master proclaims a new election, by sending an announcement message containing its own priority (function code Ann).
2. The end of the announcement message acts as a time reference point, shared among all masters, which marks the beginning of a series of np fixed-length reply slots, each *sl_t* ms long (8 ms in a typical implementation). Each master, except the active one, replies to the announcement in the slot corresponding to its priority (function code Pot or Qui, depending on whether the master is potential or quiescent); as before, the reply contains the priority of the sender.
3. After the reply slots, the active master announces the result of the election by means of an acknowledgment message, bearing the priority of the winning master (function code Ack).

All messages exchanged during the election have the same structure and are broadcast to all masters. Besides the MODBUS function code, which determines the message type, their payload contains an 8-bit priority, with the meaning discussed above.

During initialization, as well as when the notion of which is the active master is lost (e.g., due to a communication error), a timeout mechanism is triggered to force the election of a new master. In particular, if a potential master at priority cp detects that no elections have been proclaimed for *ato_t* · $(cp + 1)$ ms, it starts an election on its own. Since the timeout value is directly proportional to each master's priority value (lower values correspond to higher priorities), the unfortunate event of two or more masters starting an election at roughly the same time should be avoided. The typical value of *ato_t* used in practice is 1000 ms.

As an additional protection against the collision between two or more masters— that is, the presence of two or more masters which simultaneously believe they are the active master on the MODBUS segment—the active master monitors the segment and looks for unexpected election-related traffic generated by a higher-priority master. If such traffic is detected, the active master immediately demotes itself to potential master.

In the real protocol implementation, slot timings are slightly more complex than shown in Figure 14.1, because it is necessary to take into account the limited precision of the real-time operating system embedded in the masters to synchronize with the Ann message, in the order of 1 ms. The most obvious consequence is that the real slots are longer than the messages they are meant to contain, and masters try to send their messages in the center of the slots. The basic operating requirements the protocol should satisfy by design are:

1. It shall be correct in absence of communication errors, that is, if the whole proto-col is performed correctly, the highest-priority master shall win.
2. It shall be resilient to a *single* communication error in a protocol round and the subsequent recovery. The worst effect of such an error shall be the loss of the notion of which is the active master, a transient condition that shall be recovered in the next election round.
3. The outcome of an election shall not be affected by the *silent* failure of one or more masters. By silent failure, we mean that a master simply ceases functioning without generating any spurious traffic on MODBUS.

The occurrence of multiple communication errors, as well as the presence of *bab-bling* masters, has not been considered in the design, because their probability is deemed to be negligible with respect to other sources of failure present in the system as a whole.

The protocol and the masters' behavior have been specified by a timed finite state machine (FSM), as shown in Figure 14.2. What's more, the corresponding C-language implementation of the protocol has been directly derived from it. Although the full details of the FSM will not be discussed here, the FSM transitions have been reported in the *upper* part of Table 14.1, except those related to error counter man-agement for the sake of conciseness.

Those transitions are essential because, as will be described in Section 14.3, the bulk of the formal protocol model is heavily based on them, to the extent that they can be used to generate the model in a fully automatic way. In turn, this reduces the possi-bility of any hidden discrepancy between the protocol model and its implementation.

The most important FSM states are the *active* and *standby* states. They character-ize the active master, and the potential/quiescent masters, respectively.

For simplicity, the following discussion just focuses on the table. In Table 14.1, a transition from an old to a new state is composed of two parts: a *predicate* and an *action*. The predicate enables the transition, that is, the transition can take place only if the predicate is true. Predicates are written with a C-like syntax in which the clause rx m(p) is true if a message of type m has been received. If this is the case, variable p is bound to the priority found in the message itself. The symbol _ (underscore) matches any message type defined by the election protocol.

Actions are carried out when a certain transition is followed, before bringing the FSM into a new state. The clause tx m(p) denotes the transmission of an election message of type m, carrying priority p in the payload.

In both predicates and actions, the special variable t represents a free-running, resettable counter holding the local master time, in ms. The Boolean variable men distinguishes between potential (true) and quiescent (false) masters; mrl, when it is set to true in an active master, forces that master to relinquish its role as soon as another potential master is detected, regardless of its priority. The array pr is local

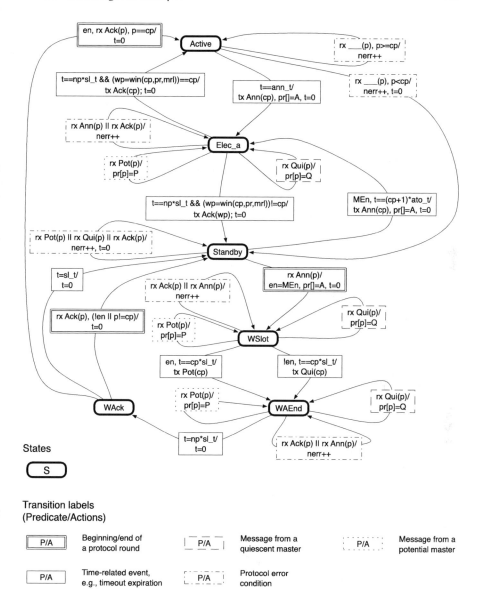

Figure 14.2   Election protocol's timed FSM.

to each master; it collects the state of each master (A-bsent, Q-uiescent, or P-otential) derived from the reply messages they sent during the election. Finally, the function win is used by the active master to determine which master won the election from cp, pr, and mrl.

Table 14.1

Original (N) and Additional (A) Transitions of the Election Protocol's Timed FSM

Id	Old State	Predicate / Action	New State
N1	Active	`t==ann_t / tx Ann(cp); pr[]=A; t=0`	Elec_a
N2	Active	`rx _(p) \&\& p<cp / t=0`	Standby
N3	Active	`rx _(p) \&\& p>=cp /` *none*	Active
N4	Elec_a	`rx Qui(p) / pr[p]=Q`	Elec_a
N5	Elec_a	`rx Pot(p) / pr[p]=P`	Elec_a
N6	Elec_a	`t==np*sl_t && (wp=win(cp,pr,mrl))!=cp /` `tx Ack(wp); t=0`	Standby
N7	Elec_a	`t==np*sl_t && (wp=win(cp,pr,mrl))==cp /` `tx Ack(cp); t=0`	Active
N8	Elec_a	`rx Ann(p) || rx Ack(p) /` *none*	Elec_a
N9	Standby	`men && t==(cp+1)*ato_t /` `tx Ann(cp); pr[]=A; t=0`	Elec_a
N10	Standby	`rx Ann(p) / en=men; pr[]=A; t=0`	WSlot
N11	Standby	`rx Qui(p) || rx Pot(p) || rx Ack(p) / t=0`	Standby
N12	WSlot	`rx Qui(p) / pr[p]=Q`	WSlot
N13	WSlot	`rx Pot(p) / pr[p]=P`	WSlot
N14	WSlot	`!en \&\& t==cp*sl_t / tx Qui(cp)`	WAEnd
N15	WSlot	`en \&\& t==cp*sl_t / tx Pot(cp)`	WAEnd
N16	WSlot	`rx Ann(p) || rx Ack(p) /` *none*	WSlot
N17	WAEnd	`rx Qui(p) / pr[p]=Q`	WAEnd
N18	WAEnd	`rx Pot(p) / pr[p]=P`	WAEnd
N19	WAEnd	`t==np*sl_t / t=0`	WAck
N20	WAEnd	`rx Ann(p) || rx Ack(p) /` *none*	WAEnd
N21	WAck	`rx Ack(p) && (!en || p!=cp) / t=0`	Standby
N22	WAck	`rx Ack(p) && en && p==cp / t=0`	Active
N23	WAck	`t==sl_t / t=0`	Standby
A1	Elec_a	`t==cp*sl_t, !sentpot /` `tx Pot(cp); sentpot=true`	Elec_a
A2	*any*	`rx Col(_) / t=0`	Standby

For example, transition N9 is enabled for a potential master (`men`), when $ato\_t \cdot (cp + 1)$ ms have elapsed since the last election (`t==(cp+1)*ato_t`). In this case, that master's FSM goes from the Standby to the Elec_a state after transmitting an announcement message (`tx Ann(cp)`), initializing all elements of the array `pr` to A (`pr[]=A`), and resetting the local timer to zero (`t=0`).

14.3 FORMAL PROMELA PROTOCOL MODEL

14.3.1 SIMPLIFIED TIME MODEL

Neither the PROMELA language, nor the SPIN model checker have any native, built-in concept of time. When necessary, the concept of time must therefore be included explicitly in the model. A few general SPIN extensions for discrete time have been developed in the past, most notably DTSPIN [22].

Although these extensions are quite powerful—they even support the automatic translation of a specification and description language (SDL) model [93] into extended PROMELA and its verification [23]—for simple protocols, like the one analyzed here, an *ad hoc* approach is simpler to deploy and, most importantly, incurs less verification overhead. Moreover, as will be described later, a proper time model is also helpful to model a broadcast channel with collisions in an efficient way.

Other tools, UPPAAL [21] in particular, directly support Timed Automata, and hence, would be good candidates to perform the analysis, too. The relative merits of SPIN and UPPAAL have been discussed at length in literature. A detailed comparison—involving a protocol similar to the one considered here—is given, for instance, in [96]. Generally speaking, the price to be paid to have a richer model is often a loss of performance in the verification.

In analogy with the protocol FSM specification presented in Section 14.2, the timer of each master has been represented as an element of the global array t, encompassing NP elements, one for each master (Figure 14.3, lines 1–2). To reduce the timer range, time is expressed as an integral multiple of the *slot length sl_t*, rather than milliseconds, in the model.

This represents a compromise between the accuracy and the complexity of the analysis because, even if the slot length is the minimum unit of time that masters are able to resolve, our approach entails the *assumption* that the underlying real-time operating system is nonetheless able to synchronize each master with respect to slot boundaries in a proper way.

It must also be remarked that doing otherwise would require the inclusion of an operating system and the MODBUS protocol stack model in the PROMELA specification to be verified. This would be a daunting task because both components lack a formal specification. On the other hand, checking masters for proper slot synchronization—a very simple timing property—can be and has been done in an immediate and exhaustive way through bench testing.

During initialization, all the timers are initialized nondeterministically with negative values in the range $[-xs, 0]$, representing the power-up skew between masters, by means of the choose_skews inline function (Figure 14.3, lines 5–19). Namely, an initial timer value of $-k$ conveys in a simple way the fact that a certain master started operating on the bus k slots after the system was powered up.

Similarly, XS represents the maximum time difference between the power-up times of different masters. It is useful to remark that, in this function, the break statement at line 12 does not prevent the statement at line 13 from being executed. Rather, it gives rise to the nondeterministic choice between abandoning the inner do loop and further decrementing t[i].

By the way, the body of an inline function is directly passed to the body of a process whenever it is invoked. In PROMELA, an inline call can appear anywhere a standalone statement can appear.

```
1   #define NP 4
2   short t[NP];
3   bool tf = true;
4
5   inline choose_skews()
6   {
7     i=0;
8     do
9     :: (i < NP) ->
10      {
11         do
12         :: break
13         :: (t[i] > -XS) -> t[i]--
14         od;
15         i++
16      }
17    :: else -> break
18    od
19  }
20
21  /* The Tick process updates global time when there is nothing else to
22     do in the system, i.e. all other processes are blocked, by using
23     the timeout special variable.
24
25     In this way, it is guaranteed that time will not elapse until no
26     processes are in the condition of carrying out any other action.
27     Implicitly, this also means that processes will always honor their
28     timing constraints.
29  */
30  active proctype Tick()
31  {
32    byte i;
33
34    do
35    :: timeout ->
36       d_step {
37         if
38         :: (tf) -> do_bcast(); tf = false
39         :: (!tf) ->
40            i=0;
41            do
42            :: (i < NP) -> t[i]++; i++
43            :: else -> break
44            od;
45            tf = true
46         fi
47       }
48    od
49  }
```

Figure 14.3 Simplified time model.

In the time model, each slot has been further divided into two sequential phases, a transmit and a receive phase. The global variable tf distinguishes between them. This subdivision has no counterpart in the real system. It has been included in the specification as part of the broadcast channel model, which will be described later.

The last noteworthy point to be discussed is how and when the local timers are incremented. This is done by a separate process, Tick (Figure 14.3, lines 30–49), in a d_step sequence enabled by the timeout guard (line 35). According to the

PROMELA language specification, this guard is true if and only if no other statements in the whole model are executable.

Therefore, informally speaking, timers are incremented when no other actions are possible in the whole system. More specifically, masters go from one slot to another when they have completely carried out all the actions they are supposed to perform in the current slot. It is important to remark that, in its simplicity, this approach contains two important assumptions that must be understood and checked for validity in the real system:

1. Within an election round, masters are perfectly synchronized at the slot level, because their timers are always updated together, albeit the time value each master uses can still differ from the others by an integral number of slots. In the system being considered here, this hypothesis has been satisfied in the design phase, by calculating the slot length and message location so that, even under the worst-case relationship among oscillator tolerances and synchronization errors among masters, they are still able to send and receive their messages in the right slot.
2. All timing constraints concerning the actions that must be accomplished by the masters within a slot are satisfied, because the model lets all masters perform all actions they have to perform in a certain slot before switching to the next one. In the real-world system, this hypothesis has been checked through real-time schedulability analysis and subsequent testing.

As remarked before, a more precise assessment of these properties through formal verification would require a formal model of the real-time operating system and the MODBUS protocol stack. Due to the complexity of both components, it would likely have been infeasible.

14.3.2 BROADCAST CHANNEL MODEL

Native PROMELA channels cannot model a broadcast transmission directly, because the ordinary receive statement is destructive. As a consequence, only *one* process will receive a given message from a channel, when using this kind of statement. Another limitation is that PROMELA channels do not contemplate the concept of message *collision*, and the subsequent garbling, when two or more message transmissions overlap. A more sophisticated channel model is needed for the system being analyzed here, because an accurate model of a TIA/EIA–485 channel requires both those features.

Masters ask for their messages to be transmitted during the transmit phase of each slot, according to their FSM actions, by using the bcast inline function (Figure 14.4, lines 12–26). This function keeps track of the senders' identity by means of the mb_s array (line 15) and checks whether or not a collision occurred (lines 20–21). The message being sent is stored in the mb_t and mb_p variables. The message type Non indicates that *no message* has been sent in the current slot yet, and therefore, it does not correspond to any "real" message. The message type Col indicates that a collision occurred, and it will be further discussed later in this section. Since each master is

```
1   mtype = { Ann, Qui, Pot, Ack, Non, Col };
2   chan mc[NP] = [1] of { mtype, byte };
3   mtype mb_t = Non;
4   byte  mb_p = 0;
5   bool  mb_s[NP];
6
7   /* Prepare the broadcast message mt(p) to all the other masters,
8      but myself.  Real transmission will be done in the do_bcast()
9      function. After the transmission, blocks until the receive
10     phase (!tf) starts, to simulate the transmission time.
11   */
12  inline bcast(myself, mt, p) {
13    d_step {
14      /* myself is a sender. */
15      mb_s[myself] = true;
16      if
17      /* If a message has been sent within the same slot, then
18         message collision happens.
19       */
20      :: (mb_t != Non) -> mb_t = Col
21      :: else -> mb_t = mt; mb_p = p
22      fi
23    }
24
25    !tf
26  }
27
28  /* Send the message in the message buffer (mb_t) to all masters, except
29     to the sender itself, which is recorded in mb_s[].
30
31     The message is sent only if there is a message in the buffer (!=
32     Non), even if there was a collision, because it is assumed that the
33     receivers can detect activity on the channel in a given slot even
34     they cannot receive the message correctly, due to a collision.  In
35     this case, a Col message is sent in the model.
36   */
37  inline do_bcast() {
38    if
39    :: (mb_t != Non) ->
40       i=0;
41       do
42       :: (i < NP) ->
43          if
44          :: !mb_s[i] -> mc[i] ! mb_t(mb_p)
45          :: else -> mb_s[i] = false /* Don't send to the sender */
46          fi;
47          i++
48       :: else -> break  /* End of loop */
49       od
50    :: else -> skip /* No message */
51    fi;
52
53    /* Clear the buffer for the next slot */
54    mb_t = Non
55  }
```

Figure 14.4 Broadcast channel model.

allowed to send only a single message in any given slot, the last statement of bcast (line 25) is a wait that lasts until the end of the transmit phase.

The actual message broadcast is performed by the do_bcast inline function (Figure 14.4, lines 37–55). This function is invoked by the Tick process at the boundary between the transmit and receive phase of each slot (Figure 14.3, line 38). If a message to be sent has been collected in mb_t and mb_p (Figure 14.4, line 39), it is sent to all masters, except the senders themselves (line 44). Each master has its own private input channel for incoming messages, namely, mc[i] for the i-th master (line 2). The message buffers are then brought back to their initial state (lines 45 and 54).

Excluding the senders from message relay is necessary to model the channel accurately, because the TIA/EIA–485 transceivers used in practice are half-duplex, that is, they are unable to receive and transmit a message at the same time. Message reception is therefore inhibited during transmission.

If a collision occurs, a special message (with message type Col) is broadcast, *only* in the model. This message makes all receivers aware of the collision and has no counterpart in the actual protocol. However, this is consistent with the real behavior of the system because all the universal asynchronous receiver transmitter (UART)/transceiver combinations of the four different microcontroller families picked for the practical implementation of the protocol [30] (namely, the Atmel ATmega 1284P [12], Atmel AT 91 [11], NXP LPC2468 [126], and NXP LPC1768 [128]) have the capability of generating an interrupt when they detect MODBUS traffic. This happens even if the message cannot be decoded due to framing, parity, or checksum errors. This quite reliable way to detect collisions, which is generally available at no additional hardware cost, is mimicked in the model by means of the Col message.

14.3.3 MASTERS MODEL

Masters have been modeled in the most straightforward way possible, even at the cost of sacrificing verification efficiency. In fact, their model descends directly from the timed FSM specification sketched in Section 14.2 and could be generated automatically from it. Since the C-language implementation of the protocol has been derived in a similar way from the same specification, this reduces the probability of discrepancy between what the protocol actually does, and what is formally checked.

Additional problems with the protocol could have been identified by using its C-language implementation as the starting point of the analysis. However, the automatic translation of C code into PROMELA has been addressed only recently in literature, for instance, in [97]. Moreover, to the best of the authors' knowledge, the existing translators only support a subset of the C language.

As shown in Figure 14.5, each master corresponds to a process, called Master (lines 1–12). All NP processes are instantiated during initialization by means of the run_agents inline function (lines 14–23). Each instance has its priority cp and initial FSM state ist as parameters. The priority value is also used to identify each master. The values of men and mrl can be individually set for each master as well, by means of the corresponding parameters.

```
1   proctype Master(byte cp; byte ist;
2                       bool men; bool mrl)
3   {
4     byte st;
5     ...other local variables...
6
7     st = ist;
8     do
9     :: ...translation of FSM transition...
10    :: ...
11    od
12  }
13
14  inline run_agents()
15  {
16    i=0;
17    do
18    :: (i < NP) ->
19       run Master(i, Standby,
20                    true, false); i++
21    :: else -> break
22    od
23  }
```

Figure 14.5   Outline of the masters model.

In this example, the agents are given a unique priority in the range $[0, np - 1]$. Moreover, none of them is active at startup (they all start in the Standby state), they are potential masters (men is true), and they are not willing to relinquish their role of active master after they acquire it (mrl is false).

The body of each master contains declarations for the local FSM variables, most importantly the FSM state st (line 4). The initial value of st is taken from the initial state of the master, held in the ist parameter (line 7). The rest of the body is an infinite loop containing one nondeterministic choice for each FSM transition listed in Table 14.1. Each nondeterministic choice is the direct translation of one FSM transition, obeying the following rules:

- The predicate of the FSM transition is encoded in the guard of the nondeterministic choice. There, the channel can be inquired for the presence of a message by means of a channel polling expression (see Section 13.3). For example, the expression mc[cp]?[Ann(p)] is true if master cp received an announcement.

  Since the polling expression does not imply the actual reception of the message from the channel, the message must be then explicitly received with an ordinary receive clause, mc[cp]?Ann(p), in a subsequent statement of the nondeterministic choice.

- The actions embedded in the FSM transition are encoded in the subsequent statements of the nondeterministic choice, after the guard. To handle the TIA/EIA–485 channel in a proper way, message transmission must be performed by means of the bcast inline function previously discussed.
- The next_state(ns) macro brings the FSM to the new state ns. Its implementation, not shown here for conciseness, it very simple and basically changes the value of st at the end of the receive phase of the current slot.

For example, transition N1 of Table 14.1 has been translated into:

```
:: (st == Active && t[cp] == ann_t) ->
   bcast(cp, Ann, cp);
   set_pr(A);
   next_state(Elec_a);
   t[cp] = 0
```

In the translation, the set_pr inline function is a shortcut to set all elements of the pr array to A. Similarly, transition N9 has been translated into:

```
:: (st == Standby && men
   && t[cp] == (cp+1)*ato_t) ->
   bcast(cp, Ann, cp);
   set_pr(A);
   next_state(Elec_a);
   t[cp] = 0
```

For the sake of completeness, the masters model also contains several additional lines of code. Their purpose is to make a few internal variables globally visible, so that they can be used in the LTL formulae presented in Section 14.4. They are neither shown nor further discussed due to their simplicity. Overall, the complete protocol specification consists of about 400 lines of PROMELA code.

14.4 FORMAL VERIFICATION RESULTS

The SPIN model checker has been used to verify three distinct properties of the protocol. In all cases, the model parameters have been set as follows: NP=4 (four masters), XS=20 (up to 20 slots of skew among masters), ann_t=1 (1 slot between the end of the previous election and the beginning of the next one), and ato_t=2 (election timeout of 2 slots for the highest-priority master).

The choice of NP and XS has been chosen based on the actual protocol configuration used in practice; the minimum meaningful value has been chosen for ann_t and ato_t, to minimize the verification time. The total execution time of SPIN, on a contemporary laptop PC, never exceeded 30 minutes for all experiments discussed in this section. The memory consumption (with the -DCOLLAPSE option enabled) was below 3 GByte.

For higher values of XS, and especially NP, both the verification time and memory consumption rise considerably. It is still possible to perform the verification within

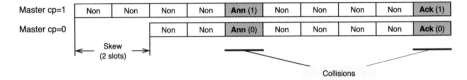

Figure 14.6 First bug found by SPIN in the original protocol. Collided messages shown in gray.

the limits stated above, up to NP=5, but only by setting the supertrace/bitstate verification option, albeit the verification is *approximate* in this case.

The properties of interest are:

- **Liveness**, written as:

```
#define p (prv[0] && prv[1] \
            && prv[2] && prv[3])
[]   ( (! p)  -> ( <>  p) )
```

where prv is an array of Booleans, set by the masters when they reach the end of an election. Informally speaking, the property states that, in any computation, the end of an election is reached infinitely often by all masters.

- **Agreement**, written as:

```
#define q (prres[0] == prres[1] \
          && prres[1] == prres[2] \
          && prres[2] == prres[3])
[]   ( p  -> q )
```

where prres is an array that collects the election winners, calculated by each master. Informally speaking, the property states that, when the masters have got a result (p is true), then they agree on the result (q is true, too).

- **Weak Agreement**, a property stating that there is *at most one* active master at any given time. It is a weaker form of the agreement property, because it contemplates the possibility of losing the notion of which is the active master. It is written as:

```
#define p ((
   ((gst[0]==St_Active) -> 1 : 0) + \
   ((gst[1]==St_Active) -> 1 : 0) + \
   ((gst[2]==St_Active) -> 1 : 0) + \
   ((gst[3]==St_Active) -> 1 : 0)) <= 1)
[] ( p )
```

where gst[i] is the globally observable state of the *i*-th master.

The original version of the protocol, presented in Figure 14.1, satisfied the *liveness* property when checked by SPIN. On the contrary, SPIN showed that the *agreement*

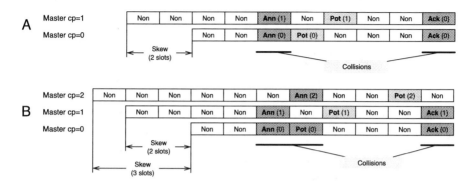

Figure 14.7   Fix for the first bug (A), and second bug found by SPIN (B), involving three masters.

property was not satisfied and provided a counterexample. The corresponding time diagram, derived from the message sequence chart (MSC) produced by SPIN as part of the counterexample, is depicted in Figure 14.6.

The time diagram shows that, when two masters start operating with a very unfortunate (and very unlikely to happen) time relationship—namely, a power-up time skewed by exactly 2 slots—both their announcement and their acknowledge messages collide on the network. This happens because the highest-priority master ($cp = 0$) waits for $ato\_t = 2$ slots before proclaiming an election, whereas the other master ($cp = 1$) waits for $2 \cdot ato\_t = 4$ slots. As a result, neither master receives the messages sent by the other, and both become active because they believe they are alone on the MODBUS segment.

The bug can apparently be fixed by observing that in the original version of the protocol, when a master proclaims an election, it sends an announcement and then stays silent during all the reply slots, waiting for replies from other masters. Hence, if the announcement is lost due to a collision, the other masters will have no clue of its presence.

Since, by design, the k-th reply slot is reserved for the master at priority k—and there is a single master at that priority—the protocol can be enhanced by making the k-th master send a Pot message in its own slot without fear of colliding with others. The additional FSM transition is marked A1 in Table 14.1. As shown in part (A) of Figure 14.7, this is enough to fix the bug in the scenario being discussed, because both masters now agree that master 0 is the winner.

However, this enhancement is not sufficient to solve the issue in the general case. When fed with the enhanced protocol model, SPIN was in fact able to provide a new, more complex counterexample, shown in part (B) of Figure 14.7. This second counterexample is particularly interesting because it involves the concurrent activity of three distinct masters and, as in the previous case, it also requires a very precise time relationship among them. The combination of those factors makes the counterexample extremely difficult to find by intuition.

Looking again at part (B) of Figure 14.7, we notice that a third master at priority 2 may send an announcement message that collides with the Pot message sent by master 0. This happens because of a combination of two factors:

- Master 1 was powered up 2 slots before master 0, and hence, its announcement message collided with the announcement sent by master 0 itself. As a result, no masters actually received any valid announcement.
- Master 2 was powered up 3 slots before master 0 and, by design, it waited for $3 \cdot ato\_t = 6$ slots before sending its announcement. It then sent its announcement because, during the wait, it did not receive any other message. In fact, the only messages sent during this period are the announcements, which were lost due to the collision.

As a consequence, master 1 did not receive the Pot message sent by master 0 and considered itself to be the winner. On the other hand, master 0 did receive the Pot message sent by master 1, but this was irrelevant for the outcome of the election, as determined by master 0 itself. The final result is that both master 0 and master 1 become active.

This second bug can be fixed by leveraging the hardware-based collision detection capabilities of TIA/EIA–485 UARTs and transceivers. Namely, all masters shall go back to the Standby state and reset their timer as soon as they detect a collision. The corresponding FSM transition is marked A2 in Table 14.1.

With this further enhancement, the master election protocol satisfied all three properties discussed above in the absence of communication errors. In the scenario shown in Figure 14.7, part (B), master 2 will no longer send its announcement, because it will detect the collision between the announcement messages sent by the other two masters. The election will therefore proceed correctly, as shown in part (A) of the same figure.

Finally, the model was further extended to introduce *single communication errors* and *silent master failures* because, according to its design, the protocol shall be resilient to them. Indeed, SPIN proved that the liveness property is still satisfied, and hence, all functioning masters still get election results within a finite number of execution steps in all cases. The agreement property is no longer satisfied—because the notion of which is the active master may be lost in the presence of communication errors—but the *weak* agreement property still is. The most dangerous condition the protocol may encounter, that is, the presence of more than one active master at the same time on the same MODBUS segment, is therefore still avoided in all failure modes foreseen in the design.

For the sake of conciseness, just the models corresponding to them are shown in the following. For what concerns single communication errors, it is assumed that there is at most one error (either a transmit error or a receive error) in a single election round and no additional errors during the subsequent recovery. Moreover, a transmit error has a *global* effect, that is to say, it affects all receivers and none of them can receive the correct message. Instead, a receive error is *local* and it just affects a single receiver.

```
1    /* Channel error: At most one error in a single election round;
2       no additional errors during recovery.
3    */
4    bool rxtxe = false;
5
6    inline bcast(myself, mt, p) {
7      if
8      /* Channel error: A transmit error affects all other masters.
9         All of them will receive a Col notification.
10     */
11     :: atomic { !rxtxe ->
12          rxtxe=true;
13          mb_s[myself] = true;
14          mb_t = Col
15        }
16     :: d_step {
17          mb_s[myself] = true;
18          if
19          :: (mb_t != Non) -> mb_t = Col
20          :: else -> mb_t = mt; mb_p = p
21          fi
22        }
23     fi;
24
25     !tf
26   }
27
28   inline do_bcast() {
29     if
30     :: (mb_t != Non) ->
31        i=0;
32        do
33        :: (i < NP) ->
34           if
35           :: !mb_s[i] ->
36              if
37              /* Channel error: When a master has a receive error,
38                 it is notified by means of a Col message.
39              */
40              :: atomic { !rxtxe -> rxtxe=true; mc[i] ! Col(mb_p) }
41              :: mc[i] ! mb_t(mb_p)
42              fi
43           :: else -> mb_s[i]=false
44           fi;
45           i++
46        :: else -> break
47        od
48     :: else -> skip
49     fi;
50
51     mb_t = Non
52   }
```

Figure 14.8   Extended broadcast channel.

More specifically, a global Boolean variable `rxtxe` is used to indicate whether a single communication error already takes place or not. In addition, the broadcast channel model has also been extended, as shown in Figure 14.8. The error message is represented as a `Col` message as well, in order not to further enlarge the message type set which may slow down the verification process.

```
1   active proctype Tick()
2   {
3       byte i;
4
5       do
6       :: timeout ->
7           /* Channel error: use atomic instead of d_step */
8           atomic {
9               if
10              :: (tf) -> do_bcast(); tf = false
11              :: (!tf) ->
12                  i=0;
13                  do
14                  :: (i < NP) -> t[i]++; i++
15                  :: else -> break
16                  od;
17                  tf = true
18              fi
19          }
20      od
21  }
```

Figure 14.9 Extended time model.

A transmit error is modeled by extending the inline function bcast with one more nondeterministic choice in the if statement (lines 11–15), which is implemented as an atomic sequence. If no communication error has been generated, a Col message could be prepared here and then transmitted by the do_bcast function. Within the same atomic sequence, rxtxe is set to true so that no more error could be introduced.

On the other hand, it is more convenient to model the receive error (line 40) in the do_bcast function because it should affect only one receiver. If neither a transmit error nor a receive error has been generated, one master could be affected by a receive error, which is indicated by receiving a Col message. Instead, if a transmit error has already been generated, only the alternative nondeterministic choice can be followed, that is to broadcast the transmit error message to all masters (except the sender).

A side effect of introducing the receive error in the model (Figure 14.8, line 40) is that execution in do_bcast can no longer be performed in a deterministic way. As a consequence, d_step in the time model is not suitable to enclose the do_bcast function any more, even though d_step leads to more efficient verification. It can be replaced with the atomic keyword, as shown in Figure 14.9.

Last but not least, in order to model *silent master failures*, a Boolean array silent[] with NP elements is introduced, with each element indicating whether the corresponding master encounters silent failure or not. Besides, the bcast function is extended, as shown in Figure 14.10.

```
 1   bool rxtxe = false;
 2   bool silent[NP];
 3
 4   inline bcast(myself, mt, p) {
 5     if
 6     :: silent[myself] == false ->
 7        if
 8        /* Master becomes silent. */
 9        :: silent[myself] = true
10        :: atomic { !rxtxe ->
11              rxtxe=true;
12              mb_s[myself] = true;
13              mb_t = Col
14           }
15        :: d_step {
16              mb_s[myself] = true;
17              if
18              :: (mb_t != Non) -> mb_t = Col
19              :: else -> mb_t = mt; mb_p = p
20              fi
21           }
22        fi;
23        /* The master failed silently in the past. */
24     :: else -> skip
25     fi;
26
27     !tf
28   }
```

Figure 14.10   Silent master failure.

As mentioned in Section 14.2, *silent master failures* mean that one or more masters cease functioning without generating any spurious traffic. As a consequence, a master just set its flag in the array when it becomes silent (line 9), and skips the collection of message and detection of collision since then (line 24). In this way, it will not affect the traffic sent by other masters within the same slot.

14.5   SUMMARY

This chapter shows how formal verification through model checking can be used to verify industrial real-time communication protocols, like the one presented in this chapter. Even though SPIN does not have built-in support for concepts such as time and broadcast channels, which is common to most popular model checkers, it is possible to write models that include these concepts explicitly, without suffering huge verification overheads.

The case study discussed in this chapter also demonstrates that model checking can be profitably used to identify low-probability issues which involves complex interleaving among multiple nodes in a concurrent system. Moreover, the counterexamples provided by the model checker when a property being verified is violated can be extremely helpful to gain further information about the issues and fix them accordingly.

15 Memory Protection Techniques

CONTENTS

Given that it is usually impossible to design and—even more importantly—develop perfect software even after extensive testing, another way to enhance software dependability is to introduce various kinds of consistency checks at runtime, in order to ensure that the software is really performing as intended and, upon error detection, initiate some form of recovery.

In this sense, preventing or at least detecting memory corruption is crucial and it will be the topic of this chapter. Memory corruption takes place when part of a memory-resident data structure is overwritten with inconsistent data. This issue is often hard to detect, because the effects of memory corruption may be subtle and become manifest a long time after the corruption actually took place.

After a short overview of hardware-based memory protection techniques, which are able to prevent memory corruption *before* it occurs, the main focus of the chapter is on software-based techniques, because many embedded microcontrollers, especially low-end ones, lack any form of hardware-based memory protection. The negative side effect of this approach is that software-based techniques are only able to detect memory corruption *after* it has occurred, rather than prevent it.

15.1 MEMORY MANAGEMENT UNITS (MMUS)

Ideally a fragment of code written in a high-level programming language—or, more generically, a task—should be able to access only its own variables, that is, variables local to the task itself or global variables voluntarily and explicitly shared with other tasks to implement inter-task communication as described in Chapter 5.

However, as discussed in Chapter 3, when the linker performs memory allocation it is possible—and even likely—that variables belonging to different tasks are stored at contiguous memory addresses.

For instance, a scalar variable belonging to a certain task A may immediately follow an array of integers belonging to task B. If, due to a programming error, task B is allowed to store a value into the array using an index whose value exceeds the maximum legal value by one, it will certainly overwrite task A's variable and very likely corrupt its contents.

Similarly, an error in pointer arithmetic, which may occur in programming languages that support this concept and make it available to programmers, may lead a task to attempt read and write operations virtually anywhere in memory. Programming languages that by default do not perform any index and pointer consistency checks at runtime for performance reasons, like the C language, are particularly prone to this issue.

In order to tackle memory corruption, most general-purpose processors implement some form of hardware-based *memory protection*, by introducing the concept of *virtual memory* and putting it in force by means of a hardware component known as memory management unit (**MMU**).

Virtual memory is a mechanism by which memory addresses generated by tasks during execution *no longer* correspond to the addresses sent to memory. Figure 15.1 A represents, in an abstract way, the path followed by memory addresses from where they are generated, that is, the processor core, until they reach memory. From this point of view, Figure 15.1 A is a simplified version of Figure 2.1 (noted in Chapter 2) focusing only on memory and memory addresses, which are the focal point of this chapter.

In particular, the figure shows that, based on what has been discussed so far, for any given memory transaction the memory address issues by the processor—called *virtual* memory address V in the following—is *identically equal* to the address given to memory—called *physical* address P. Since, in formula, $V \equiv P$, tasks have direct, total control on which addresses are sent to memory and no checks are performed on them.

Instead, the role of a MMU—when it is present—is to *translate* every virtual memory address P generated by the processor into a different physical memory address V *before* it reaches memory through the on-chip interconnection system and, possibly, the external memory controller and its associated bus.

This is possible because, as is shown in Figure 15.1 B, the MMU is located on the address path between the processor and the on-chip interconnection system, and hence, it has visibility of all addresses issued by the processor itself. Thus, one primary goal of a MMU is to implement virtual to physical address translation, or *mapping* as it is sometimes called, $V \rightarrow P$.

The second, but not less important, goal of a MMU is to perform some *consistency checks* between

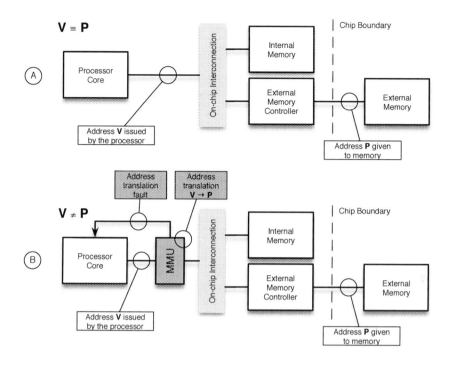

FIGURE 15.1 Memory management unit (MMU) and address translation.

- the address issued by the processor and the associated *access mode*, which specifies the kind of memory transaction the processor is requesting (for instance, instruction fetch, data read, or data write), and
- some *memory protection information* known to the MMU and that the executing task cannot modify on its own initiative.

If the check just mentioned fails, or if the translation fails for other reasons, the MMU incurs in an address translation fault. In this case, it reports the occurrence to the processor by raising an *exception* and blocks the memory transaction before it takes place.

The reaction of the processor to an exception signal is similar to the reaction to an interrupt. As better described in Chapter 8, informally speaking an exception stops the execution of the current task—in this case, the task that attempted the illegal memory access—and diverts execution to the corresponding exception handler.

Due to its criticality from the point of view of memory consistency and system integrity, the duty of handling address translation faults is most often left to the operating system. In turn, the typical action carried out by the operating system is to terminate the offending task under the assumption that it is malfunctioning.

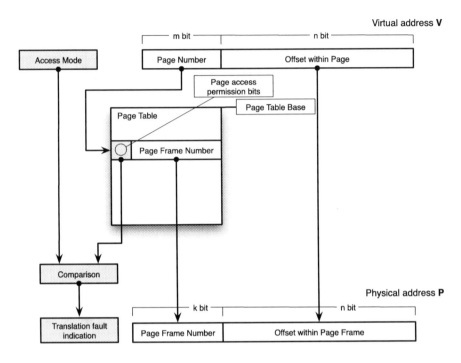

Figure 15.2   MMU address translation and memory protection principle.

The details of the translation may vary, depending on the computer architecture, but the basic mechanism always relies on a data structure called *page table*. As shown in the simplified diagram of Figure 15.2, the virtual memory address V (shown at the top of the diagram) is translated into a physical address P (bottom of the diagram) according to the following steps.

For the sake of completeness, it is also worth mentioning that virtual addresses are sometimes called *logical* addresses, especially in the context of general-purpose operating systems.

1. The virtual address V, of total width $m + n$ bits, is divided into two parts. The first one consists of the n least significant bits, while the second one is composed of the m most significant bits. These two parts are usually called *page offset* and *page number*, respectively, for reasons that will become clearer in the following.
2. The offset is not translated in any way. Instead, as shown in the right part of Figure 15.2, it goes directly into the least significant portion of the physical address P.
3. The page number is used as an *index* in the page table. Since the page number is an m-bit value, the page table must have 2^m entries, at least conceptually, although in practice several techniques are adopted to reduce the overall page table size by not storing unused elements. This is especially important when the virtual address space (that is, the set of legal values V can assume) is large, for instance when

using a 64-bit processor. In these cases m may be quite a large number, and the page table size grows exponentially with m.

4. The MMU retrieves from the page table entry corresponding to the most significant portion of V two kinds of information:

   a. Address translation information, consisting of a k-bit value called *page frame number* and shown as a white box in Figure 15.2.

   b. Memory protection information, often implemented as a set of page access permission bits and shown as a smaller, light gray box in the figure.

5. For what concerns address translation, the page frame number becomes the most significant part of the physical address P. It must also be noted that the size of the virtual address space may not coincide with the size of the physical address space (that is, the range of legal values P can assume). This happens when $m \neq k$.

6. Regarding memory protection, the MMU compares the access mode requested by the processor with the page access permission bits, in order to determine whether or not the access shall be allowed to take place. As described previously, if the result of the comparison is negative, the MMU aborts the memory transaction and raises an address translation fault.

Overall, the address translation process put in place by the MMU divides the virtual address space into 2^m pages. Each page is uniquely identified by its page number and is 2^n memory locations wide. The exact location within the page corresponding to a given virtual address V is given by its page offset. Similarly, the physical address space is also divided into 2^k page frames of 2^n locations, that is, page frames have the same size as pages.

Then, the page table can be considered as a page number translation mechanism, from the virtual page number into the physical page frame number. On the other hand, the relative position of a location within a page and within the corresponding page frame is the same, because this information is conveyed by the least-significant n bits of V and P, and they are the same.

The page table used by the MMU is associated with the running task and a pointer to it, often called *page table base*, is part of the task context managed by the operating system. Therefore, when the operating system switches from one task to another, it also switches the MMU from using one page table to another.

As a consequence, different tasks can (and usually do) have different "views" on physical memory, because the translations enforced by their page tables are different. This view comprises the notions of which physical page frames are accessible to the task and for which kind of access (address translation mechanism), as well as which page frame corresponds to, or supports, each valid virtual page.

It must also be noted that, although page tables are stored in memory, like any other data structure, the MMU itself is able to enforce the validity and consistency of its address translation and memory protection work. This is because it is possible to protect the page table of a task against accidental (or deliberate) modifications by the task itself or another task. Write access to page table contents is usually granted only to the operating system, which is responsible for, and trusted to, store correct information into them.

A full description of how a MMU works and how to use it effectively is beyond the scope of this book. Nevertheless, two simple examples shall persuade readers of its usefulness and power.

1. Let us consider two tasks that must run the same code. This is perfectly normal in general-purpose computing—for instance, it is perfectly normal that two instances of a Web browser or editor run together on the same PC—but also in embedded application.

   In the last case, a common way to design an embedded software component responsible for making a service available to other parts of the system, and increase its performance, is to organize the component as a set of "worker tasks," all identical for what concerns the code. Each worker task runs concurrently with the others and is responsible for satisfying one single service request at a time.

   If no virtual memory were supported, two instances of the same code executed concurrently by two worker tasks would interfere with each other since they would access the same data memory locations. In fact, they are running the same code and the memory addresses it contains are the same.

   This inconvenience is elegantly solved using a MMU and providing two different page tables—and hence, two different address translations—to the two tasks, so that the same virtual page referenced by the two tasks is mapped onto two different physical page frames.

2. In order to prevent memory corruption, it becomes possible to allocate all local data to be used by a task—like, for instance, its local variables and its stack—in a set of pages and map them onto a set of physical page frames.

   As long as the set of page frames accessible to the different tasks in the system are disjoint, no memory corruption can take place because any attempt made by a task to access or modify the page frames assigned to another task would be prevented and flagged by the MMU.

   Shared data, which are often needed in order to achieve effective inter-task communication, as described in Chapter 5, can still be implemented by storing them in a page frame and granting write access only to the set of tasks which are supposed, by design, to share those data.

A convenient starting point interested readers can use to gain more information about how a MMU aimed at high-end embedded systems works is Reference [6], which describes the MMU architecture optionally available in several recent ARM processor versions.

15.2 MEMORY PROTECTION UNITS (MPUS)

The main disadvantage of MMUs, which should be evident even from the terse and simplified description given in the previous section, is that the address translation process is quite complex. For this reason, a MMU also consumes a significant amount of silicon area and power. Moreover, since the MMU contributes to overall chip complexity in a significant way, it is also likely to increase the cost of the microcontroller.

The problem is further compounded by the fact that address translation must be performed in an extremely critical and very frequently used part of the execution path, that is, whenever the processor needs to access memory to fetch an instruction, read, or write data. Hence, its performance has a deep impact on the overall speed of the processor.

On the other hand, attentive readers have surely noticed that the abstract MMU address translation process, as depicted in Figure 15.2, is extremely inefficient if implemented directly. In fact, due to their size, page tables must necessarily be stored in main memory.

In this case, the interposition of a MMU between the processor and memory implies that at least *two* memory accesses are needed whenever the processor requests *one*. For instance, when the processor performs an instruction fetch:

- The MMU first consults the current page table to determine if the virtual address V provided by the processor has got a mapping in the table and whether or not that mapping permits instruction fetches depending on the current processor mode. Since the page table resides in main memory, consulting the table implies one memory access even though—as depicted in the figure—the table is implemented as a flat array of entries.

  However, as also discussed in Section 15.1, more sophisticated data structures—such as multilevel, tree-structured tables—are often needed in order to keep page table memory requirements down to an acceptable level. The price to be paid is that those data structures need more than one memory access to be consulted.

- After the MMU has confirmed that the memory access request is permissible and determined the physical address P at which it must be performed, yet another memory access is needed to retrieve the instruction that the processor wants to fetch. Only after all these memory accesses have been completed, is the instruction available to the processor.

In order to solve the problem of issuing multiple physical memory accesses to satisfy a single instruction or data request from the processor, additional hardware components are needed, further adding complexity, power consumption, and cost to the memory translation process. The adoption of a hardware translation lookaside buffer (TLB) integral to the processor architecture [83] is the most widespread method for speeding up memory accesses in a virtual memory system.

The operating principle of a TLB makes use of the memory access locality already described in Section 2.2.3 when discussing DRAM and cache memories. Informally speaking, we can say that the TLB is to the page table what cache memory is to DRAM.

In fact, if memory accesses are local and it is true that after a certain virtual address has been accessed there is a high probability that "nearby" virtual addresses will be accessed again within a short time, then it is also true that after a certain page table entry has been accessed, the probability of accessing it again within a short time is high.

This is because small variations in the virtual address will likely not modify the page number that, as shown in Figure 15.2, is the only part of the virtual address used to determine which page table entry must be used to translate it.

Therefore, it is convenient to keep in a small and fast associative memory—the TLB—the page translation information for a subset of the memory pages most recently accessed, with the goal of decreasing the overall memory access time by reducing the number of extra memory accesses needed to consult the page table. In this way:

- Before consulting the page table to translate a certain virtual address V, the MMU checks if the result of the translation is already available in the TLB because the translation has already been performed in the recent past. This check is performed without introducing any extra delay because, as mentioned previously, the TLB is based on a fast, on-chip associative memory. In this case (TLB hit), there is no need to read the page table entry from memory to get the corresponding physical page number and access permission information. The MMU simply uses this information and proceeds.

- Otherwise, the MMU consults the page table by issuing one or more additional memory address requests. Afterward, it updates the TLB and stores the result of the translation into it. Since the TLB capacity is limited—on the order of one hundred entries—in order to do this it likely becomes necessary to delete a TLB entry stored in the past.

  As happens for caches, the algorithm responsible for choosing the entry to be replaced is critical for performance. It is the result of a trade-off between the goal of deleting the entry with the minimum probability of being reused in the future versus the need for taking the decision quickly and with limited information about future task behavior.

It should also be noted that—again, with a strong analogy to what happens to caches—even though the TLB greatly improves the *average* address translation performance, it is usually hard to predict whether a particular memory access will give rise to a TLB hit or to a miss.

In the latter case, the overall memory access delay incurred by the processor is significantly higher than in the former. Moreover, TLB behavior depends in part on events external to the task under analysis. In particular, when a context switch occurs—due, for example, to task preemption—the TLB must be flushed and will be empty when task execution resumes.

This is because the page table is part of the task context and different tasks have different address translation information in them. Therefore, it would be incorrect to keep using the address translation results calculated for a certain task after switching to another. In turn, by intuition, a TLB flush "slows down" subsequent memory accesses until TLB contents have been reestablished.

For all these reasons, most microcontrollers do not embed a full-fledged MMU. Instead, they may implement a simpler component that does not perform address translation and is dedicated exclusively to the second typical function of a MMU,

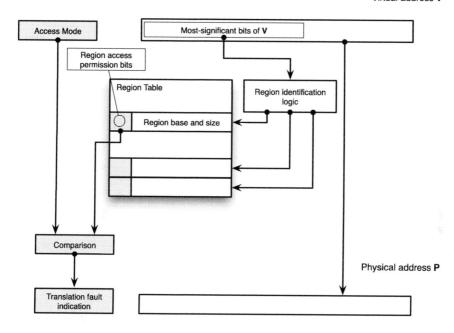

Figure 15.3 MPU memory protection principle.

that is, memory protection. To better highlight its function, this component is often called memory protection unit (MPU).

As shown in Figure 15.3, which depicts the operating principle of a typical MPU [8] in a simplified way, in systems equipped with a MPU there is always full correspondence between virtual and physical addresses (V and P in the figure) because no address translation takes place.

Memory protection information is kept in an on-chip table, often called a *region table*. Each entry of the table contains protection information pertaining to a contiguous memory region, identified by its *base address* and *size*.

The MPU architecture often imposes strict constraints on admissible values of region base addresses and sizes, in order to simplify the hardware. A typical constraint is that the size must be an integer power of two and the base address must be an integer multiple of the size.

In this way, the comparison between the address issued by the processor for a certain memory transaction and individual region table entries, to determine which region the address falls in, can be performed in a very efficient way. This is because it can be based solely on a bit-by-bit comparison between several most significant bits of the address and the most significant bits of region base addresses stored in the table.

Table 15.1

Fields of a Typical Memory Protection Unit (MPU) Region Table Entry

Field	Width	Purpose
ADDR	27 bits	Base address of the region. The number of significant bits, and hence, the alignment of the base address, depends on the region size indicated by the SIZE field.
SIZE	5 bits	Region size, constrained to be a power of two and equal to $2^{(\text{SIZE}+1)}$ bytes. The minimum region size is 32 bytes, and hence, it must be SIZE $\geq$ 4.
ENABLE	1 bit	This bit enables (ENABLE = 1) or disables (ENABLE = 0) the region table entry as a whole. See the description of SRD for a finer-grained way of enabling and disabling a region.
SRD	8 bits	Regions with SIZE $\geq$ 7, that is, with a size of at least 256 bytes are divided into eight sub-regions of size $2^{(\text{SIZE}-2)}$ bytes. Each sub-region can be individually enabled or disabled by setting the corresponding bit of SRD to 0 or 1, respectively.
AP	3 bits	Access and privilege properties of the region, encoded according to Table 15.2.
XN	1 bit	Indicates whether or not the processor is allowed to fetch and execute instructions from the region. Regardless of the setting of this bit, read access is needed anyway for instruction fetch.
TEX	3 bits	Works in concert with the C and B bits described below. It indicates the memory type of the region and which kind of ordering constraints must be obeyed when accessing it.
C	1 bit	Enables or disables caching when referring to the region.
B	1 bit	Enables or disables write buffering for the region.
S	1 bit	Indicates whether or not the region is shared among multiple processor cores.

In addition, the number of MPU table entries is usually very limited—eight is a typical value—and hence, much smaller than the typical number of MMU page table entries. For this reason, all comparisons between the processor-issued address and table entries can be, and usually are, performed in parallel by hardware. In the figure, this operation corresponds to the region identification logic block.

On the other hand, having a very small number of regions to divide the overall memory address space into, implies that memory access control granularity is much broader when using a MPU rather than a MMU. In other words, the limited number of regions provided by a MPU forces programmers to set the same access permission bits for larger areas of memory.

Regarding the detailed contents of a region table entry the ARM architecture v7m MPU [8], taken as an example, specifies the fields listed in Table 15.1. As we can see, the practical implementation of MPU region table entries closely resembles the more abstract description given previously. In particular:

- The ADDR and SIZE fields specify the position of the region in memory and its size. These are the fields that the MPU must match against the address

Table 15.2

Access Permission Field Encoding in the ARM MPU

AP	Processor mode	
Value	Privileged	Unprivileged
000	None	None
101	Read	None
001	Read, Write	None
010	Read, Write	Read
011	Read, Write	Read, Write
110	Read	Read
111	Read	Read
100	—	—

issued by the processor upon every memory access, to determine which table entry must be used for memory protection.

- The region table, as implemented by hardware, has a fixed predetermined number of entries. All entries are always consulted in parallel looking for a match against the address issued by the processor. The ENABLE bit allows programmers to enable only the table entries that are actually in use. The others are kept disabled so that their (presumably random) contents do not interfere with intended MPU operations.
- Since, as described above, memory regions are by necessity of a relatively large size this particular MPU implementation offers, as a workaround, the possibility of dividing each region into eight equally sized, smaller sub-regions. The bits in the SRD field work as a bit mask to selectively disable individual sub-regions. When a sub-region is disabled, the table entry corresponding to the whole region is not used even though the address issued by the processor matches it. In this way, the memory protection rules established by the table entry can be applied with finer granularity.
- The AP field contains the main information about access and privilege properties of the region, and determines how the MPU will enforce memory protection for the region itself. As can be seen in Table 15.2, the value of AP specifies which kinds of access to the region (read, write, or both) the MPU must allow, also depending on the current privilege mode of the processor.
- The XN bit further specifies access properties of the region for what concerns instruction fetch and execution. Namely, if XN = 1 instruction fetches within the region are not allowed. If XN = 0, instruction fetches are allowed if the AP field allows (at least) read accesses within the region.
- Finally, the TEX, C, B, and S bits convey additional information about the properties of the memory region described by the entry and the ordering constraints to be met when accessing it. This information alters the behavior of the memory access subsystem, but it is not directly related to MPU

operations and behavior. For this reason, it will not be discussed further. Similarly, a thorough description of memory access ordering strategies is beyond the scope of this book.

Another interesting aspect of the region table access algorithm adopted by this particular MPU is that table entries have different priorities. The priority ordering is fixed and is determined by the entry number, so that higher-numbered entries have higher priority.

Priorities become of importance when an address issued by the processor matches multiple region table entries. This may happen when the regions described by the entries overlap in part, which is allowed. Hence, for instance, it is possible to define a large lower-priority region in memory, with a certain access permission, and a smaller higher-priority region, with a different access permission, within it.

In this case, when the address lies within the large region but not within the smaller region, the access permission defined for the large region is used. On the contrary, when the address lies within the smaller region, and hence, matches both region table entries, the address permission set for the smaller region prevails.

This feature can be very useful to set the general access permission of a large memory area and then "dig a hole" with a more specific access permission. In turn, this helps in circumventing the previously recalled address alignment restrictions the region start address is subject to, which become more and more severe as the region size grows bigger, and also use a smaller number of table entries to put in effect the same access permissions.

As an example, even neglecting address alignment restrictions, configuring the MPU as discussed previously if region table entries overlap were not allowed, would require three entries instead of two. This is an important difference given that the total number of entries is usually very small, eight being a common value [8].

One last point about MPUs, which is worth mentioning and should emerge from the description given previously, is that they operate in a much simpler way than MMUs. Moreover, a MPU region table is much smaller than a typical MMU page table. For these reasons, the MPU region table can be kept in an on-chip memory. Moreover, all MPU operations are performed completely in hardware and do not require any acceleration mechanism that—as happens, for instance, with the TLB— may introduce non-deterministic timings into the system.

15.3 MPUS VERSUS MMUS

Table 15.3 presents a synthetic comparison between the main features and characteristics of MMUs and MPUs, as well as their implementation.

As can be seen from the table, the primary difference between MMUs and MPUs concerns their cost and complexity of implementation in terms of silicon area, stemming from the much simpler structure of MPUs. In turn, this also affects energy consumption, which is usually higher for MMUs than for MPUs.

This primary difference has several important side effects. Firstly, it limits MMU capabilities to memory protection only, barring address translation. Moreover, the

Table 15.3
Comparison between MMU and MPU Features and Their Implementation

Feature	MMU	MPU
Cost and complexity of implementation	High	Low
Energy consumption	High	Low
Address translation	Yes	No
Memory protection	Yes	Yes
Number of pages / regions	Thousands	Tens
Protection granularity	Fine	Coarse
Page table / Region table location	External memory	On-chip registers
Acceleration mechanism	Yes	No
Memory access time	Variable	Deterministic

protection granularity is much coarser in a MPU with respect to a MMU. This is because memory protection rules are applied with a granularity that corresponds to the page size for MMUs and to the region size for MPUs.

In a typical MMU the page size is a few kilobytes, whereas the typical MPU region size is several orders of magnitude larger. As a further consequence, MMUs need to handle a significantly larger number of pages, whereas MPUs support a very limited number of regions.

Due to their size, page tables must be stored in a large bank of external, off-chip memory, whereas a set of on-chip registers is usually devoted to store the current MPU region table. Since accesses to the MPU region table do not require any additonal memory transactions, MPUs do not require any acceleration mechanism and work in a fully deterministic way. In other words, they are able to implement memory protection without altering memory access timings in any way with respect to a system without a MPU.

All these MPU features—the last one in particular—are extremely important in real-time embedded systems. On the other hand, being able to perform address translation may not be as important. This explains why MPUs are generally more popular than MMUs, especially in low-cost microcontrollers.

15.4   MEMORY CHECKSUMMING

As discussed in the previous sections, many microcontrollers do not provide any form of hardware-assisted memory protection, mainly due to cost and complexity constraints.

Even when they do, the kind of memory protection granted by a MMU or MPU is effective only to protect *local* task memory from corruption due to other tasks. On the other hand, hardware has no way to detect memory corruption when it is caused by a task that incorrectly writes into its own local memory, or when it affects a memory area that is *shared* among multiple tasks and one of them behaves incorrectly. This is because, in both cases, tasks corrupt a memory area they must necessarily be granted access to.

In these cases, it is useful to resort to software-based techniques. The exact details of these techniques vary and, by necessity, they will only be summarized here, but they are all based on the same underlying principles. Namely,

- They stipulate that software must read from and write into a certain memory area M, possibly shared among multiple tasks, only by obeying a specific access algorithm, or *protocol*.
- Any write operation performed on M without following the access protocol shall be detected and taken as an indication of memory corruption.
- Detection is based on computing a short "summary" of the contents of M, generically called *checksum* C. The value of C is a mathematical function of the contents of M.
- When memory is used according to the proper protocol, the checksum is kept consistent with the contents of the memory area.
- Instead, if memory is modified without following the protocol—as a wayward task would most likely do—checksum C will likely become inconsistent with the corrupted contents of M.
- In this way, it will afterward be possible to *detect* memory corruption with *high probability*.

Before going into the details of how memory checksumming works, it is important to note one important difference with respect to what MMUs and MPUs do. Namely, MMUs and MPUs are able to *prevent* memory corruption because they are able to detect that a task is attempting to illegally access or modify a memory location *before* the memory operation is actually performed so that, in the case of an attempted write operation, memory contents are not corrupted.

Informally speaking, this makes recovery easier because the system state, which is normally held in memory, is not affected. On the other hand, software-based techniques are able to detect memory corruption with *high probability*, but not in all cases, for reasons that will be better described in the following.

Moreover, detection takes place only upon execution of a software procedure, which typically takes place when a task makes use of the shared memory area. As a consequence, memory corruption is detected only after it occurs. At this time, the system state has been irreparably lost and must be reconstructed as part of the recovery process.

Another point worth mentioning is that software-based techniques are unable to detect illegal *read* operations. In other words, they can neither detect nor prevent that a task gets access to information it has no permission for.

Figure 15.4 summarizes how memory checksumming works. In the middle of the figure, the shared memory area M to be protected is surrounded by two auxiliary data items, associated with it to this purpose:

- As described in Chapter 5, concurrent access to a shared memory area must take place in mutual exclusion, with the help of a semaphore S.

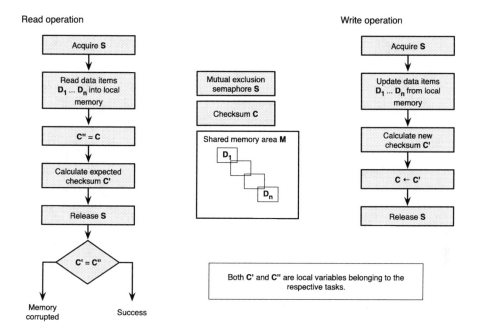

Figure 15.4 Memory protection by means of checksumming techniques.

This prevents memory corruption due to *legal* operations performed concurrently on M.

- In addition, as described in this section, checksum C is used to prevent memory corruption due to *illegal* operations performed on M.

The two flowcharts on the sides of the figure outline the protocol to be followed for read (on the left) and write (on the right) accesses to M. In order to perform a read operation:

1. Task T must first of all acquire the mutual exclusion semaphore S and it may need to wait if the semaphore is currently held by another task. After acquiring S, T is granted exclusive access to both M and C until it releases the semaphore at a later time.
2. Within the mutual exclusion region just opened, T can transfer data items $D_1, \ldots, D_n$ held in M into its own local memory.
3. Afterward, it also copies checksum C into a local variable C″.
4. Then, task T recalculates the checksum of M and stores the result into another local variable, called C′ in the figure.
5. At this point, T releases semaphore S because it will no longer use any shared data structure.
6. The final operation performed by T is to compare the two checksums C′ and C″.

Any mismatch between the two values compared in the last protocol step indicates that either the memory area M or its checksum C or both have been corrupted. This is because

- C″ is a copy of C and represents the expected checksum of M, as calculated the last time M was updated, whereas
- C′ is the actual checksum of M based on the current contents of M.

Instead, in order to perform a write operation, a task T must update both M and C to keep them consistent, by executing the following steps.

1. As for a read operation, task T must first of all acquire the mutual exclusion semaphore S.
2. At this time, T can update data items $D_1, \ldots, D_n$ held in M by means of a sequence of write operations from local task memory into M. Partial updates, involving only a subset of $\{D_1, \ldots, D_n\}$ are possible as well.
3. After completing the update, task T calculates the new checksum of M, denoted as C′ in the figure, and stores it into C.
4. Eventually, T releases S in order to allow other tasks to access M.

It should be noted that since the beginning of step 2 and until C has been updated, at the very end of step 3, M and C are indeed inconsistent even though no memory corruption occurred. However, this does not lead to any false alarm because any tasks wishing to perform a read operation—and hence, check for memory corruption— would be forced to wait on the mutual exclusion semaphore S during those steps.

Let us now consider what happens if a task T′ writes into either M or C in an uncontrolled way, that is, without following the above-specified protocol. In order to do this in a realistic way, we must also assume that T′ will not acquire semaphore S beforehand. As a consequence, T′ can perform its write operations at any time, even when the mutual exclusion rule enforced by S would forbid. A couple of distinct scenarios are possible:

1. If memory corruption occurs when neither a read nor a write operation are in progress, it will likely make M inconsistent with C. The next time a task T performs a read operation, it will detect the inconsistency during the last step of the read protocol.
2. If memory corruption occurs while a *read* operation is in progress, it may or may not be detected immediately, depending on when it takes place during the progress of the read protocol.
   Namely, any corruption that takes place after step 4 of the read protocol will not be detected in step 6. However, it will be detected upon the next read operation.
3. The most difficult scenario occurs when memory corruption occurs while a *write* operation is in progress because, in this case, there is a time window during which any corruption is *undetectable*.
   This happens if memory corruption due to T′ occurs during write protocol step 2, that is, when T is updating M, too. In this case, task T will later recalculate

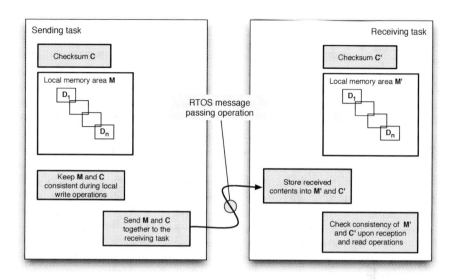

FIGURE 15.5 Checksumming technique applied to message passing.

the correct checksum of M, during protocol step 3, and store it into C "on behalf of" T'.

At the end of the day, C will indeed be consistent with the *corrupted* contents of M and it will pass the check performed during read operations.

Another reason why memory corruption may not lead to an inconsistent checksum is that, for practical reasons, C is always much smaller than M in size. In fact, the typical size of a checksum does not exceed 32 bits whereas the size of M can easily be on the order of several hundred bytes or more.

As a consequence, different contents of M may still lead to the same value of C, and hence, there is a certain probability that the expected value of C after memory corruption will still be the same as before.

In general, this probability is called *residual* error probability and this is the reason why, in the first scenario described above, we said that memory corruption will just "likely" make M inconsistent with C, rather than certainly.

It is worth remarking that, although Figure 15.4 shows how checksumming can be used to protect a shared memory area, the same method can also be applied to systems based on message passing, in order to make the system more robust against corruption occurring locally, within the communicating tasks, or during message passing itself.

For instance, as shown in Figure 15.5, if we assume that there is a unidirectional data flow between a sending task (on the left) and a receiving task (on the right), checksumming can be organized as follows:

- The sending task keeps its local memory area M and checksum C consistent while it is working on the contents of M and preparing to send them to the receiving task.

- The sending task sends M and C together to the receiving task by means of one of the RTOS message passing operations described in Chapter 5.
- The receiving task stores the message it received into its local memory, as M$'$ and C$'$.
- The receiving task checks M$'$ against C$'$ for consistency upon reception (to confirm that message passing at the RTOS level did not corrupt the message) and subsequent read operations (to safeguard against memory corruption occurring locally, within the receiving task).

Attentive readers would certainly have noticed the analogy between checksumming used in this way—that is, for message passing within tasks residing on the same microcontroller—and what is normally done in *network* communication, in which checksums are commonly used to ensure data integrity when messages travel across the network and network equipment, as mentioned in Chapter 7. This is indeed the case, and also the checksumming algorithms used in both cases overlap significantly.

15.5   CRC CALCULATION

From the descripion given in the previous section, it is clear that a central part of any memory checksumming method is checksum *calculation*, that is, the function that determines the checksum C of the contents of a memory area M.

In particular, since checksum calculation is performed very frequently, execution efficiency becomes extremely important, also with respect to the error detection capability of the algorithm. In turn, execution efficiency includes two different aspects, namely *algorithmic* efficiency in general and *implementation* efficiency, on a particular programming language and architecture.

A thorough description of checksumming methods and their properties is beyond the scope of this book. Interested readers are referred to the classic books [27, 109] for more information about this topic and error control coding in general.

In this section, we will briefly describe only one specific method of checksumming, based on cyclic redundancy check (CRC), focusing on the algorithm and, even more, on its efficient implementation using the C language. Reference [114] contains a comparison between CRC and other checksumming methods, characterized by a lower complexity but less capable of detecting errors.

Conceptually, a memory area M is a sequence of n_M bits denoted

$$m_{N-1}, \ldots, m_0 \ . \tag{15.1}$$

For what concerns CRC computation, the same area can also be seen as a polynomial $M(x)$, defined as

$$M(x) = \sum_{i=0}^{n_M-1} m_i \cdot x^i = m_0 + m_1 x + \ldots + m_{N-1} x^{n_M-1} \ , \tag{15.2}$$

Those two representations (sequence of bit and polynomial) are totally interchangeable and we will freely switch from one to the other in the following.

As shown in (15.2), if the memory area M is n_M bits wide, the degree of $M(x)$ is $n_M - 1$ and its coefficients m_i $(0 \leq i \leq N - 1)$ are the same as the bits in M.

The CRC of M, denoted as C in this chapter, is defined as the remainder of the division between $M(x)x^{n_G}$ and a *generator polynomial* $G(x)$ of degree n_G. Even though other choices are possible in theory, in most cases of practical interest related to computer engineering, coefficient arithmetic is carried out within a Galois field of two elements, denoted GF(2).

In this way, the two elements of the Galois field correspond to the two values a bit can assume, 0 and 1, as outlined above. Even more importantly, coefficient arithmetic can be performed by means of efficient bitwise operations on the strings of bits corresponding to the polynomials.

The choice of $G(x)$ heavily affects the "quality," that is, the error detection capability of the resulting CRC [109].

In formula, if it is

$$M(x)x^{n_G} = Q(x)G(x) + R(x) \ , \tag{15.3}$$

then $Q(x)$ is the quotient of the division and $R(x)$, the remainder, is the CRC of M. By definition, if the degree of $G(x)$ is n_G, then the degree of the remainder $R(x)$ is $n_G - 1$. Therefore, the polynomial $R(x)$ corresponds to a sequence of n_G bits

$$r_{n_G-1}, \ldots, r_0 \ , \tag{15.4}$$

which is the sequence of bits of C.

Obviously, actually using polynomial arithmetic to calculate the CRC would be overly expensive from the computational point of view in most cases. Fortunately, however, by means of the properties of polynomial division in GF(2), it is possible to calculate the CRC of a memory area M incrementally, considering one bit at a time, with a simple algorithm mainly based on the logical shift and the exclusive-or bitwise operation (XOR).

In particular, the CRC C of M can be calculated in the following way.

1. Initialize C by setting $C \leftarrow 0$.
2. For each bit of M, denoted as m_i, perform the three steps that follow, then go to step 6.
3. Calculate the XOR of the most significant bit of C (that is, bit $n_G - 1$) and m_i, and call it X.
4. Shift C one position to the left, dropping its most significant bit.
5. If X is 1, XOR C with the sequence of bits G corresponding to $G(x)$ and store the result into C itself.
6. At the end, C holds the result of the calculation.

The listing that follows shows, as an example, a possible C-language implementation of the bit-by-bit algorithm just presented considering a specific CRC generator polynomial used, among many other applications, for network error detection in the controller area network (CAN) [91].

In this case, $G(x)$ is of degree $n_G = 15$ and is defined as

$$G(x) = x^{15} + x^{14} + x^{10} + x^8 + x^7 + x^4 + x^3 + 1 \ , \tag{15.5}$$

and the corresponding bit string G, to be considered in the algorithm, is

$$G = 0100\ 0101\ 1001\ 1001_2 = 4599_{16} \ . \tag{15.6}$$

The most significant coefficient of $G(x)$ is not considered in the bit sequence because it is implicitly considered to be 1 by the algorithm. A value of 0 would not make sense because it would imply that $G(x)$ is not of degree n_G as assumed.

```
/* --- Slow, bit-by-bit CRC implementation.
        Used to build the lookup table
        for the faster implementation.
*/

/* Update crc considering the LSb of nxtbit */
uint16_t crc_nxtbit(uint16_t crc, uint16_t nxtbit)
{
    int crcnxt = ((crc & 0x4000) >> 14) ^ nxtbit;
    crc = (crc << 1) & 0x7FFF; /* Shift in 0 */
    if(crcnxt)  crc ^= 0x4599;
    return crc;
}

/* Update crc considering the nbits (<=16) LSb of nxtbits,
   MSb first
*/
uint16_t crc_nxtbits(uint16_t crc, uint16_t nxtbits, int nbits)
{
    int i;

    for(i = nbits-1; i >= 0; i--)
        crc = crc_nxtbit(crc, (nxtbits >> i) & 0x0001);

    return crc;
}
```

The code is organized as two distinct functions:

- The function crc_nxtbit performs algorithm steps 3 to 5. That is, it considers one single bit of M, passed to it in the nxtbit argument. It takes as input the previous value of C, in argument crc, and returns the updated value.
- The function crc_nxtbits implements the outer loop, corresponding to algorithm step 2. For the sake of simplicity, this function assumes that all the bits it should consider are held in its nxtbits argument.

Given that `nxtbits` is a 16-bit unsigned integer (as specified by its type, `uint16_t`) the maximum number of bits `nbits` it can process is limited to 16, but a larger number of bits can be handled in a similar way.

Also in this case, the function is designed to be used for incremental CRC calculations. Hence, it takes the current value of C as argument `crc` and returns the updated value.

In particular, the function `crc_nxtbit`:

- Extracts the most significant bit of `crc` (that is, bit $n_G - 1 = 14$) by masking and shifting. In this way, the value of (`crc & 0x4000) >> 14`) is either 1 or 0 according to the value of the most significant bit of `crc`.
- It performs the XOR (denoted by the `^` operator in the C language) of this bit with the next bit to be considered, `nxtbit`, and stores the result into local variable `crcnxt`.
  As described in Chapter 9, this variable is declared as an `int` so that its size matches the natural size of an architectural register of the processor in use, and the compiler has more optimization options available.
- It shifts `crc` left by one bit. Bit masking with `0x7FFF` is needed to discard the most significant bit of `crc`, which fell into bit position 15 after the shift. On the other hand, the left shift operator of the C language automatically shifts a 0 into the least significant bit position.
- Depending on the value of `crcnxt`, it conditionally performs the XOR of `crc` with the generator polynomial bit sequence (15.6). As for other arithmetic operators of the C language, it is possible to combine the XOR operation with an assignment and write it as `^=`.

The function `crc_nxtbits` is simpler. It calls `crc_nxtbit` repeatedly, within a `for` loop, for each of the `nbits` of `nxtbits`. For consistency with the correspondence between polynomial coefficients and sequences of bit, set forth in (15.1) and (15.2), the first bit to be considered is the most significant bit of `nxtbits`. The function then proceeds toward the right until it reaches the least significant bit, which is considered last.

Even without going into the details of full-fledged complexity analysis, it is clear that the execution time of the CRC algorithm just mentioned depends on the number of bits in M. In fact, the function `crc_nxtbit` accounts for most of the execution time and it is called once for every bit of M.

Fortunately, the mathematical properties of CRC calculation open the way to faster implementations that, instead of working on a bit-by-bit basis, operate on larger chunks of data with the help of information computed *in advance* once and for all, stored in a *lookup table* [149].

In the following we will examine how this calculation method works—and how it can be implemented in practice—with the additional goal of showing why, especially in embedded system software development, it may be convenient to shift as much computation load as possible from the target system to the development system.

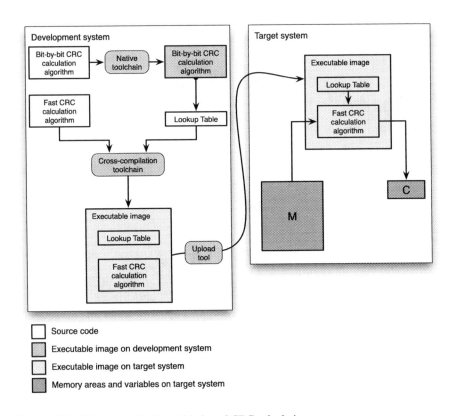

Figure 15.6   Principle of lookup table-based CRC calculation.

This is because the development system usually has much higher computational power than the target system. Moreover, the computation is performed only once, when the executable image is built, which is a time when there are no tight execution timing constraints.

Figure 15.6 illustrates the general principle of lookup table-based CRC calculation, which is designed according to the following guidelines.

- The bit-by-bit CRC calculation algorithm is used as the basis to build the lookup table used by the fast algorithm.
- To this purpose, the native toolchain is used to build an executable version of the bit-by-bit algorithm, which runs on the development system.
- The bit-by-bit algorithm is executed to generate the lookup table, which is then written in *source code* form, according to the language used to program the target system.
- The cross-compilation toolchain is used to build the executable image for the target system. The image contains, besides other modules, the compiled version of the fast CRC calculation algorithm and the lookup table.

- As described in Chapter 2, an appropriate tool, residing and running on the development system, is used to upload the executable image into the target system.
- Finally, the executable form of the fast CRC calculation algorithm runs on the target system and, with the help of the lookup table, calculates the CRC C of a memory area M in a more efficient way.

The procedure to be followed to compute the lookup table itself is fairly simple:

- If the fast CRC calculation shall work on chunks of N bits at a time, the lookup table must have 2^N entries.
- Each entry must be n_G bits wide and the i-th entry must contain the CRC of i, which can be calculated using the bit-by-bit algorithm.

It also turns out that, due to other well-known properties of CRC calculation, a table dimensioned to work on chunks of N bits can also be used to calculate the CRC using chunks smaller than N bits, still using exactly the same algorithm.

```
/* --- Lookup table for the word-by-word CRC implementation.
       For an n-bit word, the first 2^n entries of the table
       are needed.  The table is large enough for 1 <= n <= N.
*/

#define N 8

#define POW2(x)        (1<<(x))              /* 2^x */

uint16_t crc_lookup[POW2(N)];

/* Build the lookup table */
void wcrc_build_lookup(void)
{
    uint16_t i;

    for(i=0; i<POW2(N); i++)
        crc_lookup[i] = crc_nxtbits(0, i, N);
}


/* Dump the lookup table as C source,
    to be compiled and linked to the target embedded application.
*/
void wcrc_dump_lookup(void)
{
    int i;

    printf("/* %d bit wcrc lookup table */\n", N);
    printf("uint16_t\n");
    printf("crc_lookup[%d] = {\n", POW2(N));
```

```
for(i=0; i<POW2(N); i++)
    printf("     0x%04x%s /* [%d] */\n",
            crc_lookup[i],
            ((i<POW2(N)-1) ? "," : " "),
            i);
    printf("};\n");
}
```

Accordingly, the function `wcrc_build_lookup`, listed above, fills the table `crc_lookup[]` by repeatedly calling the function `crc_nxtbits(i)` and storing the result into the i-th entry of the table. The range of i goes from 0 to POW2(N)−1 included, where POW2(N) is a macro that computes 2^N efficiently—by means of a bit shift operation—and returns the result.

Then, the function `wcrc_dump_lookup` is responsible for printing out the contents of the lookup table. Namely:

- The printout consists of the source code to define an array of 16-bit unsigned integer elements, according to the `uint16_t` data type. The table is called `crc_lookup` and comprises N elements.
- After an opening brace, the initializer part of the definition contains the lookup table entries printed as hexadecimal integers according to the `printf` format specification `"0x%04x"`
- Entries are separated by commas. In order to avoid printing an extra comma after the last entry, either a comma or a blank space is printed according to the conditional expression `((i<POW2(N)-1) ? "," : " ")`.
- A closing brace followed by a semicolon concludes the printout, following the C language syntax for array definitions.

The code to use the lookup table to speed up CRC calculation is equally simple and is summarized below. Although a formal proof of correctness is beyond the scope of this book, it is useful to note its similarity with the bit-by-bit algorithm discussed previously to get at least an intuitive feeling of its correctness.

In particular, to calculate the CRC C of a memory block M working on chunks of N bits at a time, the following algorithm can be used.

1. Initialize C by setting $C \leftarrow 0$
2. For each chunk of N bits, denoted as m'_i, perform the three steps that follow, then go to step 6
3. Calculate the XOR of the N most significant bits of C and m'_i, calling the result X.
4. Shift C by N positions to the left, dropping the N most significant bits of the result.
5. Calculate the XOR of C with the contents of the X-th lookup table entry and store the result into C itself.
6. At the end, C holds the result of the calculation.

The following fragment of code implements the algorithm just presented.

```c
/* --- Parametric, lookup-table-based word-by-word CRC
       implementation.
       Works for chunks of any size nbits, 1 <= nbits <= N.
*/

#define N 8

#define POW2(x)        (1<<(x))              /* 2^x */

/* 2^x-1= Bits x-1...0 at 1 */
#define POW2M1(x)     (POW2(x)-1)

/* m bits at 1, starting @ bit 0 */
#define MASKM0(m)     (POW2M1(m))

/* m bits at 1, starting @ bit n */
#define MASKMN(m, n) (MASKM0(m)<<(n))

extern uint16_t crc_lookup[POW2(N)];

uint16_t wcrc_nxtbits(uint16_t crc, uint16_t nxtbits, int nbits)
{
    uint16_t so;

    /* Extract nbits most significant bits from crc, put them
       into so, then shift crc left by nbits bits.  Notice that
       the crc is 15 bits, so the most significant bit is
       bit 14, not bit 15.

       MASKMN and MASKM0 can be computed in advance if nbits
       is known in advance.
    */
    so = (crc & MASKMN(nbits, 15-nbits)) >> (15-nbits);
    crc <<= nbits;

    /* The lookup table index is so ^ nxtbits.  The inner & is
       necessary only if nxtbits may contain some extra
       non-zero bits besides bits 0...nbits-1
    */
    crc ^= crc_lookup[so ^ (nxtbits & MASKM0(nbits))];

    /* This bit mask is needed only once, at the end of
       the whole computation, and could be factored out of
       the computation loop.
    */
    crc &= 0x7fff;

    return crc;
}
```

At the very beginning of the code, there are a couple of macro definitions. Those macros return several values of interest, as described in the following.

- POW2(x) returns 2^x, as in the previous listing.
- POW2M1(x) returns $2^x - 1$, that is, a value whose binary representation has 1 in the x least significant bit positions and 0 elsewhere.
- This is made explicit by the definition of the macro MASKM0(m), which is defined in terms of POW2M1(x) and returns a bit mask consisting of m bits at 1 starting at bit position 0, that is, the least significant one.
- MASKMN(m, n) returns a bit mask consisting of m bits at 1 starting at bit position n, the mask is obtained by shifting the mask returned by MASKM0(m) n bits to the left.

Then, the function wcrc_nxtbits implements the lookup table-based CRC calculation algorithm proper, starting from the partial CRC crc and considering the nbits least significant bits of nxtbits. The return value of the function is the result.

In other words, this function implements steps 3 to 5 of the CRC calculation algorithm. The implementation of steps 1, 2, and 6 is not shown for conciseness. More specifically:

- With the help of the macro MASKMN(nbits, 15-nbits), which returns a bit mask with 1 in the nbits most significant positions of the CRC value (from bit 15-nbits to bit 14, included), it extracts the nbits most significant bits of the current CRC from argument crc.
- It then shifts the extracted value 15-nbits bits to the right, so that its least significant bit coincides with bit position 0, and stores the result into local variable so.
- After extraction, it shifts crc in place by nbits positions to the left, using the combined shift/assignment operator <<=.
- Then, the crc value is XORed in place (by means of the combined XOR/assignment operator ^=) with the contents of the lookup table entry at index so ^ nxtbits (that is, the XOR of local variable so and the chunk of bits to be considered).
- As a safeguard, the input parameter nxtbits is first masked with MASKM0(nbits) in order to make sure that the result of the XOR is always a valid index in the lookup table, even though nxtbits may contain some extra non-zero bits besides at bit positions 0, ... nbits-1.
- The last two statements of the function ensure that crc does not contain any bits at 1 outside bit positions from 0 to 14 included and return the result to the caller.
- The bit mask just mentioned is needed only once, at the end of the CRC computation as a whole. Therefore, it can be factored out and be performed after all chunks of M have been processed.

The function just described is *parametric* with respect to the chunk size `nbits`. In other words, the chunk size is one of the *input parameters* of the function. This makes the function more flexible and the code more compact—because the same code can be used for all legal values of `nbits`—but, as better explained in Chapter 11, gives the compiler less optimization opportunities.

As shown in the following listing, another approach, which privileges execution time with respect to code size, is possible.

```
/* --- Lookup-table-based word-by-word CRC implementation.

        Macro to instantiate a function fname that works for
        chunks of size nbits.  The code is the same as the
        parametric version but, since nbits is a constant,
        the compiler will be able to produce better code.
*/
#define instantiate_wcrc_nxtbits(fname, nbits) \
uint16_t fname(uint16_t crc, uint16_t nxtbits) \
{ \
    uint16_t so; \
    so = (crc & MASKMN(nbits, 15-nbits)) >> (15-nbits); \
    crc <<= nbits; \
    crc ^= crc_lookup[so ^ (nxtbits & MASKM0(nbits))]; \
    crc &= 0x7fff; \
    return crc; \
}
```

This fragment of code defines the macro `instantiate_wcrc_nxtbits` that, when expanded, produces the C-language definition of a function called `fname`. The body of the function contains exactly the same code as function `wcrc_nxtbits`, minus some comments, as can easily be inferred by comparing the two listings shown above.

A very important difference, though, is that `nbits` (that is, the chunk size) is a *parameter* of `wcrc_nxtbits`, whereas it is *not* a parameter of `fname`. Instead, the name `nbits` that appears in the body of `fname` is replaced during expansion with the argument of `instantiate_wcrc_nxtbits` bearing the same name. This has a couple of important consequences.

1. If, as it should, the `instantiate_wcrc_nxtbits` macro is invoked with a constant value as `nbits` argument, that constant appears in the body of `fname` in place of `nbits`.
2. At this point, all arguments of `MASKMN` and `MASKM0` become constant, too. Therefore, the results of macro expansions are expressions composed only of constant terms, which any recent compiler is able to calculate beforehand, at *compile* time rather than at *execution* time.

These facts not only give the compiler more room for optimization, but also shift part of the computation to be performed in `fname` (namely, bit mask calculation) from execution time—which occurs on the target system—to compile time—which takes place on the development system.

Therefore, the use of `instantiate_wcrc_nxtbits` repeats, on a smaller scale, the same optimization approach followed when we went from bit-by-bit CRC calculation to the lookup table-based algorithm.

It is now time to conclude the section with an example of how all the pieces just described can be put together to calculate the checksum of a whole memory area on the target system. The corresponding code is shown in the listing that follows.

```
#define POW2(x)        (1<<(x))             /* 2^x */

/* 2^x-1: Bits x-1...0 at 1 */
#define POW2M1(x)      (POW2(x)-1)

/* m bits at 1, starting @ bit 0 */
#define MASKM0(m)      (POW2M1(m))

/* m bits at 1, starting @ bit n */
#define MASKMN(m, n)  (MASKM0(m)<<(n))

/* Include the crc_lookup[] table for N=8 */
#include "crc_lookup_def.h"

/* This macro expansion defines wcrc_nxtbits_8(),
   which calculates the CRC in 8-bit chunks.
*/
instantiate_wcrc_nxtbits(wcrc_nxtbits_8, 8)

/* The following function calculates and returns the CRC of
   the k-byte memory area data[].  The calculation is performed
   byte by byte.
*/
uint16_t crc_data(int k, uint8_t data[])
{
    uint16_t crc = 0x0000;

    for(i=0; i<k; i++)
        crc = wcrc_nxtbits_8(crc, data[i]);

    return crc;
}
```

- The first part of the code defines the macros `POW2(x)`, `POW2M1(x)`, `MASKM0(m)`, and `MASKMN(m, n)`, because they are still needed to compile the target code.
- Then, the definition of `crc_lookup`, that is, the lookup table, must be included. In the code fragment above, it is assumed that the output of the `wcrc_dump_lookup` function, executed on the development system,

was saved in a file called `crc_lookup_def.h`, but the actual file name is irrelevant.

- At this point, the macro `instantiate_wcrc_nxtbits` previously described is invoked, at build time. The compiler (or more precisely, as described in Chapter 3, the C language preprocessor) expands it into a function called `wcrc_nxtbits_8`, as specified in the first argument of the macro invocation. This function has the following prototype

```
uint16_t wcrc_nxtbits_8(uint16_t crc, uint16_t nxtbits);
```

and is able to update the CRC considering 8 bits (1 byte) at a time, as specified in the second argument of the `instantiate_wcrc_nxtbits` macro invocation.

- Last, the function `crc_data` goes through a k-byte memory area represented by `data[]` and calculates its CRC. In order to do this, it first initializes local variable `crc` to `0x0000` and then invokes `wcrc_nxtbits_8`, in order to update it considering one byte at a time. Eventually, the final value of `crc` is returned to the caller.

15.6   DATA STRUCTURE MARKING

Checksumming techniques, described in the previous section, are able to detect (up to a certain extent) the corruption of memory areas as a whole. Due to the complexity of checksum calculation, those techniques are often applied only to the most critical memory areas and the ones most prone to corruption, such as the memory areas that hold the system state, which is shared among most tasks in the system.

In some cases, in order to better protect the system from software bugs and gather additional information about where they are in the code, it is also important to detect memory corruption less reliably but with a finer level of granularity—as usual, aiming at a different trade-off with respect to the price paid in terms of performance.

Structure marking [168], a concept first pioneered in the MULTICS operating system [135], and more specifically in its file system, works at the data structure level. It relies on the concept of encapsulating all data structures and adds some extra information to them.

This extra information can then be used by software to detect, with high probability, whether or not a data structure has been updated in an uncontrolled way, and hence, it has likely been damaged. As a whole, this technique works according to the more general idea of adding redundant information to data structures to improve software fault tolerance.

This is discussed, for instance, in References [162, 150, 99], to which interested readers may refer to find more detailed information about the topic.

The basic concepts of structure marking are illustrated in Figure 15.7. As shown in the middle of the figure, original data structures (depicted as gray rectangles) are encapsulated in larger structures and surrounded by several additional items of information, depicted as white rectangles and defined as follows.

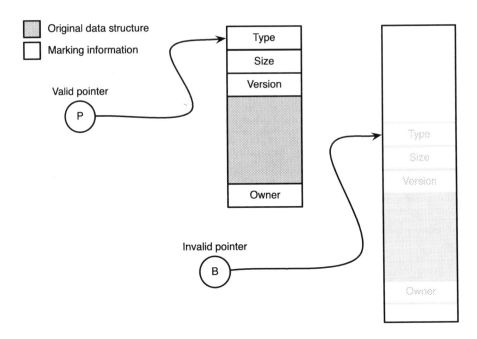

Figure 15.7   Structure marking and invalid pointer detection.

Type is a unique identifier of a certain data structure type, assigned when the data
type is first defined and introduced into the system. All data structures of the same
type contain the same value in this field.

Size represents the size of the data structure, expressed by means of a uniform unit
of measurement. Variable-length data structures, if supported by the programming
language, shall contain the actual size of the data structure.

Version is a version number that shall be changed whenever, during software development or maintenance, the definition of the data structure is changed without
changing the data type name.

Owner holds a unique identifier of the software component that "owns" the data
structure, for instance, the task that created it.

The figure also shows the expected position of the additional items of information
with respect to the original data structure. For reasons that will be better detailed in
the following, Type, Size, and Version are expected to be at the very beginning of the
extended data structure, whereas Owner shall be at the very end of it.

For variable-length data structures, Size (which can be located immediately given
a pointer to the data structure because it is at a fixed offset from the beginning of it)
can be used to determine where Owner should be and retrieve it.

In the C programming language, the possible values of Type and, for each data
type, Version numbers can, for instance, be defined in source headers as macros. In

each data structure, these fields are filled in by the function that creates it, according to its intended, initial layout.

When backward compatibility is important, the function responsible for creating a certain data structure may have an additional argument that indicates which specific version of the data structure it should create and return.

In this case, it is convenient to surround the data type definitions corresponding to all supported versions of a data structure with a `union`. In this way, a pointer to that union represents a pointer to a data structure of that kind, regardless of its version. Within the code, the Version field—which is present in all versions at the same offset—can be used to distinguish one version from another and make access to the right member of the union depending on it.

Similarly, Size and Owner are also set when the data structure is created. The Owner field is especially useful for local data structures belonging to "generic" data types, such as lists. In this case, it is important to ensure that the data structures used by a certain task are not only of the correct type, but also that they were created by the task itself and not by others.

When the owner of a data structure is a task, it is possible (and convenient) to store the unique task identifier provided by the operating system in the Owner field. However, this approach is inadequate when the concept of owner must be implemented at a finer level of granularity—for instance, at the software component or source code module level.

If the operating system itself does not provide any facilities to generate unique identifiers, a viable choice is often to initialize a free-running hardware counter clocked at a high frequency and take a snapshot of its value as a reasonable approximation of a unique identifier (neglecting counter wrap-arounds) when the component or source code module is first initialized.

As a general rule, every function that accepts a data structure as input shall first check all four additional items of information before proceeding to use the other parts of the data structure itself. If any check fails, then the data structure has likely been damaged and cannot be used safely. In particular:

- The Type field must be compared against the expected data structure type. Any mismatch prevents the function from working on the data structure, because it implies that its internal structure is unknown.
- The Size field must match the expected data structure size. In this case, a mismatch indicates that this field—and, likely, other information around it—has been corrupted, and hence, the function shall not use the data structure.
- The Owner field must match the task, component, or source code module identifier. A mismatch indicates that, even though the data structure received as input is known to the receiving function for what concerns its internal structure, it must not be used anyway because its contents belong to a different software component.
  Of course, this check must be implemented in a different way if the data structure at hand is expected to be passed from one software component to

another. If the originating component's identifier is known to the function, it may check whether or not it is correct. Otherwise, the check can be omitted.

- The kind of check to be performed on Version depends on the complexity of the function. Simple functions typically support a single version of their input data structures, and hence, any version mismatch prevents the function from working on them.

  On the other hand more complex functions, designed for backward compatibility, may support several different versions of a certain data structure and change their behavior depending on what they find in Version.

Those checks are easy to implement, even when they must be added to existing code, and do not imply a significant overhead. If all items of information can be encoded in a machine word—as is usually the case—all basic checks can be performed by means of four word comparisons against constant values plus simple address arithmetic in most cases. In any case, this is much simpler than checking the data structure content as a whole for consistency.

Structure marking also helps ensure that when a data structure is passed by reference, that is, by means of a pointer, the invoking function is indeed receiving a valid pointer. In fact, as is shown in the right part of Figure 15.7, if the function receives an invalid pointer B instead of a valid pointer P, it will basically compare the contents of several values retrieved from arbitrary memory locations with the expected values of Type, Size, Version, and Owner.

If, as is advisable, developers took care to not use common values like zero or one as expected values, there is a high probability of having a mismatch, and hence, a high probability of successfully detecting that the pointer is invalid.

Reasoning at a higher level of abstraction this is because, in general, structure marking works according to the principle of adding redundant information to data structures. With the help of this information it is possible to define additional *invariants* concerning the data structures themselves, that is, properties that they must always satisfy.

The code that accesses a data structure can then check the invariants and ensure that they hold. By intuition, the more invariants are defined and checked, the higher the probability of successfully detecting data structure corruption is.

In other words—as remarked in [168]—when using structure marking, instead of assuming that a data structure is suitable for use just because we have a pointer to it, we assume it is good because we have a pointer to it *and* some of its fields have the legal, consistent values we expect. The following kinds of issues can all be detected by structure marking.

- Write overflows from other adjacent data structures, such as arrays. Fields Type and Owner are especially useful to this purpose, because they are at the extremities of the data structure, and hence, they are overwritten first in case of overflow coming from lower or higher addresses, respectively.

  Overflow detection occurs with high probability, that is, unless the garbage written into the fields happens to match the expected values.

- Many kinds of programming errors. For instance, Type checking detects logic errors and function calls with wrong pointers. This is especially important when using a programming language (for instance, the C language) that allows programmers to freely convert pointers to a certain data type to another data type without performing any consistency check.

  Moreover, Owner checking detects programming errors that may escape pointer conversion checks performed by the compiler, for instance, passing to a function a pointer of the right type, but pointing to information intended for another function or software component.

- System integration errors, in which different parts of the software are not aligned correctly, and hence, are meant to operate on different versions of a certain data structure. This may happen, for example, when the definition of a data structure is updated, but only part of the software using it is also modified and rebuilt accordingly.

  In this case, Size and Version checking easily detects that a function is unprepared to handle the data structure it received as input.

On the contrary, structure marking is not as effective against other kinds of errors, like the ones listed in the following.

- Spurious or "random" memory writes that damage the contents of a memory word without regard to the data structure it belongs to. In fact, the probability that the spurious write ruins the original contents of the data structure (gray rectangles in Figure 15.7) without affecting the additional items of information (white rectangles) is high because their size is usually a small fraction of the total data structure size.
- Design errors. For instance, an ill-designed algorithm can easily produce incorrect results, systematically or on occasion, but still store them into perfectly formed data structures.
- Errors in the toolchain, namely the compiler because, in this case, a bug may introduce corresponding errors everywhere in the code, including the portion of code that implements structure marking and related checks.

When structure marking checks detect that a data structure has been damaged, the program actions to take depend on how the software has been designed and, in particular, on the degree of fault tolerance that has been built into the system. Some possibilities are described in the following.

- Print an error message or, more generally, report the error. Then, abort the application as a whole or halt the task that detected the damage. This kind of behavior is generally acceptable only during development and testing, especially for embedded real-time applications.
- If there is enough redundant information available, attempt to repair the damaged structure. This approach, when successful, has the advantage that data structure damage is handled transparently and the caller only notices an extra delay in function execution.

However, completely reconstructing the content of a data structure requires a significant amount of redundancy to be stored and maintained elsewhere in the system, possibly introducing overhead and consistency issues. Moreover, the extra delay that comes as a side effect of repairing the data structure may be unacceptable if it takes place in a time-critical portion of software execution.

- Put the subsystem that depends on the data structure offline, while keeping the rest of the system up and running. For instance, detecting that a data structure related to remote communication has been damaged may prevent a data-logging system from forwarding its data to a central site at the moment, but it can still keep collecting data and log them locally until the remote communication subsystem is reinitialized, either automatically or upon user intervention.

- Log the event and discard the offending data structure without further action. This approach is easy to implement and suitable for scenarios in which it is possible to attempt the same—usually, non-critical—operation again if it fails.

  For instance, a function responsible for handling remote status and diagnostic requests in a low-level control system can in some cases just not reply or return an error indication to the higher-level node if it detects that some of the data structures it needs have been damaged. The higher-level node itself will retry the request in due time and, if the issue persists, raise an exception to other parts of the control or supervision system.

- As a bare minimum and in any case, any function that detects data structure damage shall not continue using the damaged input but return an error indication to the caller.

Experience with the MULTICS operating system, in which structure marking was extensively used in most of the disk- and memory-resident file system data structures, shows that it improved system reliability noticeably [168].

At the same time, standard benchmarks showed no measurable performance cost due to the additional checks. Space cost, related to the introduction of the additional fields in every data structure, was a few percent.

Interestingly enough, the option of keeping a checksum for each structure, as described in Section 15.4, was also considered during the design phase. However, it was considered that there were not enough cases where data structures were considered to be good by structure marking even though their content was damaged, to justify the additional cost of computing a checksum at this very fine level of granularity.

15.7 STACK MANAGEMENT AND OVERFLOW DETECTION

As mentioned in Chapter 4, task stacks play a very significant role because they hold a significant part of task states and of the system state as a whole. For instance, all local task variables—that is variables declared within a function—are allocated on

Table 15.4

FREERTOS **Configuration Variables Related to Stack Overflow Checks**

Variable	Value	Meaning
INCLUDE_uxTaskGetStackHighWaterMark	0	Do not provide stack high water marking
	1	Provide stack high water marking
configCHECK_FOR_STACK_OVERFLOW	0	Disable stack overflow detection
	1	Check stack pointer
	2	Check stack pointer and guard area

the stack of the invoking task upon function entry and released upon exit, in a last-in, first-out (LIFO) fashion.

In many operating systems, including FREERTOS [18], the task stack is also used to hold the task context when it does not reside in the processor, that is, when the task is not running. As for local variables, a certain amount of space is allocated on-demand from the task stack to this purpose.

Task stacks in an embedded operating system usually have a fixed maximum size that is specified upon task creation. In the case of FREERTOS, as explained in Chapter 5, the usStackDepth argument of xTaskCreate indicates how much memory must be allocated for the task stack.

Since, as discussed above, space is dynamically allocated from and released to them dynamically during task execution, it is extremely important to prevent or at least detect any overflow, which may corrupt stack contents themselves, as well as surrounding memory areas, and cause all kinds of hard-to-predict issues in task behavior.

For this reason, many embedded operating systems implement mechanisms to assist in this detection of such an occurrence. In the case of FREERTOS, three distinct mechanisms are provided to this purpose:

1. Stack *high water* marking and detection.
2. Stack pointer check upon context switch.
3. Stack *guard area* check upon context switch, to be used in combination with the previous one.

All these mechanisms are optional because, being additional functions that the operating system performs either on demand or automatically, they introduce a certain amount of memory footprint and execution time overhead. They are selectively enabled or disabled by means of the configuration macros listed in Table 15.4. Moreover, depending on the memory model and layout, some architectures may not support them at all.

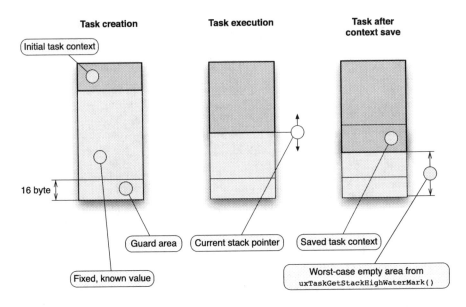

Figure 15.8 Stack overflow detection by means of a guard area.

It must also be noted that some processors may perform hardware-based stack overflow checks, and hence, they may generate an exception in response to a stack corruption before any software-based overflow check can occur. Moreover, severe stack overflows may lead the processor to address a nonexistent memory location, an occurrence that some processors are able to detect and report, usually by means of a bus fault exception.

Stack *high water marking and detection* is the simplest kind of stack overflow detection. It is also the least invasive from the point of view of execution time overhead, because the check is performed only on demand, when a certain operating system primitive is invoked. For this reason, the application has full control on the trade-off between overhead and accuracy of the check.

As shown on the left part of Figure 15.8, when a task is first created its stack is filled with a known value, depicted in light gray, and is mostly unused. Only a small portion of it, at the top, is initialized by the operating system and holds the initial task context, to be used upon the first context switch toward the task.

In this and the following figures, the used portion of the stack is colored in dark gray. Moreover, it is assumed that stacks "grow downward," that is, stack space is allocated in a LIFO fashion starting from higher addresses and going toward lower addresses. As is customary, higher addresses are depicted above lower addresses in the figures.

During task execution, as shown in the middle of Figure 15.8, the amount of stack used by the task grows and shrinks. At any time, the moving boundary between the currently used and unused portions of the stack is indicated by the current processor stack pointer.

As a consequence, when the stack grows, the known value with which the stack was filled upon task creation is overwritten by other values, for instance, the content of task local variables. On the other hand, the known value is *not* restored when the task shrinks.

Therefore, it is possible to get an approximate idea of the worst-case (minimum) amount of unused stack space reached by the task in its past history by scanning the task memory area starting from the bottom and proceeding as far as the known value is still found. This is exactly what the FREERTOS function `uxTaskGetStackHighWaterMark` does.

It it useful to remark that the check is approximate because it is possible that, by coincidence, the task overwrites the known value, with which the stack was initially filled, with the same value. As a consequence, there may be a certain amount of uncertainty about the precise location of the boundary between the light gray (still filled with the known value) and dark gray (overwritten by task) areas shown in the figure. More specifically, the function

```
UBaseType_t uxTaskGetStackHighWaterMark(
    TaskHandle_t xTask);
```

calculates and returns the worst-case amount of stack space left unused by task `xTask` since it started executing. This corresponds to the size of the light gray stack area highlighted at the extreme right side of Figure 15.8. As discussed in Chapter 5, `xTask` is a task *handle*. The special value NULL can be used as a shortcut to refer to the calling task.

By analogy with the `usStackDepth` argument of `xTaskCreate`, the return value of the `uxTaskGetStackHighWaterMark` function is not expressed in bytes, but in stack words, whose size is architecture-dependent. Refer to Chapter 5 for more information on how to determine the stack word size on the architecture at hand.

As task `xTask` approaches stack overflow, the return value becomes closer to zero. A return value of zero means that the available stack space has been completely used by the task in the past, and hence, a stack overflow likely happened.

One shortcoming of high water marking is that, by intuition, the longer the time that elapses between a stack overflow and its detection, the more likely it becomes that the system misbehaves.

From this point of view, the other two stack overflow detection mechanisms are more aggressive because the check is performed automatically, and quite frequently, by the operating system itself instead of relying on application-level code.

In order to be used, these mechanism need, first of all, a way to inform the application that a stack overflow occurred. To this purpose, the application code must define a stack overflow *hook* function whenever `configCHECK_FOR_STACK_OVERFLOW` is set to a value greater than zero. The hook function must bear a special name and adhere to the following prototype.

```
void vApplicationStackOverflowHook(
    TaskHandle_t xTask,
    signed char *pcTaskName );
```

The operating system invokes the hook function whenever it detects a task stack overflow. The arguments passed to the function both indicate the offending task in two different ways:

- The `xTask` argument is the machine-readable handle of the offending task.
- On the other hand, `pcTaskName` is the human-readable task name given to `xTaskCreate` when the task was created.

It is worth remarking that it is necessary to put extreme care in the implementation of this function. This is because, for reasons that will be better explained in the following, even in this case task stack overflow checks are not perfect. Hence, the operating system may invoke the hook function when a stack overflow already occurred and the system state has already been corrupted. The hook function arguments themselves may be invalid, too.

The first kinds of stack overflow detection provided by FREERTOS is based on checking the value of the task *stack pointer* every time the operating system saves the task context onto the stack. Statistically, as shown in the right part of Figure 15.8, it is likely that the portion of stack used by the task reaches its maximum extent at those times because, besides all task-related information—the whole task context has just been pushed onto it, too.

This method has a limited impact on performance because, even though the check is performed upon every context switch—and hence, possibly thousands of times per second—it is by itself fairly simple, since it only consists of two address comparisons. Namely, the current task stack pointer must be compared with the top and bottom addresses of the stack, to confirm that it lies within the valid space allocated to it. If this is not the case, the `vApplicationStackOverflowHook` hook is called from within the offending task context, as described above.

On the other hand, it may be unable to detect stack overflow reliably in every case. This happens, for instance, when the task overflows the stack and then the stack pointer returns within its valid range before the operating system had an opportunity to check it.

In other words, this mechanism consists of a sequence of stack pointer checks performed at very *specific points* in time, but has no *memory* about what happened to the stack pointer *between* checks. This is illustrated in Figure 15.9, which shows, from left to right, the possible evolution of a task stack state during execution, passing through a stack overflow.

This simple observation leads the way to the second method of stack detection implemented by FREERTOS. According to this method, a small portion of memory at the bottom of each task stack (16 bytes in the case of FREERTOS) is defined as a *guard area*. As shown in the left part of Figure 15.9, the guard area is filled with the same, known value as the rest of the stack during task creation.

Every time the operating system pushes the task context onto the stack, it also checks that the guard area content is still the intended one, that is, it has not been overwritten with other values. That is, the guard area content is used as a memory of past stack overflows, which remain detectable even though the stack pointer

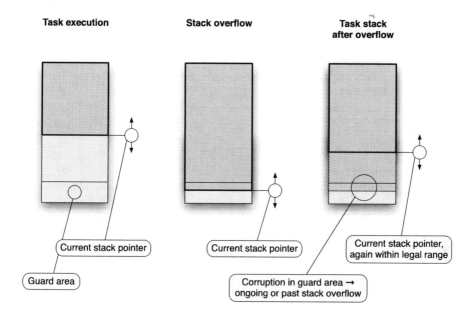

Figure 15.9 State of task stack after a stack overflow.

went back to a legal value after the overflow occurred. As in the previous case, any change in the guard area content leads the operating system to invoke the `vApplicationStackOverflowHook` hook.

Since this method requires the operating system to compare the whole content of the guard area against the known value, it is more expensive, from the execution time overhead point of view, with respect to the previous one, which only requires two address comparisons. On the other hand, it has a higher probability of detecting stack overflows successfully.

As a concluding note, it is useful to remark that automatic stack overflow checking, performed at the operating system level, may introduce a significant overhead especially on low-end microcontrollers. Therefore, it should be used only during software development and testing, whereas it can be disabled after the actual stack requirements of the various tasks present in the system become well known and stable.

15.8 SUMMARY

Internal consistency checks, especially those related to memory corruption issues, play a central role in software development, especially for embedded systems in which reliability is often an important design requirement. In addition, the same checks also speed up software development and debugging because they help programmers to quickly identify functions and code modules with questionable behavior.

As described in Sections 15.1 and 15.2, in virtually all general-purpose systems and some embedded systems as well, memory corruption issues are *prevented* by means of additional hardware components interposed between the processor and main memory. Two quite different approaches have been discussed in this chapter, and their relative merits discussed in Section 15.3.

When hardware support is unavailable, it is still possible to *detect* memory corruption with high probability and a relatively low overhead by resorting to software-based techniques, for instance, the ones presented in Sections 15.4 through 15.6.

Last, an important reason for unpredictable behavior in embedded systems is memory corruption due to stack overflow, to which none of the previously mentioned software-based techniques can be directly applied. However, as shown in Section 15.7, many operating systems offer various degrees of assistance for manual or automatic stack overflow detection, which can profitably be used during software development and testing.

16 Security and Dependability Aspects

CONTENTS

This chapter shows how to detect code mistakes and vulnerabilities, mostly related to bogus memory accesses, by means of *static code analysis* techniques. Namely, it analyzes in detail a mature open-source tool called SPLINT [60], which implements those techniques.

The goal is to enhance code reliability and, in some cases, its security. The last aspect is becoming more and more important as embedded systems are nowadays often connected to public communication networks, and hence, they are becoming more vulnerable than in the past to security attacks.

Informally speaking, static code analysis is able to infer some properties of a program by working exclusively at the *source code* level—possibly with some additional hints from the programmer, given in the form of source code *annotations*.

In any case, it is never necessary to actually *execute* the code, and hence, static code analysis does not require any kind of runtime support from the processor and the operating system.

This is especially welcome in embedded software development because, as illustrated in Chapter 2, static code analysis can be performed on the development system and does not impact the performance of the target system in any way [37].

Even more powerful methods and tools exist, based on *dynamic analysis*, of which VALGRIND [151] is a typical example in the open-source arena. However, they require various kinds of support from the target processor, operating system, and runtime libraries, which may not be available in many embedded systems. For instance, at the time of this writing VALGRIND is available only for several processor architectures and a couple of operating systems.

Moreover, even when appropriate support is indeed available, the runtime overhead that those tools introduce by permeating the target application with their instrumentation code may be unacceptable.

16.1 INTRODUCTION TO SPLINT

SPLINT [60], short for *Specification Lint* and *Secure Programming Lint*, is the evolution of an older programming tool to detect anomalies in C-language programs, called LINT.

LINT [98] was originally developed by S. Johnson for the Unix V7 operating system, to alleviate the rather severe limitations of the C-language syntax and compilers available at that time for what concerns program correctness checking. As an example, early versions of the C language did not even support function prototypes, and hence, the responsibility for calling a function with the right number of arguments and with appropriate data types fell exclusively on programmers.

Although SPLINT is able to perform many checks without help from the programmer, more advanced checks require additional information given in source code *annotations*. To avoid any interference with the C compiler and the compilation process, annotations are embedded in specially formatted comments, which are parsed by SPLINT but ignored by the compiler. A simple example of annotation is

```
/*@null@*/ int *p;
```

It indicates that pointer p may be NULL. As can be seen, an annotation is surrounded by the usual sequence of characters /* and */, which delimit C-language comments. The additional delimiter @ indicates that the comment is indeed an annotation. Within those delimiters, there is the text of the annotation proper, that is, null in this case.

It must also be noted that, unlike normal comments, the meaning of an annotation may depend on the grammatical context. For instance, the null annotation just described applies only to the pointer named in the variable definition it appears in.

Besides being used to convey additional information, special comments called *control comments* are also used to override SPLINT command line flags locally, on a file-by-file basis. For instance, the annotation

```
/*@+charint@*/
```

can be used to inform SPLINT that the code contained in a certain source file, where the annotation appears, uses data types char and int interchangeably, as was typical of legacy programs.

Clearly, assumptions like this weaken the checks that SPLINT can perform, and hence, it is better to keep their scope as narrow as possible, so that they apply only where they are strictly needed.

From the practical point of view, the tool is invoked mostly like a normal C compiler. For instance,

```
splint <flags> a.c b.c
```

Table 16.1
SPLINT **Mode Selector Flags**

weak	This flag selects a setting in which SPLINT performs only "weak" checks. These checks are suitable to analyze code without annotations without incurring nuisance warnings.
standard	This is the default operating mode of SPLINT, the one used in all examples given in this chapter, unless otherwise specified. As described in the main text, a limited amount of annotation is needed to avoid false warnings.
checks	This flag makes checks even stricter. The main difference with respect to standard is that it enables the memory management checks presented in Section 16.3.
strict	When this operating mode is selected, SPLINT performs all the checks it is able to. It can profitably be used to thoroughly scrutinize short, critical sections of code. However, as SPLINT authors themselves say [60] it is hard to produce a real program that triggers no warnings when this operating mode is in effect.

invokes SPLINT with a certain set of <flags> on source files a.c and b.c, which the tool considers to be part of the same program, and hence, analyzes together.

The tool supports a relatively high number of flags, which control different aspects of the checks it performs. For this reason several special flags, called *mode selector* flags and listed in Table 16.1, set the other flags to predefined values, in order to make the checks weaker or stricter. In other words, they provide a convenient, coarse-grained way of generally controlling which classes of errors the tool reports.

After setting a mode selector flag, it is still possible to set or reset individual flags to further configure the tool in greater detail. Instead, doing the opposite triggers a warning.

Another peculiar aspect of SPLINT is how flags are set and reset. In most "unix-style" tools, a flag is set by putting its name on the command line, preceded by either – or –– to distinguish it from other kinds of command-line arguments. For instance, in order to ask a tool to print out some help information, it is pretty common to invoke the tool with the ––help command-line option. Flags are usually reset by putting no– before their name.

Instead, SPLINT flags are preceded by a single character that determines the action to be performed on that flag. Namely:

- The + character turns the flag on.
- The – character turns the flag off.
- The = character can be used in a control comment to reset a flag to the value specified on the command line or to the default value.

Due to space limitations, only an overview of the main checks that SPLINT can perform and a short introduction on how to take advantage of it, by means of examples, will be given here. Readers interested in the fine details or aiming at adopting the tool for production use should refer to the SPLINT manual [60] for further information.

It must also be remarked that, with the steady progress of compiler technology, many of these checks can nowadays be performed by the compiler itself. Chapter 9 lists some flags that can profitably be used for this purpose. However, in many cases, SPLINT checks still have greater precision also thanks to programmers' annotations, which give it additional valuable information that is otherwise unavailable to the compiler.

16.2 BASIC CHECKS

A common source of trouble in programs are mistakes related to pointer handling. Among those, dereferencing a NULL pointer is surely a very likely cause of program failure, along with attempts to gain access, through a pointer, to memory areas that have not been correctly initialized.

In order to illustrate which kinds of pointer check SPLINT can perform unaided, let us start from the following source code module, without annotations. Before proceeding further, it is useful to remark that all the examples presented in this chapter are extremely simplified in order to make them as clear and easy to follow as possible.

As a consequence, at first sight the warnings produced by SPLINT may look obvious and the programming mistakes they point out may seem quite easy to find out, by simply looking at the code. It shall be noted, however, that automatic tools do not get tired and work equally well regardless of how large the source code size is, whereas human beings usually do not.

```c
#include <stdio.h>
#include <stdlib.h>

int *alloc_int(void)
{
    int *p = (int *)malloc(sizeof(int));
    return p;
}

void use_intp(int *p)
{
    printf("%d", *p);
}

void alloc_and_use(void)
{
    int *p = alloc_int();
    use_intp(p);
}
```

The module defines three very simple functions.

1. The function `alloc_int` dynamically allocates an integer variable (of type `int`), by means of the `malloc` C library function, and returns a pointer to it. Local variable p is used to temporarily store the pointer before returning.
2. The function `use_intp`, given a pointer to an integer, makes (trivial) use of it by printing out the integer it points to on the standard output stream.
3. Finally, the function `alloc_and_use` invokes the two previous functions in sequence, passing to the second one the return value of the first one. As before, local variable p is used as temporary storage.

As shown in the following listing, which contains the SPLINT output slightly edited by adding some blank lines for clarity, although the code is very simple, SPLINT still produces 5 warnings when invoked on the module with default flags.

```
splint basic_checks_1.c
Splint 3.1.2

basic_checks_1.c: (in function alloc_int)
basic_checks_1.c:7:12: Possibly null storage p returned as non-null: p
  Function returns a possibly null pointer, but is not declared using
  /*@null@*/ annotation of result.  If function may return NULL,
  add /*@null@*/ annotation to the return value declaration.
  (Use -nullret to inhibit warning)
   basic_checks_1.c:6:14: Storage p may become null

basic_checks_1.c:7:12: Returned storage p not completely defined (*p is
                       undefined): p
  Storage derivable from a parameter, return value or global is not
  defined.  Use /*@out@*/ to denote passed or returned storage which
  need not be defined. (Use -compdef to inhibit warning)
   basic_checks_1.c:6:41: Storage *p allocated

basic_checks_1.c: (in function alloc_and_use)
basic_checks_1.c:19:2: Fresh storage p not released before return
  A memory leak has been detected. Storage allocated locally is not
  released before the last reference to it is lost.
  (Use -mustfreefresh to inhibit warning)
   basic_checks_1.c:17:26: Fresh storage p created

basic_checks_1.c:4:6: Function exported but not used outside
   basic_checks_1:
                        alloc_int
  A declaration is exported, but not used outside this module.
  Declaration can use static qualifier.
  (Use -exportlocal to inhibit warning)
   basic_checks_1.c:8:1: Definition of alloc_int

basic_checks_1.c:10:6: Function exported but not used outside
   basic_checks_1:
                        use_intp
   basic_checks_1.c:13:1: Definition of use_intp

Finished checking --- 5 code warnings
```

Of these, the last two pertain to other kinds of checks, unrelated to pointer handling, which will be better detailed later. Concerning the first three:

1. The first one informs the programmer that function `alloc_int` may return a NULL pointer. If this is deemed to be acceptable because, for instance, the caller

or some other functions are responsible for ensuring that the pointer they receive is indeed not NULL before using it, then the programmer should annotate the return value in an appropriate way, else the code should be modified to address the potential issue.

2. The second warning highlights that, in the same function, the storage pointed by p—that becomes accessible to the caller after the function return—may have undefined content. This is because the library function malloc does not initialize the content of memory areas it allocates. As before, it is possible to disable the warning if this is the intended behavior of the function.

3. The third and last warning indicates that there is a memory leak in function alloc_and_use. This is because the function dynamically allocates some storage, by means of alloc_int. Then, it returns without releasing the storage and without making a valid reference to it available in some other ways, for instance, by returning a pointer to the storage to the caller or storing the pointer into a global variable. As a consequence, all references to the storage are lost and it becomes impossible to release it appropriately.

As shown in the following, modified listing, all three warnings can be addressed in a straightforward way, either by improving the quality of the code or by providing further information about the intended behavior of the code itself.

```c
1   #include <stdio.h>
2   #include <stdlib.h>
3
4   /*@null@*/ int *alloc_int(void)
5   {
6       int *p = (int *)malloc(sizeof(int));
7       if(p != (int *)NULL)   *p = 0;
8       return p;
9   }
10
11  void use_intp(/*@null@*/ int *p)
12  {
13      if(p != (int *)NULL)   printf("%d", *p);
14  }
15
16  void alloc_and_use(void)
17  {
18      int *p = alloc_int();
19      use_intp(p);
20      free(p);
21  }
```

In particular:

- For the sake of this example, we are willing to accept that alloc_int may return a NULL pointer because our intention is to check the pointer

value before *using* it. As a consequence, we annotate the return value with
`/*@null@*/` to indicate that it may be NULL.

- The input argument of `use_intp` has been annotated in the same way to remark that we accept it to be NULL, because the function itself will check it before use.
- In order to address the second warning, in `alloc_int` we explicitly initialize to zero the storage pointed by p after allocation. In order to avoid further warnings related to dereferencing a possible NULL pointer in `alloc_int`, initialization is performed only after checking that allocation was successful, that is, `p != (int *)NULL`.
- In function `use_intp` we ensure that p is not NULL before passing it to `printf`. This avoids further warnings about NULL pointer dereferencing in this function.
- Last, we `free` the storage pointed by p before returning from `alloc_and_use`, thus avoiding the memory leak spotted in the third warning discussed previously.

Going back to the last two original warnings, their meaning is that both `alloc_int` and `use_intp` are globally visible functions but—as far as the tool can tell—they are not used elsewhere in the program, that is, outside the source code module where they are defined. Those warnings can be addressed in two different ways:

1. If this is indeed the case, then it is advisable to use the `static` qualifier in the function definition, to avoid cluttering the global name space without reason.
2. If the warning depends on the fact that the tool has insufficient knowledge about how functions are used—for instance, because it is running only on some modules of a bigger program—it can be suppressed by means of the `+partial` flag.

In addition, SPLINT performs a variety of sophisticated checks aimed at detecting type mismatches, which are stricter than what average compilers do. Of these, probably the most interesting ones concern enumerated and Boolean data types.

The standard C language considers enumerated data types, defined by means of the `enum` keyword, to be equivalent to integers in many respects. As a consequence, it is possible to assign an arbitrary integral value to an `enum` variable, even though that value was not mentioned as an enumerator member. Even assigning a member defined for a certain enumerated data type to another enumerated data type merely triggers a compiler warning, and not in all cases.

Let us consider the following fragment of code as an example.

```
1  /*@ -enumint @*/
2
3  enum a {
4      A_ONE = 1,
5      A_TWO,
6      A_THREE
7  };
```

```
 8
 9   enum b {
10       B_ONE = 1,
11       B_TWO
12   };
13
14   void work_with_enum(void)
15   {
16       enum b x = A_THREE;
17       enum b y = 8;
18   }
```

The code defines two enumerated data types (enum a and enum b), along with their members (A_ONE to A_THREE for enum a, and B_ONE to B_TWO for enum b). According to the C-language specification, the numeric values of these members corresponds to the numbers spelled out in their name.

Then, function work_with_enum defines two local variables (x and y) and assigns a value to them. Both assignments are questionable, for two different reasons.

- The first one, at line 16, assigns to x (which is an enum b) a valid enumeration member A_THREE. However, that member has been defined for the enum a data type, not enum b, and its numeric value does not correspond to any member of enum b.
- The second one, at line 17, assigns to y (which is again an enum b) the integer value 8, which does not correspond to the numeric value of any enum b members.

Since it conforms to the C language standard, neither of those assignments triggers an error from the GCC compiler, not even when it is configured for strict error checks. In particular, it generates a warning about the first assignment and nothing about the second one.

On the contrary, SPLINT produces the following four warnings, when it is configured to consider enumerated data types to be distinct from integers. This is done by means of the control comment at line 1 of the previous listing.

```
splint basic_checks_3.c
Splint 3.1.2 --- 19 Apr 2015

basic_checks_3.c: (in function work_with_enum)
basic_checks_3.c:16:16: Variable x initialized to type enum a
   { A_ONE, A_TWO, A_THREE }, expects enum b: A_THREE
   Types are incompatible. (Use -type to inhibit warning)

basic_checks_3.c:17:16: Variable y initialized to type int,
   expects enum b: 8

basic_checks_3.c:16:12: Variable x declared but not used
   A variable is declared but never used. Use /*@unused@*/ in front of
   declaration to suppress message. (Use -varuse, to inhibit warning)

basic_checks_3.c:17:12: Variable y declared but not used

Finished checking --- 4 code warnings
```

- The first two warnings highlight that the two assignments are questionable, for the reasons explained previously.
- The last two warnings are of little interest because they merely remark that local variables x and y are never used. This is a condition that most compilers can indeed detect and report to the user.

Before the adoption of C99, the informal name of the ISO/IEC 9899:1999 international standard [89], the C programming language did not foresee an explicit Boolean data type, and used integers in its place. In particular:

- The result of a comparison operator was an integer.
- Test expressions accepted integers and pointers as operands.
- Any non-zero integer and any non-NULL pointer were considered to be true, and false otherwise.

Therefore, it was possible to make a confusion between Boolean and integer values, and hence, introduce errors in the code, without receiving any warning from the compiler. C99 introduced a Boolean data type, revolving around the stdbool.h header, but did not specify any stronger type checking.

The following fragment of code declares several bool variables, the Boolean data type in C99, and performs some checks on them by means of the if statement. The control comment at the very beginning of the listing conveys two pieces of information to SPLINT.

1. -booltype bool specifies that the name of the Boolean data type is indeed bool. The option of using a different name for this data type is useful when the program defines its own Boolean data type, as happens, for instance, when the program was written before C99 came into effect.
2. +predboolptr enables one additional check that forbids a test expression from being a pointer.

```
 1    /*@ -booltype bool +predboolptr @*/
 2
 3    #include <stdbool.h>
 4
 5    void work_with_bool(int a, int b)
 6    {
 7        bool x = (a == b);
 8        bool *px = &x;
 9
10        if (px)   return;
11
12        if (a = b)   return;
13
14        if (a + b)   return;
15    }
```

When run on the code fragment, the tool produces the following output.

```
splint basic_checks_4.c
Splint 3.1.2 --- 19 Apr 2015

basic_checks_4.c: (in function work_with_bool)
basic_checks_4.c:10:8: Test expression for if not bool, type bool *: px
  Test expression type is not boolean.
  (Use -predboolptr to inhibit warning)

basic_checks_4.c:12:8: Test expression for if is assignment
  expression: a = b
  The condition test is an assignment expression. Probably, you mean
  to use == instead of =. If an assignment is intended, add an extra
  parentheses nesting (e.g., if ((a = b)) ...) to suppress this message.
  (Use -predassign to inhibit warning)

basic_checks_4.c:12:8: Test expression for if not bool, type int: a = b
  Test expression type is not boolean or int. (Use -predboolint to inhibit
  warning)

basic_checks_4.c:14:8: Test expression for if not bool, type int: a + b

Finished checking --- 4 code warnings
```

- As expected, no warnings are generated about variable declaration plus initialization at line 7. In fact, the result of the comparison a == b is a Boolean and the result is correctly assigned to a Boolean variable. The tool infers that x is indeed a Boolean variable because its type is exactly the one specified by means of -booltype.
- The if statement at line 10 triggers a warning because the test expression is a pointer (that is, px) instead of a Boolean. In this case, according to the C language standard, the pointer is implicitly compared with NULL. Namely, the result of the test expression is true if and only if the pointer was not NULL. Since the comparison is not syntactically evident, this kind of test expression, although perfectly legal from the language point of view, is sometimes considered poor programming practice.
- The next two warnings concern the if statement at line 11. They highlight that the test expression is an *assignment* and its value is an integer instead of a Boolean. This is a well-formed test expression but, since = and == are quite similar to each other, it is also quite likely that its actual meaning is not what was intended by the programmer.
- The fourth and last warning indicates that the test expression of the if statement at line 14 is an integer instead of a Boolean. In fact, it is the result of the arithmetic expression a + b. As before, even though this is legal, there is also a high probability that it is the outcome of a programmer's mistake.

For what concerns standard compilers behavior, at the time of this writing GCC only warns about the if(a = b) statement. This is not surprising because, as re-marked previously, all those statements are totally legal as far as the language is concerned.

The last group of basic checks performed by SPLINT is about questionable *control flow*. Examples of checks belonging to this group are:

- Detection and proper handling of unreachable code.
- Code behavior that depends on expression evaluation order that is not defined by the standard.
- Likely infinite loops.
- Execution fall through within a `switch` statement.
- `break` statements nested within more than one loop or switch.
- Statements with no effect.
- Ignored function return values.

Even though these mistakes seem trivial and easy to spot by code inspection, they are indeed responsible for a fairly large share of programming errors that, as pointed out for instance in [167], may be hard and time-consuming to find.

Regarding automatic checks, it is important to remark that, since the C-language syntax, by itself, does not provide detailed flow control information, many checks are effective and do not produce false warnings only if the program contains a sufficient number of annotations. To better illustrate this point, let us consider the following excerpt of code, which addresses in a different way the issues pointed out by SPLINT about the code shown on page 470.

```
1   #include <stdio.h>
2   #include <stdlib.h>
3
4   void report_error(void);
5
6   int *alloc_int(void)
7   {
8       int *p = (int *)malloc(sizeof(int));
9       if (p == NULL)  report_error();
10      *p = 0;
11      return p;
12  }
13
14  void use_intp(int *p)
15  {
16      printf("%d", *p);
17  }
18
19  void alloc_and_use(void)
20  {
21      int *p = alloc_int();
22      use_intp(p);
23      free(p);
24  }
```

In this case, instead of checking if pointer p is NULL before *using* it, we would like to check it immediately after attempting to *allocate* the dynamic memory it must point to.

To this purpose, at line 9 of the listing—within the `alloc_int` function that is responsible for allocating memory for p—we added a statement to perform the above-mentioned check and call the function `report_error` if memory allocation was unsuccessful.

When called, the function `report_error` reports the error in some ways, and then aborts the program without ever returning to the caller. In this way, the program behaves correctly because, if memory allocation fails, it never makes use of p.

Nonetheless, as shown in the listing that follows, SPLINT warns that it is indeed possible to dereference a possibly NULL pointer.

```
splint +partial basic_checks_5.c
Splint 3.1.2 --- 19 Apr 2015

basic_checks_5.c: (in function alloc_int)
basic_checks_5.c:10:6: Dereference of possibly null pointer p: *p
  A possibly null pointer is dereferenced.  Value is either the result of a
  function which may return null (in which case, code should check it is not
  null), or a global, parameter or structure field declared with the null
  qualifier. (Use -nullderef to inhibit warning)
   basic_checks_5.c:8:14: Storage p may become null

Finished checking --- 1 code warning
```

This false warning is due to the fact that—unless it can prove this is not the case—SPLINT assumes that all functions eventually return to the caller, and hence, program execution always continues after a function call.

Therefore, it is strongly recommended that functions that never return are annotated to improve the quality of the analysis, by adding `/*@ noreturn @*/` before the function prototype.

To conclude this section, Tables 16.2 and 16.3 summarize the main flags and annotations that configure SPLINT basic checks, respectively. Due to lack of space, not all flags listed in the tables have been mentioned and thoroughly described in the text. However, the short description given in the tables ought to provide readers with a starting point for further research on the subject.

16.3 MEMORY MANAGEMENT

Memory management issues are a major source of bugs in C programs. Those bugs are often hard to detect because they may show their effect only sporadically, especially when the program comprises multiple, concurrent tasks, and a long time after the problem occurred.

For instance, using a dynamically-allocated memory area after it has been freed may or may not cause a malfunction depending on whether or not the same memory area, or part of it, has already been reused for other purposes in the meantime.

If dynamically allocated memory is drawn from the same memory pool for all tasks in the system—as often happens in small embedded systems—this depends in

Table 16.2
SPLINT Flags Related to Basic Checks

`usedef`	When set, it enables checks concerning the use of a location before it has been initialized.
`mustdefine`	When this flag is turned on, the tool emits a warning if a function parameter annotated as `out` (see Table 16.3) has not been defined before the function returns.
`impouts`	When this flag is set, unannotated function parameters are implicitly assumed to be annotated as `out`.
`charint`	When set, this flag makes the `char` data type indistinguishable from integers, to avoid false warnings in legacy programs.
`charindex`	When set, the tool allows array indexes to be of type `char` without warnings.
`enumint`	Like `charint`, but for enumerated (`enum`) data types.
`enumindex`	Line `charindex`, but for enumerated data types.
`booltype`	Informs the tool about the name used for the Boolean data type. Flags `booltrue` and `boolfalse` can be used to specify the (symbolic) names of true and false.
`predboolptr`	When set, enables a warning when a pointer is used as a test expression.
`predboolint`	When set, enables a warning when an integer, rather than a Boolean, is used as a test expression.
`predboolothers`	When set, enables a warning when any data type (other than Boolean, pointer, or integer) is used as a test expression. The flag `predbool` can be used to set the three flags just described all together.
`eval-order`	When set, the tool checks and warns the programmer when it finds an expression whose result is unspecified, and may be implementation-dependent, because it depends on its sub-expression evaluation order, which is not defined by the standard.
`casebreak`	When set, the tool warns about cases in which control flow falls through a `switch` statement in a doubtful way.
`misscase`	When set, the tool emits a warning if not all members of an enumerated data type appear in a `switch` statement concerning that data type.
`noeffect`	When set, statements which have no effect are flagged with a warning.
`retvalint`	When set, ignoring an integer function return value triggers a warning.
`retvalbool`	When set, ignoring a Boolean function return value triggers a warning.
`retvalother`	When set, ignoring a function return value whose type is neither integer nor Boolean triggers a warning. The flag `predbool` can be used to set the three flags just described all together.

Table 16.3
SPLINT **Annotations Related to Basic Checks**

`null`	Applies to a pointer and states the assumption that it *may be* NULL.
`notnull`	Applies to a pointer and states the assumption that it *cannot be* NULL.
`relnull`	Relaxes NULL checks for the pointer it refers to.
`nullwhentrue`	This annotation is used for functions that check whether or not a pointer is NULL. This annotation indicates that when the return value of the function is true, then its first argument *was* a NULL pointer.
`falsewhennull`	Analogous to the previous annotation, but it specifies that if the return value of the function is true, then its first argument *was not* a NULL pointer.
`in`	This annotation applies to a pointer and indicates that the storage reachable from it must be defined.
`out`	This annotation applies to a pointer and indicates that the storage reachable from it need not be defined.
`partial`	This annotation indicates the storage reachable from the pointer it applies to, typically a `struct`, may be partially defined, that is, it may have undefined fields.
`reldef`	Relaxes definition checks for the pointer it applies to.
`noreturn`	This annotation denotes a function that never returns to the caller.
`noreturnwhentrue`	This annotation denotes that a function never returns if its first argument is true.
`noreturnwhenfalse`	Analogous to the previous one, but it denotes that a function never returns if its first argument is false.
`fallthrough`	Within a `switch`, it indicates that flow control was left to fall through a `case` statement on purpose, and hence, the occurrence shall not be flagged by the tool.

a complex and hard to predict way not only on the offending task, but also on the memory-related activities of all other tasks in the system.

At the beginning of Section 16.2, we already saw an informal example of how SPLINT can effectively track the value of a pointer and generate a warning if the pointer is NULL or the memory location it points to may be used while it has undefined contents.

In order to proceed further, it is first of all necessary to give some more precise definitions about the storage model adopted by SPLINT [58]. In this model, an *object* is a typed region of storage that holds a C-language variable, for instance, an array. The assignment of storage to objects is a crucial part of memory management. In the C language, it can be performed in two different ways.

1. The storage assigned to, and used by, some objects is *implicitly* managed by the compiler without the programmer's intervention. For instance, storage for local variables is automatically allocated from the task stack when the function is called, and released when it returns.
2. In other cases, storage must be *explicitly* managed by means of appropriate program statements. In other words, the programmer becomes responsible for allocating and releasing storage at the right time. In the C language, this must be done for dynamic memory allocated by `malloc` or similar functions.

A certain region of storage and, by analogy, the object it holds, can be in several different states. Its state evolves according to the operations that the program performs on it.

- A region of storage is *undefined* if it has been allocated to an object but has not (yet) been assigned a value.
- The region becomes *defined* after it has been assigned a value.
- When considering complex objects, like structures or objects containing pointers to other objects, we say that the object is *completely defined* if all objects that can be reached from it are defined.

As shown in Figure 16.1, the matter becomes more complex when we also consider object *pointers*. Even though they are depicted as separate entries in the figure, for the sake of clarity, pointers are themselves objects and are stored in memory like any other object is. It is therefore possible, for instance, to have a pointer Q to another pointer P to an object O. In this case, pointer P assumes a double role, because

1. P is the *object* pointed by Q, and
2. P is a *pointer* to O, too.

Therefore, a pointer shall be viewed as an object containing a memory address that references another object in memory. The operation of taking the value of a pointer to get access to the referenced object is usually called *dereference*.

A pointer may be in several different states depending on its own value, but also on the state of the storage it points to.

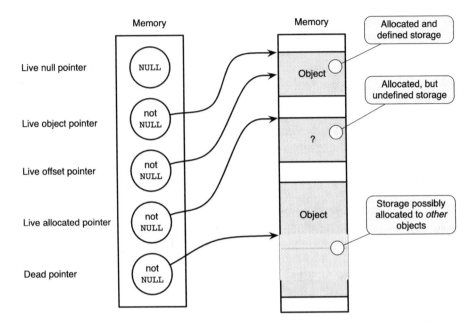

Figure 16.1   Nomenclature of object and object pointer states.

- A first, rough distinction is between *live* pointers and *dead* pointers. Live pointers are often also called *valid* or *legal* pointers, whereas dead pointers are called *dangling*, *invalid*, or *illegal*.
- A null pointer is a pointer that has a special reserved value, corresponding to the macro NULL in the C language. This reserved value indicates that the pointer does not point to any object in memory, and hence, it cannot be dereferenced.
- A live pointer is either a null pointer or a pointer to an allocated area of storage. As shown in the figure, in the second case, the pointer value cannot be NULL by definition.
- Live pointers belong to two different categories, depending on the state of the storage they point to. A pointer to storage that has been allocated to an object and is defined is called an *object* pointer. Object pointers can be dereferenced and the result is to get access to a valid object.
- A special type of object pointer is a pointer that points *within* an object and is called an *offset* pointer. For example, if the object is an array of 5 elements, the address of the third element is a valid offset pointer.

  Offset pointers have the same basic properties as object pointers. However, as better detailed in Section 16.4, extra checks are required to make sure they stay within the underlying object boundaries, especially when they are generated by means of pointer arithmetic.

Table 16.4

Kinds of Pointers and Their Usage

| Kind of pointers | Operation on referenced object | | Allowed |
	Read	Write	in programs
Live null pointer	No	No	Yes
Live object pointer	Yes	Yes	Yes
Live offset pointer	Yes*	Yes*	Yes
Live allocated pointer	No	Yes	Yes
Dead pointer	No	No	No

* Subject to extra checks, see Section 16.4

- Instead, a pointer to an allocated, but undefined area of storage (denoted as a question mark in Figure 16.1) is an *allocated* pointer. For instance, the C library function `malloc` returns a live, allocated pointer.

  Allocated pointers can be dereferenced but the result is an object with undefined contents. Hence, performing this operation and then using object contents (that is, *read* from the object) indicates a memory management anomaly in the program. On the other hand, using an allocated pointer to define object contents (that is, *write* into the object) is correct.
- Finally, a *dead* pointer points to storage that is not allocated. This includes the case of storage that had been formerly allocated to an object with `malloc`, but has then been released by means of `free`.

  As shown at the bottom of Figure 16.1, a dead pointer does not point to a valid object. Furthermore, the storage it points to may have been reused and allocated to other objects. As a consequence, dead pointers should never be used in the program.

Table 16.4 summarizes the kinds of pointers just discussed and their properties. Namely, the second and third columns of the table manifest which operations can be performed on the object they reference, and the rightmost column says whether or not that kind of pointer is allowed at all in the program.

From another point of view, informally speaking, the table summarizes the checks SPLINT is able to perform on pointer usage. In particular, any usage not marked with "yes" in the table triggers a warning.

At the same time, the table highlights a crucial distinction that must be made clear between null and dead pointers. As can be seen, neither of them can be dereferenced because they do not point to a valid object.

However, the null pointer has a fixed, well-defined value (NULL) and it is both possible and easy to determine if a pointer is null or not. For this reason, null pointers are allowed in programs and are quite frequently used in complex data structures—

in which components are bound together by means of pointers—as markers that indicate a special condition.

For instance, in a linked list an element is usually a data structure of which one member is a pointer to the next element of the list. The last element of the list contains a null pointer to indicate that it is indeed the last one and there is no "next" element.

On the contrary, a dead pointer does not have a well-defined value, and hence, it is impossible to determine whether a pointer is live or dead by looking at it. Indeed, the *same value* may correspond to a live pointer at present, and then to a dead one in the future, if the underlying storage is released in the meantime.

For this reason, the mere presence of a dead pointer in a program is a likely indication of an anomaly and it is extremely important to determine if a program action may turn live pointers into dead ones, for instance, by releasing storage (either implicitly or explicitly) while keeping one or more pointers to it. On the other hand, the opposite program error is also possible, that is, the program may fail to release storage even after the last pointer to it has been lost.

This kind of error, often called *memory leak*, is not as dangerous as the previous one—because it cannot cause memory corruption—but it must still be avoided in order to avoid depleting memory in the long term by keeping objects that the program can no longer use in any way. This is especially important in embedded systems, in which memory is usually in scarce supply right from the beginning.

Memory management problems can be solved in an easier way when it can be assumed that there is only one pointer to each object at any given time. This is indicated by the `only` annotation attached to the pointer. In this case, that pointer carries with it the *obligation* to release storage before it is destroyed or assigned a new value.

In C-language programs, all pointers to dynamically allocated storage—as well as all the release obligations that come with them—originate from the `malloc` library function and similar ones. In fact, `malloc` is implicitly annotated by the tool as

```
/*@only@*/ /*@null@*/ void *malloc (size_t s);
```

The `null` annotation, as explained in Section 16.2, indicates that the return value of `malloc` may be a null pointer. This happens when memory allocation fails. On the other hand, `only` indicates that the return value of `malloc` is the only reference to the storage it points to. This makes sense since the storage was just allocated by `malloc` itself and no other references to it are available elsewhere.

After an `only` pointer has been obtained, the obligation attached to it can be transferred to another pointer in several different ways:

- By passing it to a function, as an argument that corresponds to a parameter annotated with `only` in the function prototype.
- By assigning its value to another pointer annotated with `only`.
- By using it as the return value of a function, provided the return value was annotated with `only` in the function prototype.

In all cases, after the obligation has been transferred to a new pointer, the original pointer shall no longer be dereferenced in order to preserve pointer uniqueness, and also because it may have become a dead pointer.

Eventually, the obligation to release storage is satisfied by calling the library function `free`, which is implicitly annotated as

```
void free (/*@only@*/ /*@out@*/ /*@null@*/ void *p);
```

Regarding this function, the annotations bear the following meaning.

- `null` indicates that it is acceptable to call `free` with a null pointer as argument.
- `out` denotes that it is correct to call `free` passing it a pointer to storage that has not been completely defined.
- Finally, `only` indicates that `free` accepts the only pointer to some storage as an argument.

Let us now illustrate, by taking the code fragment that follows as an example, how SPLINT flags memory management mistakes according to Table 16.4.

```
1    #include <stdlib.h>
2    #include <assert.h>
3
4    /*@ -exportheader @*/
5
6    /*@unused@*/ static int *create_stack(void)
7    {
8        int v;
9
10       return &v;
11   }
12
13   static int *create_heap(void)
14   {
15       int *p = (int *)malloc(sizeof(int));
16
17       assert(p != NULL);
18       *p = 0;
19       return p;
20   }
21
22   static void destroy_heap(/*@ only @*/ int *p)
23   {
24       free(p);
25   }
26
27   void use_heap(void)
28   {
29       int *p = create_heap();
30
31       *p = 1;
32
```

```
33      p = create_heap();
34      *p = 2;
35
36      destroy_heap(p);
37      *p = 3;
38  }
```

- The function `create_stack` defines an integer variable v, which is implicitly allocated by the compiler on the calling task stack and returns a
  pointer to it.
  The annotation `/*@unused@*/` suppresses a warning related to the fact
  that this function is defined but never used. The warning would certainly
  be pertinent when examining a real program but it is expected in this case,
  since we are examining only a fragment of example code.
- The function `create_heap` has the same goal as `create_stack`, but it
  explicitly allocates an integer variable from the heap by calling the library
  function `malloc`. After ensuring that memory allocation succeeded, by
  means of `assert`, it initializes the variable to zero and returns a pointer to
  the caller.
- The function `destroy_heap` destroys a variable created by `create_heap`
  by means of the library function `free`. A pointer to the variable to be destroyed must be passed as argument p.
  As explained above, the annotation `/*@ only @*/` passes the obligation
  of releasing the storage associated with p from the caller to the function
  `destroy_heap` itself.
- At the end of the listing, the function `use_heap` makes use of the functions
  `create_heap` and `destroy_heap` to perform several operations on an
  integer variable pointed by pointer p.
  - It first creates the variable by means of `create_heap` at line 29.
  - It assigns the value 1 to the variable at line 31.
  - It creates a new variable at line 33 and reuses p to store a pointer to it.
  - It assigns the value 2 to the new variable at line 34.
  - It destroys the variable at line 36.
  - Nevertheless, it dereferences p to assign the value 3 to the variable.

It should already be clear from the description that this code fragment contains
several serious memory management errors, which SPLINT is able to detect, as highlighted by the tool output that is listed in the following.

```
splint +checks mem_checks_1.c
Splint 3.1.2 --- 19 Apr 2015

mem_checks_1.c: (in function create_stack)
mem_checks_1.c:10:12: Stack-allocated storage &v reachable from return value:
                      &v
  A stack reference is pointed to by an external reference when the function
  returns. The stack-allocated storage is destroyed after the call, leaving a
  dangling reference. (Use -stackref to inhibit warning)
```

```
mem_checks_1.c:10:12: Returned storage &v not completely defined: &v
  Storage derivable from a parameter, return value or global is not defined.
  Use /*@out@*/ to denote passed or returned storage which need not be defined.
  (Use -compdef to inhibit warning)

mem_checks_1.c:10:12: Immediate address &v returned as implicitly only: &v
  An immediate address (result of & operator) is transferred inconsistently.
  (Use -immediatetrans to inhibit warning)

mem_checks_1.c: (in function use_heap)
mem_checks_1.c:33:5: Fresh storage p (type int *) not released before
                    assignment: p = create_heap()
  A memory leak has been detected. Storage allocated locally is not released
  before the last reference to it is lost. (Use -mustfreefresh to inhibit
  warning)
    mem_checks_1.c:29:28: Fresh storage p created

mem_checks_1.c:37:6: Variable p used after being released
  Memory is used after it has been released (either by passing as an only param
  or assigning to an only global). (Use -usereleased to inhibit warning)
    mem_checks_1.c:36:18: Storage p released

Finished checking --- 5 code warnings
```

In particular, the tool produces 5 distinct warnings.

1. The first warning states that `create_stack` returns a dead pointer to the caller. In fact, it is a pointer to local variable v and the storage allocated to it is automatically released when the function itself returns.

2. Furthermore, in the second warning the tool remarks that the storage made accessible to the caller by `create_stack` is not completely defined. In fact, v has not been assigned any value since it was defined. According to the nomenclature introduced previously, this makes a pointer to it an allocated pointer.

   As shown in Table 16.4, allocated pointers are allowed in programs, but are subject to usage restrictions. In particular, the value of the object they reference must not be used.

   For pointers of this kind, the tool requires the annotation /*@out@*/, as mentioned in the warning itself, in order to check them appropriately. This annotation has already been introduced informally in Section 16.2 and Table 16.3.

3. Since there is no explicit annotation regarding it, the return value of `create_stack` is implicitly assumed to be marked with `only`. This is inconsistent with returning an immediate address of a local variable as the responsibility for managing the allocation and release of storage for local variables falls onto the compiler and should never be performed explicitly in the program.

4. In the fourth warning, SPLINT warns that a new value was assigned to p at line 33, without first releasing the storage it was pointing to. Since p contained the only reference to that storage, according to the `only` annotations, this indicates a memory leak.

   In fact, the storage previously pointed by p is still allocated but it will no longer be possible to access it, and not even release it, in the future. As a debugging aid, the tool also points out that the leaked storage was allocated at line 29.

5. The last warning remarks that the program dereferences pointer p after the storage it points to has been released and p became a dead pointer. The line numbers

cited in the warning are the line at which storage was released (line 36) and then dereferenced (line 37).

As was already noted in the previous examples, the checks performed by a standard compiler like GCC are much more limited, also due to the inability to provide annotations. In this case, it is able to detect only the very first issue in the previous list, and not the others.

In real programs, even when working with relatively simple data structures, it is often necessary to have more than one pointer to the same object. Those additional pointers are called *aliases* in the SPLINT documentation.

A typical example is a circular, linked list in which the first element of the list must be accessible by means of a pointer to the head of the list, but it is also referenced by the last element of the list, in order to make the list circular.

In this case, the `only` annotation must be replaced with other, weaker annotations, to be described in the following.

- The annotations `owned` and `dependent` are used together. Of them, `owned` is similar to `only` because it indicates that a pointer carries with it the obligation to release the storage allocated to the object it points to.

  Unlike `only`, though, it does not prevent additional pointers to the same object from being created, provided that they are annotated as `dependent` pointers. When the original pointer is assigned to a `dependent` pointer, the dependent pointer does *not* take the obligation to release storage, which therefore remains with the `owned` pointer.

  In return for additional flexibility in pointer handling, the use of `owned` and `dependent` instead of `only` weakens the checks made by the tool. In particular, it becomes impossible to verify automatically that `dependent` pointers are no longer dereferenced after storage has been released by means of the `owned` pointer, because the tool is unable to keep track of the program control flow accurately enough to this purpose.

- The annotation `keep` can be attached to a function parameter that is a pointer. The meaning is similar to `only` and the main difference lies in how the transfer of obligation to release and the right to use the original pointer after the transfer are handled when the function is called.

  - When a pointer is passed as an `only` parameter, the caller transfers the obligation to release to the called function and loses the right to use the pointer afterward.
  - When a pointer is passed as a `keep` parameter, the transfer of obligation still occurs, but the caller keeps the right to use the pointer afterward.

  Using the `keep` annotation relaxes the automatic checks that the tool performs and makes them weaker. In particular, it leaves open possible memory management issues related to two conflicting behaviors.

  - After the call, the called function—or some other code module called directly or indirectly by it—becomes responsible for freeing the storage referenced by the pointer.

- At the same time, the caller can keep using the original pointer for as long as it wants.

  Therefore, after the call, the storage may be released at any time by the called function, without informing the caller, whereas the caller can keep using a (possibly dead) pointer to it. It is up to the programmer to ensure that this does not happen, because the condition cannot be checked automatically.

- A pointer annotated with shared gives even more freedom to the programmer and further relaxes the checks the tool performs on its usage. Multiple shared pointers can point to the same object and there are no limits on how pointers can be aliased. The only constraint enforced by the tool is that storage should never be released explicitly. This annotation is adequate, for instance, to model a program that makes use of garbage-collected memory management.

- The temp annotation can be attached to a function parameter to indicate that the function uses the corresponding argument, which must be a pointer, only temporarily. When the function is called, the obligation to release storage is *not* transferred to the function. Therefore, SPLINT outputs a warning if the function releases the storage referenced by the pointer or creates aliases of the pointer that are still visible after the function returns.

The difference between only, keep, and temp for what concerns the transfer of obligation to release storage is best illustrated by means of an example.

```
1    #include <stdio.h>
2    #include <stdlib.h>
3    #include <assert.h>
4
5    /*@ -exportheader @*/
6
7    void f(/*@only@*/ int *p);
8    void g(/*@keep@*/ int *p);
9    void h(/*@temp@*/ int *p);
10
11   /*@only@*/ static int *a(void)
12   {
13       int *p = malloc(sizeof(int));
14       assert(p != (int *)NULL);
15       *p = 0;
16       return p;
17   }
18
19   void m(void)
20   {
21       int *pf = a();
22       int *pg = a();
23       int *ph = a();
```

```
24
25        f(pf); g(pg); h(ph);
26
27        printf("%d, %d, %d\n", *pf, *pg, *pg);
28    }
```

Basically, the function m shown above first allocates three integer variables by means of function a and stores the corresponding pointers into variables pf, pg, and ph (lines 21–23).

Function a, in turn, allocates memory by calling malloc (line 13). After verifying that the allocation succeeded (assert at line 14), it initializes the variable to zero (line 15) and eventually returns a pointer to it. The return value is annotated with only because it is the only existing pointer to the variable just allocated.

Then, m calls the three functions f, g, and h, passing pointers pf, pg, and ph to them, respectively (line 25). The prototypes of the three functions (declared at lines 7–9) are identical except for what concerns the annotations of their parameters, which are only for f, keep for g, and temp for h.

Finally, b dereferences the three pointers to print the values of the corresponding variables (line 27).

When executed on the code fragment above, SPLINT outputs two warnings, shown and commented on in the following. The tool has been invoked with the +partial flag set, to prevent spurious warnings about the fact that functions f, g, and h are declared (because we give their prototype), but not defined.

```
splint +checks +partial mem_checks_2.c
Splint 3.1.2 --- 19 Apr 2015

mem_checks_2.c: (in function m)
mem_checks_2.c:27:29: Variable pf used after being released
   Memory is used after it has been released (either by passing as an only param
   or assigning to an only global). (Use -usereleased to inhibit warning)
   mem_checks_2.c:25:7: Storage pf released

mem_checks_2.c:28:2: Fresh storage ph not released before return
   A memory leak has been detected. Storage allocated locally is not released
   before the last reference to it is lost. (Use -mustfreefresh to inhibit
   warning)
   mem_checks_2.c:23:19: Fresh storage ph created

Finished checking --- 2 code warnings
```

1. The first warning highlights that pointer pf has been dereferenced after the storage it points to has been released. In fact, pf was passed as argument to f and the corresponding parameter is annotated as only.

   According to the annotation, the called function took the obligation to release storage and the caller lost the right of using the pointer after the call.

2. The second warning reveals that the storage pointed by ph has not been released correctly before m returns to the caller. This indicates a memory leak because ph is lost upon return and it is the only available reference to that storage.

   In this case, even though function h was invoked with ph as argument, it did not take the obligation to release storage because its parameter is annotated with temp.

Thus, the caller m retained the right of using ph after the call, but also the obligation to release storage, which it did not fulfill.

Instead, no warnings are raised about pointer pg. This is because the parameter of function g is annotated with keep. As a consequence, g took the obligation to release storage (and no warnings are raised about this aspect) but m kept the right of using the pointer after the call (and hence, the use of pg after the call is considered correct).

Besides being often associated with memory management issues, aliasing may also trigger problems within some functions, when their implementation somewhat assumes that arguments do not alias each other, but they do.

A well-known example is represented by the strcpy library function, which copies a ' \0' -terminated source string, pointed by its second argument, into a destination string pointed by the first argument. The behavior of this function when source and destination overlap even partially is *undefined*, because the source—and, even more importantly, its termination character—may be overwritten while the copy is in progress.

In addition to the ones discussed previously, SPLINT supports two annotations to provide information and constrain aliasing of function parameters and return values. In particular:

- When a function parameter is annotated with unique, it shall not be aliased by any other storage that can be reached from within the function body, through either other parameters or global variables. The tool generates a warning when aliasing takes place.
  The meaning of unique is therefore similar to only, but it does not convey any additional information about how the obligation to release storage is transferred.
- When a function parameter is annotated by returned, the information conveyed to the analysis tool is that that parameter may be aliased by the function return value. In turn, this improves the quality of the checks that the tool performs concerning improper use of aliasing.

As an example of how these two annotations work, let us consider the following fragment of code.

```
1   /*@ -exportheader @*/
2
3   int *f(/*@returned@*/ int *p);
4   void g(/*@unique@*/ int *p, /*@unique@*/ int *q);
5
6
7
8   void m(void)
9   {
10          int a;
11          int *b, *c;
```

```
12
13        a = 0;
14        b = &a;
15        c = f(b);
16
17        g(b, c);
18    }
```

In this code, function m performs several operations on local variable a. In particular:

- It defines the variable and initializes it to zero (line 13).
- It defines a pointer b and sets it to the address of a (line 14).
- It defines another pointer c and sets it to the return value of function f, which takes b as argument (line 15).
- It calls function g passing both pointers as arguments.

Taking into account the annotations of f and g, the tool emits the following two warnings:

```
splint +checks +partial mem_checks_3.c
Splint 3.1.2 --- 19 Apr 2015

mem_checks_3.c: (in function m)
mem_checks_3.c:17:7: Parameter 1 (b) to function g is declared unique but is
                aliased externally by parameter 2 (c) through alias b
  A unique or only parameter is aliased by some other parameter or visible
  global. (Use -aliasunique to inhibit warning)
mem_checks_3.c:17:10: Parameter 2 (c) to function g is declared unique but is
                aliased externally by parameter 1 (b) through alias c

Finished checking --- 2 code warnings
```

As can be inferred from the warning messages above, these two warnings are actually symmetric. They remark that pointers b and c, passed to function g, may be aliases of each other.

This is because, due to the presence of the returned annotation for parameter p of f, the return value of f may be an alias of p. Therefore, when f is called by m using b as argument, c may become an alias of b, and both may therefore point to variable a.

Then, m passes pointers b and c to function g upon calling it. However, the corresponding parameters have a unique annotation attached to them, and this triggers the warning.

Tables 16.5 and 16.6 summarize the most important SPLINT flags and annotations related to memory management. It should be noted that not all of them have been discussed in this book for brevity.

16.4   BUFFER OVERFLOWS

Most programs make use of fixed-length memory areas, called *buffers*, to store information. In the C language, they are usually implemented as arrays whose size is

Table 16.5

SPLINT Flags Related to Memory Management

`stackref`	When this flag is turned off, warnings about returning the address of variables allocated on the stack are suppressed.
`compdef`	When this flag is off, the tool does not warn the programmer when a function returns a pointer to storage that is not completely defined.
`immediatetrans`	When this flag is off, no warnings are given when an immediate object address, obtained with the & operator, is used in an inconsistent way.
`mustfree`	When off, the tool does not produce any warning about memory leaks.
`usereleased`	When off, using storage after it has been released does not trigger a warning.

constant and is determined when they are defined. Then, buffer contents are accessed by referring to array elements by means of an integer index.

As any other variable, buffers are surrounded by other objects in memory. Hence, accessing an array element with an invalid index (either lower than zero or higher than the number of array elements minus one) is illegal because it references storage that is outside the region allocated to the array.

In particular, read operations may cause part of the program to use seemingly random information and malfunction. Write operations usually result in memory corruption that, in some cases, can be exploited to make the program execute arbitrary, malicious code. In fact, it has been estimated that buffer overflows are responsible for about 50% of all security attacks [107].

As happens for memory corruption in general, buffer overflows are often difficult to detect because their effects, especially in a concurrent system, may be different from one program execution to another. Moreover, they are inherently data-dependent and may not show up during testing.

The general techniques used by static analysis tools—and SPLINT in particular—to detect buffer overflows are quite complex and thoroughly discussing them is beyond the scope of this book. Here, they will mainly be described in an intuitive, rather than formal way, and the results they can achieve will be shown by means of simple examples. Interested readers should refer to more specialized literature, for instance [59, 107], for further information about this topic.

Informally speaking, in order to perform buffer overflow analysis, SPLINT tags buffers with two properties. If b is a buffer, then:

- The property maxSet(b) represents the highest index of b that can legally be set, by using it as the target of an assignment.

Table 16.6

SPLINT **Annotations Related to Memory Management**

only	This annotation indicates that a pointer is the *only* reference to the object it points to. Therefore, the pointer has attached to it the obligation of releasing the storage associated to the object.
owned	This annotation indicates that a pointer has attached to it the obligation of releasing the storage pointed by it. Unlike only, it is however possible to have other pointers to the same storage, provided they are annotated as dependent (see below).
dependent	This annotation indicates that a pointer references storage to which other pointers refer, too. One of those other pointers, annotated with owned, has the obligation to release the storage.
keep	This annotation applies to a function parameter that is a pointer. Like only, it indicates that the function takes the obligation of releasing the storage associated to the referenced object. Unlike only, the caller can however keep using the original pointer.
shared	This annotation indicates that a pointer points to storage that has one or more pointers to it and it is never explicitly released, as happens in garbage-collected memory management systems.
temp	This annotation applies to a function parameter that is a pointer, like keep does. It indicates that the function uses the pointer only temporarily, and hence, it does not take the obligation of releasing the storage associated to the referenced object.
unique	This annotation is attached to a function parameter and denotes that the parameter shall not be aliased by any storage reachable from within the function body. Unlike only, this annotation does not imply any obligation to release storage.
returned	This annotation, when attached to a function parameter, indicates that the parameter may be aliased by the function return value, and hence, the call should be checked for correctness accordingly.

- The property maxRead(b) denotes the highest index of b that can legally be read. Namely, it is considered illegal to read a buffer element that is beyond the highest-index element that has been initialized or, for character strings, any element beyond the ' \0' terminating character.

In general, the execution of a statement involving a certain buffer b requires certain Boolean predicates called *preconditions*—related to the properties of b—to be satisfied in order to be legal. In turn, statement execution may make true additional predicates, called *postconditions*, concerning b.

In other words, preconditions can be seen as constraints that must be satisfied for a statement to be legal. Then, postconditions made true by the execution of a certain legal statement can be leveraged to prove that the preconditions of subsequent statements are also satisfied, and so on. Informally speaking, static verification proceeds in this way, trying to prove that all statements belonging to a block of code—for instance, the body of a function—are indeed legal.

The tool is able to establish preconditions and generate preconditions for a variety of C statements. In addition, functions can be annotated to specify their preconditions and postconditions, by means of the `requires` and `ensures` annotations, respectively. C library functions are implicitly annotated in this way, too, and no programmer's intervention is needed for them.

For instance, when analyzing the statement

```
static int b[10];
```

the tool establishes that there are no preconditions for it to be legal, and it generates the postcondition

$$maxSet(b) == 9 \tag{16.1}$$

because the last allocated element of the array is at index 9.

As a further example, let us consider the implicit `requires` and `ensures` annotations attached automatically by the tool to the well-known `strcpy` library function:

```
void strcpy(char *s1, char *s2)
/*@requires maxSet(s1) >= maxRead(s2) @*/
/*@ensures maxRead(s1) == maxRead (s2) @*/;
```

- The `requires` annotation specifies the precondition, regarding the arguments passed to `strcpy`, which shall be true when the function is called. Namely, the precondition

$$maxSet(s1) >= maxRead(s2) \tag{16.2}$$

  indicates that the index of the highest element that can legally be written into the argument passed to `strcpy` as parameter s1 (the destination of the string copy) must be at least as high as the highest element that can legally be read from the argument passed as parameter s2 (the source of the string copy).
- The `ensures` annotation specifies the postcondition made true by the execution of `strcpy` after the function returns to the caller.
  In this example, the postcondition

$$maxRead(s1) == maxRead (s2) \tag{16.3}$$

  indicates that after string s2 has been copied into s1 the highest element that can legally be read from s1 is the same as s2, because the contents of both strings are now identical.

Furthermore, SPLINT is also able to determine automatically the values of maxRead for literals. Hence, if the literal string "abc" is passed as argument s2 to strcpy, the tool adds

$$maxRead(s2) \ == \ 3 \qquad (16.4)$$

to the list of postconditions that are true at the function call point and can be used to satisfy the function's preconditions. It must be noted that the value of maxRead is 3 and not 4 because the literal is terminated by a ' \0' character, which can be legally read, but arrays are indexed from zero in the C language.

As a consequence, the three characters "abc" are at indexes from 0 to 2, inclusive, and the terminating ' \0' character is at index 3.

Let us now consider the following fragment of code:

```
1   /*@ -exportheader @*/
2
3   static char b[10];
4
5   void m(void)  /*@modifies b@*/
6   {
7       char x;
8       int i;
9
10      strcpy(b, "abc");
11
12      x = b[2];
13
14      i = 4;
15      x = b[i];
16
17      strcpy(b, "abcde");
18      x = b[i];
19
20      b[2*i] = 'x';
21      b[3*i] = 'x';
22  }
```

- In this fragment of code, function m operates on three variables:
  1. The global array of characters b, with 10 elements.
  2. The local character variable x.
  3. The local integer variable i.

- The function first initializes b by copying a 3-character string (plus the terminating ' \0' character) into it. The copy is performed by means of the strcpy library function (line 10).
- Afterward, the function reads two elements from the character string, one after another, and assigns them to x.
  1. First, it reads element 2, using an integer literal as index (line 12).

2. Then, it reads element 4, but this time is uses variable i as index, after setting it to 4 (lines 14–15).

- At line 17, the function changes the contents of b by copying another literal string into it. This time, the string is 5 characters long (plus the terminating `'\0'` character).
- At this point, the function reads again element 4 of b, as before (line 18).
- The last two statements of m (lines 20–21) store a character into b at two different positions determined by performing a simple calculation on i.
  1. Position 2 * i, that is, 8.
  2. Position 3 * i, that is, 12.

When invoked on the fragment of code just illustrated, the tool produces the following output.

```
splint +checks +bounds +partial buffer_overflow_1.c
Splint 3.1.2 --- 19 Apr 2015

buffer_overflow_1.c: (in function m)
buffer_overflow_1.c:15:9: Likely out-of-bounds read: b[i]
    Unable to resolve constraint:
    requires 3 >= 4
     needed to satisfy precondition:
    requires maxRead(b @ buffer_overflow_1.c:15:9) >= i @ buffer_overflow_1.c:15
    :11
 A memory read references memory beyond the allocated storage. (Use
 -likelyboundsread to inhibit warning)

buffer_overflow_1.c:21:5: Possible out-of-bounds store: b[3 * i]
    Unable to resolve constraint:
    requires 3 * i @ buffer_overflow_1.c:21:7 <= 9
     needed to satisfy precondition:
    requires maxSet(b @ buffer_overflow_1.c:21:5) >= 3 * i @
    buffer_overflow_1.c:21:7
 A memory write may write to an address beyond the allocated buffer. (Use
 -boundswrite to inhibit warning)

Finished checking --- 2 code warnings
```

Let us now informally follow the procedure followed by SPLINT to emit the two warnings that appear in the output and justify why they indeed indicate issues in the code itself.

- Upon analyzing the definition of b, the tool asserts the postcondition maxSet(b) == 9, as in (16.1).
- When the strcpy function is called at line 10, the tool tries to prove that its precondition (16.2) is true. This can be done because:
  - From (16.1) it is possible to derive the postcondition maxSet(s1) == 9 because b corresponds to parameter s1 in the function call.
  - From (16.4), it is maxRead(s2) == 3.
  By substituting these postconditions in the precondition (16.2), we obtain

$$9 >= 3 \qquad\qquad (16.5)$$

that is obviously true. From this, the tool concludes that the call to `strcpy` is legal and adds

$$maxRead(b) \ == \ 3 \tag{16.6}$$

to the list of postconditions. This postcondition is derived from (16.3) by back-substituting parameter and argument names, and replacing `maxRead(s2)` with its known value.

- In order to verify that the assignment at line 12 is valid, the tool consults the current postconditions concerning `maxRead(b)` to prove the precondition of the assignment, that is,

$$maxRead(b) \ >= \ 2 \ . \tag{16.7}$$

Due to (16.6) this is true because $3 \ >= \ 2$ and hence, the assignment is considered to be legal. The assignment statement does not generate any further postcondition.

- The tool follows the same procedure to analyze the next operation on `b`, at line 15. In this case, the precondition to be satisfied is

$$maxRead(b) \ >= \ i \ . \tag{16.8}$$

By inspecting the code, the tool concludes that the current value of `i` is `i == 4` due to the assignment at line 14. Moreover, (16.6) is still true. By substitution, the precondition becomes

$$3 \ >= \ 4 \tag{16.9}$$

and it is therefore not satisfied. For this reason, SPLINT emits the first warning shown in the above listing.

- When analyzing line 17, the tool proceeds exactly like it did for the previous call to `strcpy`. The only difference is that, due to the different length of the literal passed to `strcpy` as `s2` (5 characters instead of 3), the preconditions and postconditions are modified accordingly.

  The conclusion is that the function call is legal and establishes the postcondition

$$maxRead(b) \ == \ 5 \ . \tag{16.10}$$

- The statement at line 18 is syntactically the same as the one at line 15, which was flagged as illegal. However, the statement at line 18 shall be analyzed according to the *current* set of postconditions, which leads to a different result.

  In particular, the most recent postcondition concerning `maxRead(b)` is now (16.10) instead of (16.6) and precondition (16.8) is satisfied in this case.

- The two assignments to elements of `b` at lines 20 and 21 are analyzed in a similar way. In this case, the preconditions to be satisfied concern `maxSet(b)` and are

$$maxSet(b) >= 2*i \; , \quad and \quad (16.11)$$

$$maxSet(b) >= 3*i \; , \quad (16.12)$$

respectively.

The value of `maxSet(b)` can readily be assessed from the postcondition established by the definition of `b` (16.1). On the other hand, it is possible to calculate the values of $2*i$ and $3*i$ from the current value of `i`, obtaining the values 8 and 12, respectively. By substituting these values back into (16.11) and (16.12), the result is

$$9 >= 8 \; , \quad and \quad (16.13)$$

$$9 >= 12 \; . \quad (16.14)$$

Of these preconditions, the second one is clearly not satisfied and triggers a warning from the tool.

As may already be clear from the example just illustrated, the checks SPLINT performs are extremely strict and may lead to a high number of false warnings unless all the code is thoroughly and properly annotated as required by the tool.

For this reason, the tool offers the ability to classify warnings into two different categories—according to a heuristic that tries to determine their likelihood of indicating a real flaw in the program—and selectively suppress them. The two categories correspond to different conclusions that can be drawn about constraints.

1. Unresolved constraints that can be directly reduced to a plain numerical inconsistency by substitution, like the first one discussed in the example, are considered to be "likely bound errors" and always trigger a warning when buffer overflow checks are enabled, unless one of the flags starting with `likelybounds` is reset.
2. Constraints that cannot be directly reduced to a numerical inconsistency like the second one discussed in the example—because it also requires an index calculation concerning variable `i` rather than a simple substitution—are considered to be "less likely" to indicate a real error. Accordingly, they trigger a warning only if bound checks are completely enabled by means of one of the flags starting with `bound`.

To conclude this section, Tables 16.7 and 16.8 summarize the main flags and annotations related to buffer overflow checks, respectively.

16.5 FUNCTION INTERFACE ANNOTATIONS

In modern high-level programming languages, functions communicate with their caller through a well-defined *interface*. The interface specifies the number and type of the arguments that the caller must pass to the function, as well as the type of the function return value. Further information exchange between the caller and the called

Table 16.7
SPLINT **Flags Related to Buffer Overflow Checks**

bounds	This flag sets both boundsread and boundswrite. Hence, it enables all kinds of buffer overflow checks that the tool is able to perform.
boundsread	When set, the tool produces a warning for any attempt to *read* from a buffer at a position that may lie beyond the bounds of allocated storage, because the buffer access constraints determined by the tool cannot be proven true.
boundswrite	When set, the tool produces a warning for any attempt to *write* into a buffer at a position that may lie beyond the bounds of allocated storage, because the buffer access constraints determined by the tool cannot be proven true.
likelybounds	This flag sets both likelyboundsread and likelyboundswrite. It enables a less stringent form of buffer overflow checks with respect to bounds, which is weaker but also less likely to produce false warnings.
likelyboundsread	When set, the tool produces a warning for any attempt to *read* from a buffer at a position that is likely to lie beyond the bounds of allocated storage, because it induces a numerical inconsistency in the buffer access constraints determined by the tool.
likelyboundswrite	When set, the tool produces a warning for any attempt to *write* into a buffer at a position that is likely to lie beyond the bounds of allocated storage, because it induces a numerical inconsistency in the buffer access constraints determined by the tool.

function may take place through global variables and, more in general, any storage reachable from both parties.

In the C language, the interface to a function is defined by its function *prototype*. In early versions of the language the prototype was extremely simple and only described the type of the function return value, but it was later extended to also define the number and type of its arguments.

By means of annotations, SPLINT provides programmers a way to express further information about the following aspects of a function interface, not covered by its prototype but still very important.

- Which parts of the storage accessible through its arguments a function may use or modify, and how.
- Whether or not the function has a persistent internal state and modifies it.
- Which global variables the function modifies.

Table 16.8

SPLINT Annotations Related to Buffer Overflow Checks

`requires`	Attached to a function, establishes a precondition on the function's arguments that must be satisfied when the function is called. The precondition may contain references to the `maxSet` and `maxRead` properties of the arguments.
`ensures`	Attached to a function, states a postcondition on the function's arguments and, possibly, return value, which the function makes true upon return. The postcondition may assert facts about the `maxSet` and `maxRead` properties of the arguments and the return value.

- Whether or not the function indirectly modifies the system state as a whole.
- Specify any assumptions made by the function about the state of its arguments and global variables when it is called.
- Assert which predicates—concerning arguments, return value, and global variables—the function makes true at the caller site when it returns.

The main annotations concerning function interfaces are summarized in Table 16.9. The first kind of annotation, introduced by the keyword `modifies` lists which variables visible to the caller the function *may* modify. Implicitly, the tool assumes that any variables *not* mentioned in the list cannot be modified.

For example, if function `f` is annotated in this way:

```
1   struct a
2   {
3       int x;
4       int y;
5   };
6
7   void f(struct a *p) /*@modifies p->y @*/
8   {
9       p->x = 3;
10      p->y = 5;
11  }
```

it means that `f` may modify field `y` of the structure passed as argument ans pointed by `p` but it cannot modify any other field, for instance `x`. In fact, when run on the code fragment above, SPLINT produces the following warning:

```
splint +checks -exportheader int_checks_1.c
Splint 3.1.2 --- 19 Apr 2015

int_checks_1.c: (in function f)
int_checks_1.c:9:5: Undocumented modification of p->x: p->x = 3
```

Table 16.9

SPLINT **Annotations for Function Interfaces**

Argument and return value handling

modifies	Specifies which variables visible to the caller the function may modify.
special	Indicates that an argument or return value is further annotated using the more specific *state clauses* listed below.
uses	The function makes use of a variable, and hence, it must be completely defined when the function is called.
sets	The function assigns a value to a variable. Storage for that variable must have already been allocated at the time of the call and the variable must be completely defined when the function returns.
allocates	The function allocates storage for a variable. It is an error if storage for the same variable has already been allocated at the time of the call.
defines	The function allocates storage and assigns a value to a variable.
releases	The function releases the variable. The variable is therefore assumed to be defined when the function is called and it can no longer be used after the function returns.

State modification (special modifies names)

internalState	The function has an internal state and, when called, may modify it in a way that may affect its future computation.
fileSystem	The function may modify the "file system." This term shall be interpreted in a broad sense, and refers to any modifications to system state items external to the program.
nothing	The function does not modify any variable, that is, it has no side effects.

Global variables

globals	This annotation is attached to a function and contains the list of global variables that the function may use.
checkedstrict	This one and the following three annotations are attached to a global variable, in order to control how strictly SPLINT shall check accesses to it. This annotation selects the strictest checks, so that all undocumented accesses to the global variable are flagged.
checked	Any undocumented accesses to the variable are reported only for functions annotated with a globals or modifies list.
checkmod	Undocumented *modifications* of the global variable trigger a warning, but undocumented *uses* do not.
unchecked	This annotation disables all checks on the global variable it refers to.

```
An externally-visible object is modified by a function, but not
listed in its modifies clause. (Use -mods to inhibit warning)

Finished checking --- 1 code warning
```

It should be noted that this mechanism is more powerful than the `const` qualifier foreseen by the C language. In fact, it is possible to declare `void f(const struct a *p)` to indicate that `f` does not modify the structure pointed by `p` at all, but not on a field-by-field basis.

In the same way, it is possible to attach the `const` qualifier to the `struct` type definition as a whole or referring to individual fields, but then it applies to *all* functions using the structure. Hence, it is not possible to override it on a function-by-function basis.

More specific annotations, listed in Table 16.9 but not discussed here due to lack of space, allow programmers to specify how the function handles its arguments and the return value at an even finer level of detail, also in relation to memory management, discussed in Section 16.3. They are especially useful when the function receives references to variables by means of pointers.

As outlined previously an important aspect of a function interface, which is not mentioned at all in its standard C-language prototype, is whether or not the function modifies other parts of the program state besides those reachable through its arguments. To this purpose, SPLINT supports three special names, to be used with the `modifies` annotation.

- The name `internalState` specifies that a function has got a hidden internal state and uses/modifies it. For instance, it defines some `static` variables, whose value persists from one function call to another and may affect its future computations.

  In other words, since the result of the function depends not only on its arguments, but also on its internal state, the results of two calls to the same function may differ even though the function was given exactly the same arguments.

  This item of information is therefore important when SPLINT checks whether or not the result of an expression may depend on the order of evaluation of its sub-expressions, and also when it tries to prove that a certain statement is side-effect free.
- The name `fileSystem` indicates that a function modifies the file system state or, more in general, change the system state. For instance, a function may create a new file, write new contents into a file, or delete a file. The consequences on the analysis performed by SPLINT are the same as for `internalState`.
- A function annotated with `modifies nothing` is completely side-effect free, and hence, it can affect the caller's computation only through its return value.

Last, but not least, a sometimes unintended interaction between a function and the outside world is through *global variables*. Checking global variables for correct, intended use involves adding two kinds of annotations to the code:

1. Functions using the value of some global variables shall be accompanied by a `globals` annotation containing the list of global variables they use. If they also modify some global variables, those variables must be cited in a `modifies` annotation, as described earlier. Stating that a function modifies a certain global variable implies that the function uses it, too. Hence, if a variable is in the `modifies` list, it is unnecessary to mention it in `globals` as well. Functions for which neither `globals` nor `modifies` annotations are provided are checked in a special way, as described in the following.
2. Global variables may be annotated to declare in which way accesses must be checked. When no annotations are given for a certain variable, an annotation is added implicitly depending on flag settings.

The possible annotations that can be attached to global variables are summarized at the bottom of Table 16.9 and discussed in more detail in the following.

- All undocumented accesses to a global variable annotated with `checkedstrict` trigger a warning. An access can be either a *use* of the variable or a *modification*. It is considered to be undocumented when it is performed either by a function that does not reference that variable in its `globals` or `modifies` list, or by a function for which neither `globals` nor `modifies` annotations are provided at all.
- Global variables annotated with `checked` are checked in a more relaxed way. Namely, they are *not checked* when they are used or modified within a function for which neither `globals` nor `modifies` annotations are provided.
- When a global variable is accompanied by a `checkmod` annotation, only *modifications* to the variable are checked, whereas use checks are disabled.
- Global variables annotated with `unchecked` are not checked at all. Hence, both uses and modifications of those variables are allowed from anywhere in the code, without restrictions.

In order to show how global variable checks work, let us consider the following fragment of code.

```
1   /*@ checkedstrict @*/ int a;
2   /*@ checkedstrict @*/ int b;
3   /*@ unchecked @*/    int c;
4
5   int bad_use(void) /*@globals a @*/
6   {
7       return a + b + c;
8   }
9
```

```
10   void bad_mod(void) /*@globals a, b @*/
11   {
12         c = b;
13         a = b;
14   }
15
16   void good_mod(void) /*@modifies a @*/ /*@globals b @*/
17   {
18         c = b;
19         a = b;
20   }
```

For the sake of the example, the code fragment defines three global variables (a, b, and c). According to their annotations, a and b shall be checked in the strictest way, whereas c shall not be checked at all. Moreover, it also defines three functions.

- bad_use makes use of all three global variables, because it calculates their sum and returns it to the caller. On the other hand, its annotation states that it only uses global variable a.
- bad_mod copies the value of b into a and c. The annotation states that it uses variables a and b.
- good_mod is identical to bad_mod except for the annotation. In this case, the annotation states that the function uses b and modifies a.

As can be seen from the listing that follows, SPLINT identifies and reports two issues with the code just discussed. The tool was invoked with the −exportheader and −exportlocal command-line options to avoid spurious warning related to the fact that the fragment of code, being just an example, defines globally visible functions without declaring them in an appropriate header and defines global variables that are not used elsewhere.

```
splint +checks -exportheader -exportlocal int_checks_2.c
Splint 3.1.2 --- 19 Apr 2015

int_checks_2.c: (in function bad_use)
int_checks_2.c:7:16: Undocumented use of global b
  A checked global variable is used in the function, but not listed in its
  globals clause. By default, only globals specified in .lcl files are checked.
  To check all globals, use +allglobals. To check globals selectively use
  /*@checked@*/ in the global declaration. (Use -globs to inhibit warning)

int_checks_2.c: (in function bad_mod)
int_checks_2.c:13:5: Undocumented modification of a: a = b
  An externally-visible object is modified by a function, but not listed in its
  modifies clause. (Use -mods to inhibit warning)

Finished checking --- 2 code warnings
```

1. The first message remarks that, contrary to what has been stated in its annotation, function bad_use makes use of global variable b, which is subject to strict access checks according to the annotation checkedstrict.

2. The second message highlights that function `bad_mod` modifies the value of global variable `a`, whereas the function annotation states that it should only use it. In fact, `a` is mentioned in the `globals` annotation, but not in the `modifies` annotation. Actually, there is no `modifies` annotation at all for `bad_mod`.

On the other hand, no warnings are raised for the `good_mod` function, because its annotation correctly states that it modifies `a`.

Futhermore, no warnings are given concerning variable `c`, even though all functions use or modify it, because it has been annotated as `unchecked`.

16.6 SUMMARY

Static code analysis tools are nowadays able to perform a wide variety of checks, ranging from basic ones (like the ones discussed in Section 16.2) to very complex ones (for instance, buffer overflow checks described in Section 16.4).

In addition, as shown in Section 16.5, when the source code is properly annotated, they can also verify that different source code modules (possibly written by different groups of programmers at different times) interface in a consistent way. This aspect is becoming more and more important every day, as it becomes more and more common to build new applications (especially when they are based on open-source software) starting from existing components that are migrated from one project to another.

Static code analysis is therefore a useful technique to improve source code quality, reliability, and alleviate security concerns. Another important feature of this kind of analysis is that it does not require any kind of runtime support and—perhaps even more importantly for embedded software developers—it does not introduce any overhead on program execution on the target system.

Last, but not least, the availability of free, open-source tools, like the one discussed in this chapter, further encourages the adoption of static code analysis even when developing very low-cost applications, in which acquiring a commercial tool may adversely affect project budget.

References

1. Alfred V. Aho, Monica S. Lam, Ravi Sethi, and Jeffrey D. Ullman. *Compilers: Principles, Techniques, and Tools*. Pearson Education Ltd., Harlow, England, 2nd edition, September 2006.

2. James H. Anderson and Mark Moir. Universal constructions for large objects. *IEEE Transactions on Parallel and Distributed Systems*, 10(12):1317–1332, 1999.

3. James H. Anderson and Srikanth Ramamurthy. A framework for implementing objects and scheduling tasks in lock-free real-time systems. In *Proc. 17th IEEE Real-Time Systems Symposium*, pages 94–105, December 1996.

4. James H. Anderson, Srikanth Ramamurthy, and Kevin Jeffay. Real-time computing with lock-free shared objects. In *Proc. 16th IEEE Real-Time Systems Symposium*, pages 28–37, December 1995.

5. ANSI/INCITS. *ANSI/INCITS 408-2005 – Information Technology – SCSI Primary Commands – 3 (SPC–3)*, 2005.

6. ARM Ltd. *ARM System Memory Management Unit Architecture Specification — SMMU architecture version 2.0*, January 2-15. IHI 0062D.a.

7. ARM Ltd. *ARM PrimeCell^{TM} Vectored Interrupt Controller (PL192) — Technical Reference Manual*, December 2002. DDI 0273A.

8. ARM Ltd. *ARMv7-M Architecture Reference Manual*, February 2010. DDI 0403D.

9. ARM Ltd. *Cortex^{TM}-M3 Technical Reference Manual, rev. r2p0*, February 2010. DDI 0337H.

10. ARM Ltd. *Cortex^{TM}-M4 Devices, Generic User Guide*, December 2010. DUI 0553A.

11. Atmel Corp. *AT 91 ARM Thumb-based Microcontrollers*, 2008.

12. Atmel Corp. *8-bit AVR Microcontroller with 128K bytes In-System Programmable Flash – ATmega 1284P*, 2009.

13. Neil C. Audsley, Alan Burns, Mike Richardson, and Andy J. Wellings. Hard real-time scheduling: The deadline monotonic approach. In *Proc. 8th IEEE Workshop on Real-Time Operating Systems and Software*, pages 127–132, 1991.

14. Neil C. Audsley, Alan Burns, and Andy J. Wellings. Deadline monotonic scheduling theory and application. *Control Engineering Practice*, 1(1):71–78, 1993.

15. Christel Baier and Joost-Pieter Katoen. *Principles of Model Checking*. The MIT Press, Cambridge, MA, 2008.

16. Theodore P. Baker and Alan Shaw. The cyclic executive model and Ada. In *Proc. IEEE Real-Time Systems Symposium*, pages 120–129, December 1988.

17. Richard Barry. The FreeRTOS.org project. Available online, at http://www.freertos.org/.

18. Richard Barry. *Using the FreeRTOS Real Time Kernel – Standard Edition*. Lulu Press, Raleigh, North Carolina, 1st edition, 2010.

19. Richard Barry. *The FreeRTOS^{TM} Reference Manual*. Real Time Engineers Ltd., 2011.

20. Mordechai Ben-Ari. *Principles of the Spin Model Checker*. Springer-Verlag, London, 2008.

21. Johan Bengtsson, Kim Larsen, Fredrik Larsson, Paul Pettersson, and Wang Yi. Uppaal – a tool suite for automatic verification of real-time systems. In *Hybrid Systems III, LNCS 1066*, pages 232–243. Springer-Verlag, 1995.

22. Dragan Bosnacki and Dennis Dams. Integrating real time into Spin: A prototype implementation. In *Proc. FIP TC6 WG6.1 Joint International Conference on Formal Description Techniques for Distributed Systems and Communication Protocols and Protocol Specification, Testing and Verification*, pages 423–438, 1998.

23. Dragan Bosnacki, Dennis Dams, Leszek Holenderski, and Natalia Sidorova. Model checking SDL with Spin. In *Proc. 6th International Conference on Tools and Algorithms for Construction and Analysis of Systems*, pages 363–377, 2000.

24. Robert Braden, editor. *Requirements for Internet Hosts — Communication Layers, RFC 1122*. Internet Engineering Task Force, October 1989.

25. Alan Burns and Andy Wellings. *Real-Time Systems and Programming Languages*. Pearson Education, Harlow, England, 3rd edition, 2001.

26. Giorgio C. Buttazzo. *Hard Real-Time Computing Systems. Predictable Scheduling Algorithms and Applications*. Springer-Verlag, Santa Clara, CA, 2nd edition, 2005.

27. William Cary Huffman and Vera Pless. *Fundamentals of Error-Correcting Codes*. Cambridge University Press, February 2010.

28. Per Cederqvist et al. *Version Management with CVS, for CVS 1.12.13*. Free Software Foundation, Inc., 2005.

29. Gianluca Cena, Marco Cereia, Ivan Cibrario Bertolotti, and Stefano Scanzio. A Modbus extension for inexpensive distributed embedded systems. In *Proc. 8th IEEE International Workshop on Factory Communication Systems*, pages 251–260, May 2010.

30. Gianluca Cena, Ranieri Cesarato, and Ivan Cibrario Bertolotti. An RTOS-based design for inexpensive distributed embedded system. In *Proc. IEEE International Symposium on Industrial Electronics*, pages 1716–1721, July 2010.

31. Gianluca Cena, Ivan Cibrario Bertolotti, Tingting Hu, and Adriano Valenzano. Fixed-length payload encoding for low-jitter Controller Area Network communication. *IEEE Transactions on Industrial Informatics*, 9(4):2155–2164, 2013.

32. Gianluca Cena, Ivan Cibrario Bertolotti, Tingting Hu, and Adriano Valenzano. A mechanism to prevent stuff bits in CAN for achieving jitterless communication. *IEEE Transactions on Industrial Informatics*, 11(1):83–93, February 2015.

33. Steve Chamberlain and Cygnus Support. *Libbfd — The Binary File Descriptor Library*. Free Software Foundation, Inc., 2008.

34. Steve Chamberlain and Ian Lance Taylor. *The GNU linker ld (GNU binutils) Version 2.20*. Free Software Foundation, Inc., 2009.

35. ChaN. *FatFs Generic FAT File System Module*, 2012. Available online, at `http://elm-chan.org/fsw/ff/00index_e.html`.

36. Joachim Charzinski. Performance of the error detection mechanisms in CAN. In *Proc. 1st International CAN Conference*, pages 20–29, September 1994.

37. Ben Chelf and Christof Ebert. Ensuring the integrity of embedded software with static code analysis. *IEEE Software*, 26(3):96–99, 2009.

38. Ivan Cibrario Bertolotti and Tingting Hu. Real-time performance of an open-source protocol stack for low-cost, embedded systems. In *Proc. 16th IEEE International Conference on Emerging Technologies and Factory Automation*, pages 1–8, September 2011.

39. Ivan Cibrario Bertolotti and Gabriele Manduchi. *Real-Time Embedded Systems: Open-Source Operating Systems Perspective*. CRC Press, Taylor & Francis Group, Boca Raton, FL, 1st edition, January 2012.

40. Compaq Computer Corp., Hewlett-Packard Company, Intel Corp., Lucent Technologies Inc., Microsoft Corp., NEC Corp., Koninklijke Philips Electronics N.V. *Universal Serial Bus Specification*, 2000. Revision 2.0.

41. Compaq Computer Corp., Microsoft Corp., National Semiconductor Corp. *OpenHCI Open Host Controller Interface Specification for USB*, 1999. Release 1.0a.

42. John L. Connell and Linda Isabell Shafer. *Object-Oriented Rapid Prototyping*. Prentice Hall, Englewood Cliffs, NJ, October 1994.

43. Ron Cytron, Jeanne Ferrante, Barry K. Rosen, Mark N. Wegman, and F. Kenneth Zadeck. Efficiently computing static single assignment form and the control dependence graph. *ACM Transactions on Programming Languages and Systems*, 13(4):451–490, October 1991.

44. Alan M. Davis. *Software Requirements: Analysis and Specification*. Prentice Hall, Englewood Cliffs, NJ, December 1989.

45. Stephen Deering and Robert Hinden. *Internet Protocol, Version 6 (IPv6) Specification, RFC 2460*. The Internet Society, December 1998.

46. DENX Software Engineering GmbH. *The DENX U-Boot and Linux Guide (DULG) for canyonlands*, March 2015. The documentation is also available online, at `http://www.denx.de/`.

47. Raymond Devillers and Joël Goossens. Liu and Layland's schedulability test revisited. *Information Processing Letters*, 73(5-6):157–161, 2000.

48. Edsger W. Dijkstra. Cooperating sequential processes. Technical Report EWD-123, Eindhoven University of Technology, 1965. Published as [49].

49. Edsger W. Dijkstra. Cooperating sequential processes. In F. Genuys, editor, *Programming Languages: NATO Advanced Study Institute*, pages 43–112. Academic Press, Villard de Lans, France, 1968.

50. Joseph D. Dumas II. *Computer Architecture: Fundamentals and Principles of Computer Design*. Taylor & Francis Group, Boca Raton, FL, November 2005.

51. Adam Dunkels. lwIP—a lightweight TCP/IP stack. Available online, at `http://savannah.nongnu.org/projects/lwip/`.

52. Adam Dunkels. Design and implementation of the lwIP TCP/IP stack. Available online, at `http://www.sics.se/~adam/lwip/doc/lwip.pdf`, 2001.

53. Adam Dunkels. Full TCP/IP for 8-bit architectures. In *Proc. 1st International Conference on Mobile Applications, Systems and Services*, pages 1–14, 2003.

54. Eclipse Foundation, Inc. *Eclipse Luna (4.4) Documentation*, 2015. Full documentation available in HTML format at `http://www.eclipse.org/`.

55. Dean Elsner, Jay Fenlason, and friends. *Using as — The GNU Assembler (GNU binutils) Version 2.20*. Free Software Foundation, Inc., 2009.

56. Embedded Artists AB. *LPC2468 OEM Board User's Guide, EA2-USG-0702 v1.2 Rev. C*, 2008. Available online, at `http://www.embeddedartists.com/`.

57. Embedded Solutions. Modbus master. Available online, at `http://www.embedded-solutions.at/`.

58. David Evans. Static detection of dynamic memory errors. In *Proc. ACM SIGPLAN Conference on Programming Language Design and Implementation*, pages 44–53, New York, NY, USA, 1996. ACM.

59. David Evans and David Larochelle. Improving security using extensible lightweight static analysis. *IEEE Software*, 19(1):42–51, January 2002.

60. David Evans and David Larochelle. *Splint Manual, Version 3.1.1-1*. Secure Programming Group, University of Virginia, Department of Computer Science, June 2003.

61. Max Felser. Real time Ethernet: standardization and implementations. In *Proc. IEEE International Symposium on Industrial Electronics*, pages 3766–3771, 2010.

62. Free Software Foundation, Inc. *GCC, the GNU compiler collection*, 2012. Available online, at `http://gcc.gnu.org/`.

63. Free Software Foundation, Inc. *GDB, the GNU project debugger*, 2012. Available online, at `http://www.gnu.org/software/gdb/`.

64. Free Software Foundation, Inc. *GNU binutils*, 2012. Available online, at `http://www.gnu.org/software/binutils/`.

65. Free Software Foundation, Inc. *GNU Make*, 2014. Available online, at `http://www.gnu.org/software/make/`.

66. Freescale Semiconductor, Inc. *ColdFire® Family Programmer's Reference Manual*, March 2005.

67. Philippe Gerum. *Xenomai—Implementing a RTOS emulation framework on GNU/Linux*, 2004. Available online, at `http://www.xenomai.org/`.

68. Ian Graham. *Requirements Engineering and Rapid Development: An object-oriented approach*. Addison-Wesley Professional, June 1999.

69. Laurie J. Hendren, Chris Donawa, Maryam Emami, Guang R. Gao, Justiani, and Bhama Sridharan. Designing the McCAT compiler based on a family of structured intermediate representations. In *Proc. 5th International Workshop on Languages and Compilers for Parallel Computing*, number 757 in LNCS, pages 406–420. Springer-Verlag, 1992.

70. Maurice P. Herlihy. A methodology for implementing highly concurrent data objects. *ACM Trans. on Programming Languages and Systems*, 15(5):745–770, November 1993.

71. Maurice P. Herlihy and Nir Shavit. *The Art of Multiprocessor Programming*. Morgan Kaufmann, June 2012.

72. Maurice P. Herlihy and Jeannette M. Wing. Axioms for concurrent objects. In *Proc. 14th ACM SIGACT-SIGPLAN Symposium on Principles of Programming Languages*, pages 13–26, New York, 1987.

73. Gerard J. Holzmann. *Verifying Multi-threaded Software with Spin*. Available online, at `http://spinroot.com/`.

74. Gerard J. Holzmann. The model checker SPIN. *IEEE Transactions on Software Engineering*, 23:279–295, 1997.

75. Gerard J. Holzmann. An analysis of bitstate hashing. *Formal Methods in System Design*, 13(3):289–307, November 1998.

76. Gerard J. Holzmann. *The Spin Model Checker: Primer and Reference Manual*. Pearson Education, Boston, MA, 2003.

77. Gerard J. Holzmann and Dragan Bošnački. The design of a multicore extension of the SPIN model checker. *IEEE Transactions on Software Engineering*, 33(10):659–674, 2007.

78. Gerard J. Holzmann and Doron Peled. An improvement in formal verification. In *Proc. 7th IFIP WG6.1 International Conference on Formal Description Techniques*, pages 197–211, Berne, Switzerland, 1994.

79. *IEEE Std 1003.13^{TM}-2003, IEEE Standard for Information Technology—Standardized Application Environment Profile (AEP)—POSIX® Realtime and Embedded Application Support*. IEEE, 2003.

80. *IEEE Std 1149.1-2013 — IEEE Standard for Test Access Port and Boundary-Scan Architecture*. IEEE Computer Society, May 2013.

81. Intel Corp. *Universal Host Controller Interface (UHCI) Design Guide*, 1996. Revision 1.1.

82. Intel Corp. *Enhanced Host Controller Interface Specification for Universal Serial Bus*, 2002. Revision 1.0.

83. Intel Corp. *Intel® 64 and IA-32 Architectures Software Developer's Manual*, 2007.

84. Intel Corp. *eXtensible Host Controller Interface for Universal Serial Bus*, 2010. Revision 1.0.

85. *International Standard ISO/IEC 7498-1, Information Technology—Open Systems Interconnection—Basic Reference Model: The Basic Model.* ISO/IEC, 1994.

86. *International Standard ISO/IEC/IEEE 9945, Information Technology—Portable Operating System Interface (POSIX®) Base Specifications, Issue 7.* IEEE and The Open Group, 2009.

87. International Organization for Standardization and International Electrotechnical Commission. *ISO/IEC 9899, Programming Languages — C,* 1st edition, December 1990.

88. International Organization for Standardization and International Electrotechnical Commission. *ISO/IEC 9899:1990/Amd 1, Programming Languages — C Integrity,* 1st edition, March 1995.

89. International Organization for Standardization and International Electrotechnical Commission. *ISO/IEC 9899, Programming Languages — C,* 2nd edition, December 1999.

90. International Organization for Standardization and International Electrotechnical Commission. *ISO/IEC 9899, Programming Languages — C,* 3rd edition, December 2011.

91. *ISO 11898-1—Road vehicles—Controller area network (CAN)—Part 1: Data link layer and physical signalling.* International Organization for Standardization, 2003.

92. *ISO 17356-1 — Road vehicles — Open interface for embedded automotive applications — Part 1: General structure and terms, definitions and abbreviated terms.* International Organization for Standardization, January 2005.

93. ITU-T. *Recommendation Z.100, Specification and Description Language (SDL).* International Telecommunication Union, 2002.

94. ITU-T. *Recommendation Z.120 – Message Sequence Chart (MSC),* 2004.

95. Rex Jaecshke. *Portability and the C Language.* Hayden Books C library. Sams Technical Publishing, Inc., Indianapolis, IN, 1st edition, 1988.

96. Henrik Ejersbo Jensen, Kim G. Larsen, and Arne Skou. Modelling and analysis of a collision avoidance protocol using Spin and Uppaal. In *Proc. 2nd International Workshop on the SPIN Verification System,* pages 33–49, 1996.

97. Ke Jiang and Bengt Jonsson. Using Spin to model check concurrent algorithms, using a translation from C to Promela. In *Proc. 2nd Swedish Workshop on Multi-Core Computing,* pages 1–9, 2009.

98. Stephen C. Johnson. Lint, a C program checker. Technical report, Bell Laboratories Computer Science Technical report 65, December 1977.

99. Krishna Kant and A. Ravichandran. Synthesizing robust data structures — an introduction. *IEEE Transactions on Computers,* 39(2):161–173, February 1990.

100. Max Khiszinsky. *CDS: Concurrent Data Structures library.* Available online, at `http://libcds.sourceforge.net/`.

101. Jan Kiszka. The real-time driver model and first applications. In *7th Real-Time Linux Workshop, Lille, France,* 2005.

102. Sambasiva Rao Kosaraju. Limitations of Dijkstra's semaphore primitives and Petri nets. *Operating Systems Review,* 7(4):122–126, January 1973.

103. Karthik Lakshmanan and Ragunathan Rajkumar. Scheduling self-suspending real-time tasks with rate-monotonic priorities. In *Proc. 16th IEEE Real-Time and Embedded Technology and Applications Symposium,* pages 3–12, April 2010.

104. Leslie Lamport. Concurrent reading and writing. *Communications of the ACM,* 20(11):806–811, November 1977.

105. Leslie Lamport. The mutual exclusion problem: part I—a theory of interprocess communication. *Journal of the ACM,* 33(2):313–326, 1986.

106. Leslie Lamport. The mutual exclusion problem: part II—statement and solutions. *Journal of the ACM,* 33(2):327–348, 1986.

107. David Larochelle and David Evans. Statically detecting likely buffer overflow vulnerabilities. In *Proc. 10th USENIX Security Symposium*, pages 1–13, Berkeley, CA, USA, 2001. USENIX Association.

108. John A. N. Lee. Howard Aiken's third machine: the Howard Mark III calculator or Aiken-Dahlgren electronic calculator. *IEEE Annals of the History of Computing*, 22(1):62–81, January 2000.

109. Shu Lin and Daniel J. Costello. *Error Control Coding*. Prentice Hall, 2nd edition, June 2004.

110. Chung L. Liu and James W. Layland. Scheduling algorithms for multiprogramming in a hard-real-time environment. *Journal of the ACM*, 20(1):46–61, 1973.

111. Jane W. S. Liu. *Real-Time Systems*. Prentice Hall, Upper Saddle River, NJ, 2000.

112. Robert Love. Kernel korner: Kernel locking techniques. *Linux Journal*, August 2002.

113. Zohar Manna and Amir Pnueli. *The Temporal Logic of Reactive and Concurrent Systems – Specification*. Springer-Verlag, New York, NY, 1992.

114. Theresa C. Maxino and Philip J. Koopman. The effectiveness of checksums for embedded control networks. *IEEE Transactions on Dependable and Secure Computing*, 6(1):59–72, January 2009.

115. Marshall Kirk McKusick, Keith Bostic, Michael J. Karels, and John S. Quarterman. *The Design and Implementation of the 4.4BSD Operating System*. Addison-Wesley, Reading, MA, 1996.

116. Mentor Graphics Corp. *Sourcery CodeBench*, 2013. Available online, at `http://www.mentor.com/embedded-software/codesourcery/`.

117. C. Michael Pilato, Ben Collins-Sussman, and Brian W. Fitzpatrick. *Version Control with Subversion*. O'Reilly Media, Sebastopol, CA, 2nd edition, September 2008.

118. Microsoft Corp. *Microsoft Extensible Firmware Initiative, FAT32 File System Specification – FAT: General Overview of On-Disk Format*, 2000. Version 1.03.

119. Modbus-IDA. *MODBUS Application Protocol Specification V1.1b*. Modbus Organization, Inc., 2006. Available online, at `http://www.modbus-ida.org/`.

120. Modbus-IDA. *MODBUS Messaging on TCP/IP Implementation Guide V1.0b*. Modbus Organization, Inc., 2006. Available online, at `http://www.modbus-ida.org/`.

121. Modbus-IDA. *MODBUS over Serial Line Specification and Implementation Guide V1.02*. Modbus Organization, Inc., 2006. Available online, at `http://www.modbus-ida.org/`.

122. Bryon Moyer. *Real World Multicore Embedded Systems*. Newnes, May 2013.

123. Netlabs. *FAT32*, 2012. Available online, at `http://svn.netlabs.org/fat32`.

124. NXP B.V. *AN10703 — NXP USB host lite*, 2008. Rev. 01.

125. NXP B.V. *LPC2468 Product data sheet, rev. 4*, October 2008. Available online, at `http://www.nxp.com/`.

126. NXP B.V. *LPC24xx User manual, UM10237 rev. 2*, December 2008. Available online, at `http://www.nxp.com/`.

127. NXP B.V. *LPC1769/68/67/66/65/64/63 Product data sheet, rev. 6*, 2010. Available online, at `http://www.nxp.com/`.

128. NXP B.V. *LPC17xx User manual, UM10360 rev. 2*, August 2010. Available online, at `http://www.nxp.com/`.

129. NXP B.V. *LPCOpen Software Development Platform for NXP LPC Microcontrollers*, June 2012. Available online, at `http://www.lpcware.com/lpcopen`.

130. On-line Applications Research Corp. *RTEMS C User's Guide*, December 2011. Available online, at `http://www.rtems.com/`.

131. On-line Applications Research Corp. *RTEMS Documentation*, December 2011. Available online, at `http://www.rtems.com/`.

132. On-line Applications Research Corp. *RTEMS ITRON 3.0 User's Guide*, December 2011. Available online, at `http://www.rtems.com/`.

133. On-line Applications Research Corp. *RTEMS POSIX API User's Guide*, December 2011. Available online, at `http://www.rtems.com/`.

134. The OpenOCD Project. *Open On-Chip Debugger: OpenOCD User's Guide*, March 2015.

135. Elliott I. Organick. *The Multics System: An Examination of Its Structure*. MIT Press, Cambridge, MA, USA, 1972.

136. OSEK/VDX. *OSEK/VDX Operating System Specification*. Available online, at `http://www.osek-vdx.org/`.

137. Sergio Pérez, Joan Vila, Jose A. Alegre, and Josep V. Sala. A CORBA based architecture for distributed embedded systems using the RTLinux-GPL platform. In *Proc. IEEE International Symposium on Object Oriented Real-Time Distributed Computing*, pages 285–288, 2004.

138. Roland H. Pesch, Jeffrey M. Osier, and Cygnus Support. *The GNU Binary Utilities (GNU binutils) Version 2.20*. Free Software Foundation, Inc., October 2009.

139. Amir Pnueli. The temporal logic of programs. In *Proc. 18th Annual Symposium on Foundations of Computer Science*, pages 46–57, November 1977.

140. Jon Postel. *User Datagram Protocol, RFC 768*. Information Sciences Institute (ISI), August 1980.

141. Jon Postel. *Internet Control Message Protocol—DARPA Internet Program Protocol Specification, RFC 792*. Information Sciences Institute (ISI), September 1981.

142. Jon Postel, editor. *Internet Protocol—DARPA Internet Program Protocol Specification, RFC 791*. USC/Information Sciences Institute (ISI), September 1981.

143. Jon Postel, editor. *Transmission Control Protocol—DARPA Internet Program Protocol Specification, RFC 793*. USC/Information Sciences Institute (ISI), September 1981.

144. Ragunathan Rajkumar, L. Sha, and John P. Lehoczky. Real-time synchronization protocols for multiprocessors. In *Proc. 9th IEEE Real-Time Systems Symposium*, pages 259–269, December 1988.

145. Red Hat, Inc. *The Red Hat Newlib C Library*, 2012. Available online, at `http://sourceware.org/newlib/`.

146. Red Hat Inc. *eCos User Guide*, 2013. Available online, at `http://ecos.sourceware.org/`.

147. W. Richard Stevens and Gary R. Wright. *TCP/IP Illustrated (3 Volume Set)*. Addison-Wesley Professional, Boston, MA, USA, November 2001.

148. Ken Sakamura. *ITRON3.0: An Open and Portable Real-Time Operating System for Embedded Systems*. IEEE Computer Society Press, Los Alamitos, CA, April 1998.

149. Dilip V. Sarwate. Computation of cyclic redundancy checks via table look-up. *Communications of the ACM*, 31(8):1008–1013, August 1988.

150. Sharad C. Seth and R. Muralidhar. Analysis and design of robust data structures. In *Proc. 15th Annual International Symposium on Fault-Tolerant Computing (FTCS)*, pages 14–19, June 1985.

151. Julian Seward, Nicholas Nethercote, Josef Weidendorfer, and The Valgrind Development Team. *Valgrind 3.3 — Advanced Debugging and Profiling for Gnu/Linux Applications*. Network Theory Ltd., Bristol, United Kingdom, March 2008.

152. Lui Sha, Tarek Abdelzaher, Karl-Erik Årzén, Anton Cervin, Theodore P. Baker, Alan Burns, Giorgio C. Buttazzo, Marco Caccamo, John Lehoczky, and Aloysius K. Mok. Real time scheduling theory: A historical perspective. *Real-Time Systems*, 28(2):101–155, 2004.

153. Lui Sha, Ragunathan Rajkumar, and John P. Lehoczky. Priority inheritance protocols: an approach to real-time synchronization. *IEEE Transactions on Computers*, 39(9):1175–1185, September 1990.

154. Sajjan G. Shiva. *Computer Organization, Design, and Architecture*. Taylor & Francis Group, Boca Raton, FL, 5th edition, December 2013.

155. Muzaffer A. Siddiqi. *Dynamic RAM: Technology Advancements*. Taylor & Francis Group, Boca Raton, FL, December 2012.

156. Richard M. Stallman et al. *GNU Emacs Manual*. Free Software Foundation, Inc., 17th edition, 2014.

157. Richard M. Stallman, Roland McGrath, and Paul D. Smith. *GNU Make — A Program for Directing Recompilation, for GNU make Version 4.0*. Free Software Foundation, Inc., October 2013.

158. Richard M. Stallman, Roland Pesch, Stan Shebs, et al. *Debugging with GDB — The GNU Source-Level Debugger*. Free Software Foundation, Inc., 10th edition, 2015.

159. Richard M. Stallman and the GCC Developer Community. *GNU Compiler Collection Internals, for GCC Version 4.3.4*. Free Software Foundation, Inc., 2007.

160. Richard M. Stallman and the GCC Developer Community. *Using the GNU Compiler Collection, for GCC Version 4.3.4*. Free Software Foundation, Inc., 2008.

161. Andrew S. Tanenbaum and Todd Austin. *Structured Computer Organization*. Pearson Education Ltd., 6th edition, August 2012.

162. David J. Taylor, David E. Morgan, and James P. Black. Redundancy in data structures: Improving software fault tolerance. *IEEE Transactions on Software Engineering*, SE-6(6):585–594, November 1980.

163. TIA. *Electrical Characteristics of Generators and Receivers for Use in Balanced Digital Multipoint Systems (ANSI/TIA/EIA-485-A-98) (R2003)*. Telecommunications Industry Association, 1998.

164. USB Implementers Forum. *Universal Serial Bus Mass Storage Class UFI Command Specification*, 1998. Revision 1.0.

165. USB Implementers Forum. *Universal Serial Bus Mass Storage Class Bulk-Only Transport*, 1999. Revision 1.0.

166. USB Implementers Forum. *Universal Serial Bus Mass Storage Class Specification Overview*, 2008. Revision 1.3.

167. Peter van der Linden. *Expert C Programming: Deep C Secrets*. SunSoft Press, a Prentice Hall Title, Mountain View, CA, June 1994.

168. Tom Van Vleck. Structure marking. http://www.multicians.org/thvv/marking.html, 2015 (retrieved on April 14, 2015).

169. William Von Hagen. *The definitive guide to GCC*. Apress, Berkeley, CA, 2006.

170. Christian Walter. FreeMODBUS - a Modbus ASCII/RTU and TCP implementation, 2007. Available online, at http://freemodbus.berlios.de/.

171. Joseph Yiu. *The Definitive Guide to ARM Cortex-M3 and Cortex-M4 Processors, Third Edition*. Newnes, Newton, MA, USA, 3rd edition, 2013.

172. Lennart Ysboodt and Michael De Nil. *EFSL – Embedded Filesystems Library*, 2012. Available online, at http://sourceforge.net/projects/efsl/.

Index

Printed in the United States
by Baker & Taylor Publisher Services